AN INTRODUCTION TO GEOPHYSICAL EXPLORATION

An Introduction to Geophysical Exploration

Philip Kearey
Department of Earth Sciences
University of Bristol

Michael Brooks
Ty Newydd, City
Near Cowbridge
Vale of Glamorgan

Ian Hill
Department of Geology
University of Leicester

THIRD EDITION

Blackwell
Publishing

© 2002 by Blackwell Science Ltd,
a Blackwell Publishing company

BLACKWELL PUBLISHING
350 Main Street, Malden, MA 02148-5020, USA
9600 Garsington Road, Oxford OX4 2DQ, UK
550 Swanston Street, Carlton, Victoria 3053, Australia

First published 1984
Second edition 1991
Third edition 2002

9   2009

Library of Congress Cataloging-in-Publication Data has been applied for.

ISBN 978-0-632-04929-5 (paperback)

A catalogue record for this title is available from the British Library.

Set by SNP Best-set Typesetters Ltd, Hong Kong
Printed and bound in Singapore
by Markono Print Media Pte Ltd

The publisher's policy is to use permanent paper from mills that operate a sustainable forestry policy, and
which has been manufactured from pulp processed using acid-free and elementary chlorine-free practices.
Furthermore, the publisher ensures that the text paper and cover board used have met acceptable
environmental accreditation standards.

For further information on
Blackwell Publishing, visit our website:
www.blackwellpublishing.com

Contents

Preface

This book provides a general introduction to the most important methods of geophysical exploration. These methods represent a primary tool for investigation of the subsurface and are applicable to a very wide range of problems. Although their main application is in prospecting for natural resources, the methods are also used, for example, as an aid to geological surveying, as a means of deriving information on the Earth's internal physical properties, and in engineering or archaeological site investigations. Consequently, geophysical exploration is of importance not only to geophysicists but also to geologists, physicists, engineers and archaeologists. The book covers the physical principles, methodology, interpretational procedures and fields of application of the various survey methods. The main emphasis has been placed on seismic methods because these represent the most extensively used techniques, being routinely and widely employed by the oil industry in prospecting for hydrocarbons. Since this is an introductory text we have not attempted to be completely comprehensive in our coverage of the subject. Readers seeking further information on any of the survey methods described should refer to the more advanced texts listed at the end of each chapter.

We hope that the book will serve as an introductory course text for students in the above-mentioned disciplines and also as a useful guide for specialists who wish to be aware of the value of geophysical surveying to their own disciplines. In preparing a book for such a wide possible readership it is inevitable that problems arise concerning the level of mathematical treatment to be adopted. Geophysics is a highly mathematical subject and, although we have attempted to show that no great mathematical expertise is necessary for a broad understanding of geophysical surveying, a full appreciation of the more advanced data processing and interpretational techniques does require a reasonable mathematical ability. Our approach to this problem has been to keep the mathematics as simple as possible and to restrict full mathematical analysis to relatively simple cases. We consider it important, however, that any user of geophysical surveying should be aware of the more advanced techniques of analysing and interpreting geophysical data since these can greatly increase the amount of useful information obtained from the data. In discussing such techniques we have adopted a semiquantitative or qualitative approach which allows the reader to assess their scope and importance, without going into the details of their implementation.

Earlier editions of this book have come to be accepted as the standard geophysical exploration textbook by numerous higher educational institutions in Britain, North America, and many other countries. In the third edition, we have brought the content up to date by taking account of recent developments in all the main areas of geophysical exploration. We have extended the scope of the seismic chapters by including new material on three-component and 4D reflection seismology, and by providing a new section on seismic tomography. We have also widened the range of applications of refraction seismology considered, to include an account of engineering site investigation.

1 The principles and limitations of geophysical exploration methods

1.1 Introduction

This chapter is provided for readers with no prior knowledge of geophysical exploration methods and is pitched at an elementary level. It may be passed over by readers already familiar with the basic principles and limitations of geophysical surveying.

The science of geophysics applies the principles of physics to the study of the Earth. Geophysical investigations of the interior of the Earth involve taking measurements at or near the Earth's surface that are influenced by the internal distribution of physical properties. Analysis of these measurements can reveal how the physical properties of the Earth's interior vary vertically and laterally.

By working at different scales, geophysical methods may be applied to a wide range of investigations from studies of the entire Earth (global geophysics; e.g. Kearey & Vine 1996) to exploration of a localized region of the upper crust for engineering or other purposes (e.g. Vogelsang 1995, McCann *et al.* 1997). In the geophysical exploration methods (also referred to as geophysical surveying) discussed in this book, measurements within geographically restricted areas are used to determine the distributions of physical properties at depths that reflect the local subsurface geology.

An alternative method of investigating subsurface geology is, of course, by drilling boreholes, but these are expensive and provide information only at discrete locations. Geophysical surveying, although sometimes prone to major ambiguities or uncertainties of interpretation, provides a relatively rapid and cost-effective means of deriving areally distributed information on subsurface geology. In the exploration for subsurface resources the methods are capable of detecting and delineating local features of potential interest that could not be discovered by any realistic drilling programme. Geophysical surveying does not dispense with the need for drilling but, properly applied, it can optimize explo-

ration programmes by maximizing the rate of ground coverage and minimizing the drilling requirement. The importance of geophysical exploration as a means of deriving subsurface geological information is so great that the basic principles and scope of the methods and their main fields of application should be appreciated by any practising Earth scientist. This book provides a general introduction to the main geophysical methods in widespread use.

1.2 The survey methods

There is a broad division of geophysical surveying methods into those that make use of natural fields of the Earth and those that require the input into the ground of artificially generated energy. The natural field methods utilize the gravitational, magnetic, electrical and electromagnetic fields of the Earth, searching for local perturbations in these naturally occurring fields that may be caused by concealed geological features of economic or other interest. Artificial source methods involve the generation of local electrical or electromagnetic fields that may be used analogously to natural fields, or, in the most important single group of geophysical surveying methods, the generation of seismic waves whose propagation velocities and transmission paths through the subsurface are mapped to provide information on the distribution of geological boundaries at depth. Generally, natural field methods can provide information on Earth properties to significantly greater depths and are logistically more simple to carry out than artificial source methods. The latter, however, are capable of producing a more detailed and better resolved picture of the subsurface geology.

Several geophysical surveying methods can be used at sea or in the air. The higher capital and operating costs associated with marine or airborne work are offset by the increased speed of operation and the benefit of

Table 1.1 Geophysical methods.

Method	Measured parameter	Operative physical property
Seismic	Travel times of reflected/refracted seismic waves	Density and elastic moduli, which determine the propagation velocity of seismic waves
Gravity	Spatial variations in the strength of the gravitational field of the Earth	Density
Magnetic	Spatial variations in the strength of the geomagnetic field	Magnetic susceptibility and remanence
Electrical		
Resistivity	Earth resistance	Electrical conductivity
Induced polarization	Polarization voltages or frequency-dependent ground resistance	Electrical capacitance
Self-potential	Electrical potentials	Electrical conductivity
Electromagnetic	Response to electromagnetic radiation	Electrical conductivity and inductance
Radar	Travel times of reflected radar pulses	Dielectric constant

being able to survey areas where ground access is difficult or impossible.

A wide range of geophysical surveying methods exists, for each of which there is an 'operative' physical property to which the method is sensitive. The methods are listed in Table 1.1.

The type of physical property to which a method responds clearly determines its range of applications. Thus, for example, the magnetic method is very suitable for locating buried magnetite ore bodies because of their high magnetic susceptibility. Similarly, seismic or electrical methods are suitable for the location of a buried water table because saturated rock may be distinguished from dry rock by its higher seismic velocity and higher electrical conductivity.

Other considerations also determine the type of methods employed in a geophysical exploration programme. For example, reconnaissance surveys are often carried out from the air because of the high speed of operation. In such cases the electrical or seismic methods are not applicable, since these require physical contact with the ground for the direct input of energy.

Geophysical methods are often used in combination. Thus, the initial search for metalliferous mineral deposits often utilizes airborne magnetic and electromagnetic surveying. Similarly, routine reconnaissance of continental shelf areas often includes simultaneous gravity, magnetic and seismic surveying. At the interpretation stage, ambiguity arising from the results of one survey method may often be removed by consideration of results from a second survey method.

Geophysical exploration commonly takes place in a number of stages. For example, in the offshore search for oil and gas, an initial gravity reconnaissance survey may reveal the presence of a large sedimentary basin that is subsequently explored using seismic methods. A first round of seismic exploration may highlight areas of particular interest where further detailed seismic work needs to be carried out.

The main fields of application of geophysical surveying, together with an indication of the most appropriate surveying methods for each application, are listed in Table 1.2.

Exploration for hydrocarbons, for metalliferous minerals and environmental applications represents the main uses of geophysical surveying. In terms of the amount of money expended annually, seismic methods are the most important techniques because of their routine and widespread use in the exploration for hydrocarbons. Seismic methods are particularly well suited to the investigation of the layered sequences in sedimentary basins that are the primary targets for oil or gas. On the other hand, seismic methods are quite unsuited to the exploration of igneous and metamorphic terrains for the near-surface, irregular ore bodies that represent the main source of metalliferous minerals. Exploration for ore bodies is mainly carried out using electromagnetic and magnetic surveying methods.

In several geophysical survey methods it is the local variation in a measured parameter, relative to some normal background value, that is of primary interest. Such variation is attributable to a localized subsurface zone of

Table 1.2 Geophysical surveying applications.

Application	Appropriate survey methods*
Exploration for fossil fuels (oil, gas, coal)	S, G, M, (EM)
Exploration for metalliferous mineral deposits	M, EM, E, SP, IP, R
Exploration for bulk mineral deposits (sand and gravel)	S, (E), (G)
Exploration for underground water supplies	E, S, (G), (Rd)
Engineering/construction site investigation	E, S, Rd. (G), (M)
Archaeological investigations	Rd, E, EM, M, (S)

* G, gravity; M, magnetic; S, seismic; E, electrical resistivity; SP, self-potential; IP, induced polarization; EM, electromagnetic; R, radiometric; Rd, ground-penetrating radar. Subsidiary methods in brackets.

distinctive physical property and possible geological importance. A local variation of this type is known as a *geophysical anomaly*. For example, the Earth's gravitational field, after the application of certain corrections, would everywhere be constant if the subsurface were of uniform density. Any lateral density variation associated with a change of subsurface geology results in a local deviation in the gravitational field. This local deviation from the otherwise constant gravitational field is referred to as a gravity anomaly.

Although many of the geophysical methods require complex methodology and relatively advanced mathematical treatment in interpretation, much information may be derived from a simple assessment of the survey data. This is illustrated in the following paragraphs where a number of geophysical surveying methods are applied to the problem of detecting and delineating a specific geological feature, namely a salt dome. No terms or units are defined here, but the examples serve to illustrate the way in which geophysical surveys can be applied to the solution of a particular geological problem.

Salt domes are emplaced when a buried salt layer, because of its low density and ability to flow, rises through overlying denser strata in a series of approximately cylindrical bodies. The rising columns of salt pierce the overlying strata or arch them into a domed form. A salt dome has physical properties that are different from the surrounding sediments and which enable its detection by geophysical methods. These properties are: (1) a relatively low density; (2) a negative magnetic susceptibility; (3) a relatively high propagation velocity for seismic waves; and (4) a high electrical resistivity (specific resistance).

1. The relatively low density of salt with respect to its surroundings renders the salt dome a zone of anomalously low mass. The Earth's gravitational field is perturbed by subsurface mass distributions and the salt dome therefore gives rise to a gravity anomaly that is negative with respect to surrounding areas. Figure 1.1 presents a contour map of gravity anomalies measured over the Grand Saline Salt Dome in east Texas, USA. The gravitational readings have been corrected for effects which result from the Earth's rotation, irregular surface relief and regional geology so that the contours reflect only variations in the shallow density structure of the area resulting from the local geology. The location of the salt dome is known from both drilling and mining operations and its subcrop is indicated. It is readily apparent that there is a well-defined negative gravity anomaly centred over the salt dome and the circular gravity contours reflect the circular outline of the dome. Clearly, gravity surveys provide a powerful method for the location of features of this type.

2. A less familiar characteristic of salt is its negative magnetic susceptibility, full details of which must be deferred to Chapter 7. This property of salt causes a local decrease in the strength of the Earth's magnetic field in the vicinity of a salt dome. Figure 1.2 presents a contour map of the strength of the magnetic field over the Grand Saline Salt Dome covering the same area as Fig. 1.1. Readings have been corrected for the large-scale variations of the magnetic field with latitude, longitude and time so that, again, the contours reflect only those variations resulting from variations in the magnetic properties of the subsurface. As expected, the salt dome is associated with a negative magnetic anomaly, although the magnetic low is displaced slightly from the centre of the dome. This example illustrates that salt domes may be located by magnetic surveying but the technique is not widely used as the associated anomalies are usually very small and therefore difficult to detect.

3. Seismic rays normally propagate through salt at a higher velocity than through the surrounding sediments. A consequence of this velocity difference is that

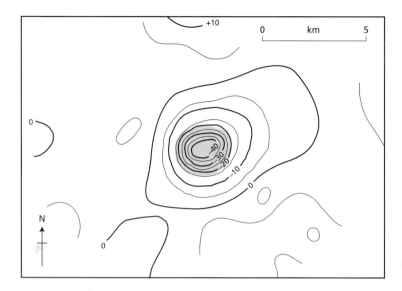

Fig. 1.1 The gravity anomaly over the Grand Saline Salt Dome, Texas, USA (contours in gravity units — see Chapter 6). The stippled area represents the subcrop of the dome. (Redrawn from Peters & Dugan 1945.)

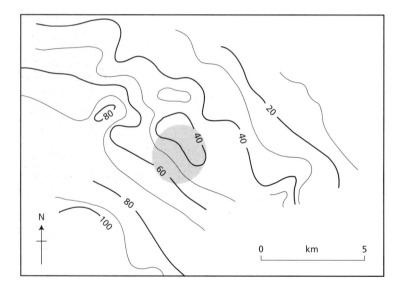

Fig. 1.2 Magnetic anomalies over the Grand Saline Salt Dome, Texas, USA (contours in nT — see Chapter 7). The stippled area represents the subcrop of the dome. (Redrawn from Peters & Dugan 1945.)

any seismic energy incident on the boundary of a salt body is partitioned into a refracted phase that is transmitted through the salt and a reflected phase that travels back through the surrounding sediments (Chapter 3). These two seismic phases provide alternative means of locating a concealed salt body.

For a series of seismic rays travelling from a single shot point into a fan of seismic detectors (see Fig. 5.21), rays transmitted through any intervening salt dome will travel at a higher average velocity than in the surrounding medium and, hence, will arrive relatively early at the recording site. By means of this 'fan-shooting' it is possible to delineate sections of ground which are associated with anomalously short travel times and which may therefore be underlain by a salt body.

An alternative, and more effective, approach to the seismic location of salt domes utilizes energy reflected off the salt, as shown schematically in Fig. 1.3. A survey

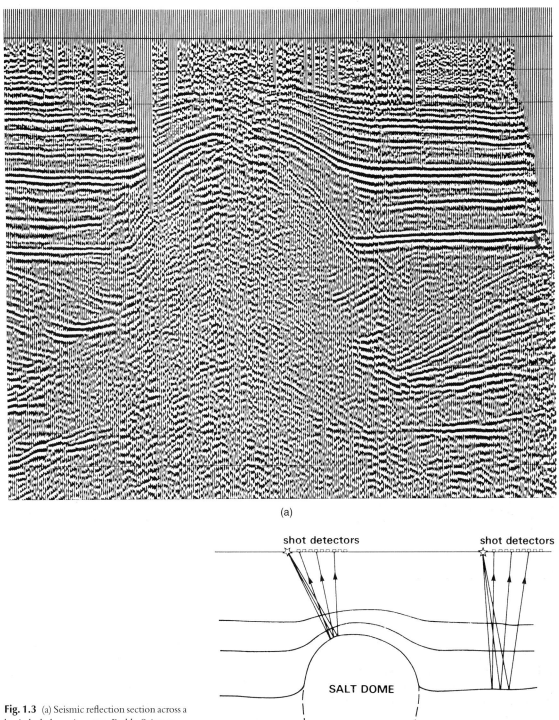

(a)

Fig. 1.3 (a) Seismic reflection section across a buried salt dome (courtesy Prakla–Seismos GmbH). (b) Simple structural interpretation of the seismic section, illustrating some possible ray paths for reflected rays.

shot detectors

shot detectors

SALT DOME

(b)

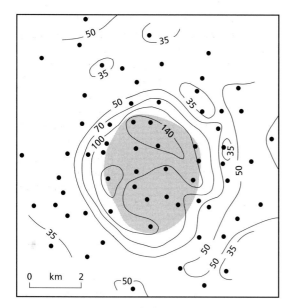

Fig. 1.4 Perturbation of telluric currents over the Haynesville Salt Dome, Texas, USA (for explanation of units see Chapter 9). The stippled area represents the subcrop of the dome. (Redrawn from Boissonas & Leonardon 1948.)

configuration of closely-spaced shots and detectors is moved systematically along a profile line and the travel times of rays reflected back from any subsurface geological interfaces are measured. If a salt dome is encountered, rays reflected off its top surface will delineate the shape of the concealed body.

4. Earth materials with anomalous electrical resistivity may be located using either electrical or electromagnetic geophysical techniques. Shallow features are normally investigated using artificial field methods in which an electrical current is introduced into the ground and potential differences between points on the surface are measured to reveal anomalous material in the subsurface (Chapter 8). However, this method is restricted in its depth of penetration by the limited power that can be introduced into the ground. Much greater penetration can be achieved by making use of the natural Earth currents (telluric currents) generated by the motions of charged particles in the ionosphere. These currents extend to great depths within the Earth and, in the absence of any electrically anomalous material, flow parallel to the surface. A salt dome, however, possesses an anomalously high electrical resistivity and electric currents preferentially flow around and over the top of such a

structure rather than through it. This pattern of flow causes distortion of the constant potential gradient at the surface that would be associated with a homogeneous subsurface and indicates the presence of the high-resistivity salt. Figure 1.4 presents the results of a telluric current survey of the Haynesville Salt Dome, Texas, USA. The contour values represent quantities describing the extent to which the telluric currents are distorted by subsurface phenomena and their configuration reflects the shape of the subsurface salt dome with some accuracy.

1.3 The problem of ambiguity in geophysical interpretation

If the internal structure and physical properties of the Earth were precisely known, the magnitude of any particular geophysical measurement taken at the Earth's surface could be predicted uniquely. Thus, for example, it would be possible to predict the travel time of a seismic wave reflected off any buried layer or to determine the value of the gravity or magnetic field at any surface location. In geophysical surveying the problem is the opposite of the above, namely, to deduce some aspect of the Earth's internal structure on the basis of geophysical measurements taken at (or near to) the Earth's surface. The former type of problem is known as a *direct* problem, the latter as an *inverse* problem. Whereas direct problems are theoretically capable of unambiguous solution, inverse problems suffer from an inherent ambiguity, or non-uniqueness, in the conclusions that can be drawn.

To exemplify this point a simple analogy to geophysical surveying may be considered. In *echo-sounding*, high-frequency acoustic pulses are transmitted by a transducer mounted on the hull of a ship and echoes returned from the sea bed are detected by the same transducer. The travel time of the echo is measured and converted into a water depth, multiplying the travel time by the velocity with which sound waves travel through water; that is, $1500 \, \text{m} \, \text{s}^{-1}$. Thus an echo time of $0.10 \, \text{s}$ indicates a path length of $0.10 \times 1500 = 150 \, \text{m}$, or a water depth of $150/2 = 75 \, \text{m}$, since the pulse travels down to the sea bed and back up to the ship.

Using the same principle, a simple seismic survey may be used to determine the depth of a buried geological interface (e.g. the top of a limestone layer). This would involve generating a seismic pulse at the Earth's surface and measuring the travel time of a pulse reflected back to the surface from the top of the limestone. However, the conversion of this travel time into a depth requires knowledge of the velocity with which the pulse travelled along the reflection path and, unlike the velocity of sound in water, this information is generally not known. If a velocity is assumed, a depth estimate can be derived but it represents only one of many possible solutions. And since rocks differ significantly in the velocity with which they propagate seismic waves, it is by no means a straightforward matter to translate the travel time of a seismic pulse into an accurate depth to the geological interface from which it was reflected.

The solution to this particular problem, as discussed in Chapter 4, is to measure the travel times of reflected pulses at several offset distances from a seismic source because the variation of travel time as a function of range provides information on the velocity distribution with depth. However, although the degree of uncertainty in geophysical interpretation can often be reduced to an acceptable level by the general expedient of taking additional (and in some cases different kinds of) field measurements, the problem of inherent ambiguity cannot be circumvented.

The general problem is that significant differences from an actual subsurface geological situation may give rise to insignificant, or immeasurably small, differences in the quantities actually measured during a geophysical survey. Thus, ambiguity arises because many different geological configurations could reproduce the observed measurements. This basic limitation results from the unavoidable fact that geophysical surveying attempts to solve a difficult inverse problem. It should also be noted that experimentally-derived quantities are never exactly determined and experimental error adds a further degree of indeterminacy to that caused by the incompleteness of the field data and the ambiguity associated with the inverse problem. Since a unique solution cannot, in general, be recovered from a set of field measurements, geophysical interpretation is concerned either to determine properties of the subsurface that all possible solutions share, or to introduce assumptions to restrict the number of admissible solutions (Parker 1977). In spite of these inherent problems, however, geophysical surveying is an invaluable tool for the investigation of subsurface geology and occupies a key role in exploration programmes for geological resources.

1.4 The structure of the book

The above introductory sections illustrate in a simple way the very wide range of approaches to the geophysical investigation of the subsurface and warn of inherent limitations in geophysical interpretations.

Chapter 2 provides a short account of the more important data processing techniques of general applicability to geophysics. In Chapters 3 to 10 the individual survey methods are treated systematically in terms of their basic principles, survey procedures, interpretation techniques and major applications. Chapter 11 describes the application of these methods to specialized surveys undertaken in boreholes. All these chapters contain suggestions for further reading which provide a more extensive treatment of the material covered in this book. A set of problems is given for all the major geophysical methods.

2 Geophysical data processing

2.1 Introduction

Geophysical surveys measure the variation of some physical quantity, with respect either to position or to time. The quantity may, for example, be the strength of the Earth's magnetic field along a profile across an igneous intrusion. It may be the motion of the ground surface as a function of time associated with the passage of seismic waves. In either case, the simplest way to present the data is to plot a graph (Fig. 2.1) showing the variation of the measured quantity with respect to distance or time as appropriate. The graph will show some more or less complex waveform shape, which will reflect physical variations in the underlying geology, superimposed on unwanted variations from non-geological features (such as the effect of electrical power cables in the magnetic example, or vibration from passing traffic for the seismic case), instrumental inaccuracy and data collection errors. The detailed shape of the waveform may be uncertain due to the difficulty in interpolating the curve between widely spaced stations. The geophysicist's task is to separate the 'signal' from the 'noise' and interpret the signal in terms of ground structure.

Analysis of waveforms such as these represents an essential aspect of geophysical data processing and interpretation. The fundamental physics and mathematics of such analysis is not novel, most having been discovered in the 19th or early 20th centuries. The use of these ideas is also widespread in other technological areas such as radio, television, sound and video recording, radioastronomy, meteorology and medical imaging, as well as military applications such as radar, sonar and satellite imaging. Before the general availability of digital computing, the quantity of data and the complexity of the processing severely restricted the use of the known techniques. This no longer applies and nearly all the techniques described in this chapter may be implemented in standard computer spreadsheet programs.

The fundamental principles on which the various methods of data analysis are based are brought together in this chapter. These are accompanied by a discussion of the techniques of digital data processing by computer that are routinely used by geophysicists. Throughout this chapter, waveforms are referred to as functions of time, but all the principles discussed are equally applicable to functions of distance. In the latter case, frequency (number of waveform cycles per unit time) is replaced by spatial frequency or *wavenumber* (number of waveform cycles per unit distance).

2.2 Digitization of geophysical data

Waveforms of geophysical interest are generally continuous (analogue) functions of time or distance. To apply the power of digital computers to the task of analysis, the data need to be expressed in digital form, whatever the form in which they were originally recorded.

A continuous, smooth function of time or distance can be expressed digitally by sampling the function at a fixed interval and recording the instantaneous value of the function at each sampling point. Thus, the analogue function of time $f(t)$ shown in Fig. 2.2(a) can be represented as the digital function $g(t)$ shown in Fig. 2.2(b) in which the continuous function has been replaced by a series of discrete values at fixed, equal, intervals of time. This process is inherent in many geophysical surveys, where readings are taken of the value of some parameter (e.g. magnetic field strength) at points along survey lines. The extent to which the digital values faithfully represent the original waveform will depend on the accuracy of the amplitude measurement and the intervals between measured samples. Stated more formally, these two parameters of a digitizing system are the sampling precision (dynamic range) and the sampling frequency.

Dynamic range is an expression of the ratio of the largest measurable amplitude A_{max} to the smallest measurable amplitude A_{min} in a sampled function. The higher the

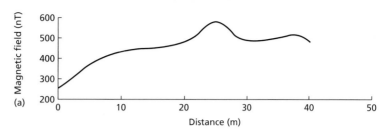

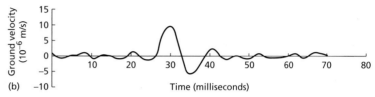

Fig. 2.1 (a) A graph showing a typical magnetic field strength variation which may be measured along a profile. (b) A graph of a typical seismogram, showing variation of particle velocities in the ground as a function of time during the passage of a seismic wave.

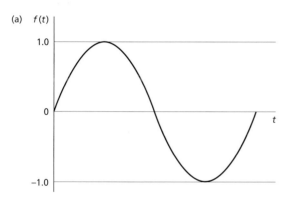

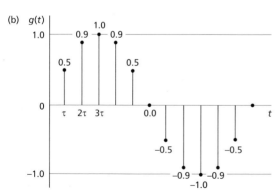

Fig. 2.2 (a) Analogue representation of a sinusoidal function. (b) Digital representation of the same function.

fine electrical power ratios: the ratio of two power values P_1 and P_2 is given by $10\log_{10}(P_1/P_2)$ dB. Since *power* is proportional to the square of *signal amplitude A*

$$10\log_{10}(P_1/P_2) = 10\log_{10}(A_1/A_2)^2$$
$$= 20\log_{10}(A_1/A_2) \qquad (2.1)$$

Thus, if a digital sampling scheme measures amplitudes over the range from 1 to 1024 units of amplitude, the dynamic range is given by

$$20\log_{10}(A_{max}/A_{min}) = 20\log_{10}1024 \approx 60\,dB$$

In digital computers, digital samples are expressed in binary form (i.e. they are composed of a sequence of digits that have the value of either 0 or 1). Each binary digit is known as a *bit* and the sequence of bits representing the sample value is known as a *word*. The number of bits in each word determines the dynamic range of a digitized waveform. For example, a dynamic range of 60 dB requires 11-bit words since the appropriate amplitude ratio of 1024 ($= 2^{10}$) is rendered as 10000000000 in binary form. A dynamic range of 84 dB represents an amplitude ratio of 2^{14} and, hence, requires sampling with 15-bit words. Thus, increasing the number of bits in each word in digital sampling increases the dynamic range of the digital function.

Sampling frequency is the number of sampling points in unit time or unit distance. Intuitively, it may appear that the digital sampling of a continuous function inevitably leads to a loss of information in the resultant digital function, since the latter is only specified by discrete values at a series of points. Again intuitively, there will be no

dynamic range, the more faithfully the amplitude variations in the analogue waveform will be represented in the digitized version of the waveform. Dynamic range is normally expressed in the *decibel* (dB) scale used to de-

significant loss of information content as long as the frequency of sampling is much higher than the highest frequency component in the sampled function. Mathematically, it can be proved that, if the waveform is a sine curve, this can always be reconstructed provided that there are a minimum of two samples per period of the sine wave.

Thus, if a waveform is sampled every two milliseconds (sampling interval), the sampling frequency is 500 samples per second (or 500 Hz). Sampling at this rate will preserve all frequencies up to 250 Hz in the sampled function. This frequency of half the sampling frequency is known as the *Nyquist frequency* (f_N) and the *Nyquist interval* is the frequency range from zero up to f_N

$$f_N = 1/(2\Delta t) \tag{2.2}$$

where Δt = sampling interval.

If frequencies above the Nyquist frequency are present in the sampled function, a serious form of distortion results known as *aliasing*, in which the higher frequency components are 'folded back' into the Nyquist interval. Consider the example illustrated in Fig. 2.3 in which sine waves at different frequencies are sampled. The lower frequency wave (Fig. 2.3(a)) is accurately reproduced, but the higher frequency wave (Fig. 2.3(b), solid line) is rendered as a fictitious frequency, shown by the dashed line, within the Nyquist interval. The relationship between input and output frequencies in the case of a sampling frequency of 500 Hz is shown in Fig. 2.3(c). It is apparent that an input frequency of 125 Hz, for example, is retained in the output but that an input frequency of 625 Hz is folded back to be output at 125 Hz also.

To overcome the problem of aliasing, the sampling frequency must be at least twice as high as the highest frequency component present in the sampled function. If the function does contain frequencies above the Nyquist frequency determined by the sampling, it must be passed through an *antialias filter* prior to digitization. The antialias filter is a low-pass frequency filter with a sharp cut-off that removes frequency components above the Nyquist frequency, or attenuates them to an insignificant amplitude level.

2.3 Spectral analysis

An important mathematical distinction exists between *periodic waveforms* (Fig. 2.4(a)), that repeat themselves at a fixed time period T, and *transient waveforms* (Fig. 2.4(b)),

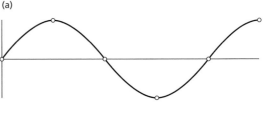

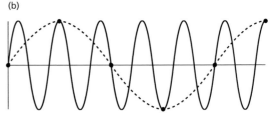

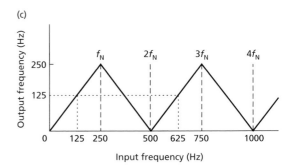

Fig. 2.3 (a) Sine wave frequency less than Nyquist frequency. (b) Sine wave frequency greater than Nyquist frequency (solid line) showing the fictitious frequency that is generated by aliasing (dashed line).
(c) Relationship between input and output frequencies for a sampling frequency of 500 Hz (Nyquist frequency f_N = 250 Hz).

that are non-repetitive. By means of the mathematical technique of *Fourier analysis* any periodic waveform, however complex, may be decomposed into a series of sine (or cosine) waves whose frequencies are integer multiples of the basic repetition frequency $1/T$, known as the *fundamental frequency*. The higher frequency components, at frequencies of n/T ($n = 1, 2, 3, \ldots$), are known as harmonics. The complex waveform of Fig. 2.5(a) is built up from the addition of the two individual sine wave components shown. To express any waveform in terms of its constituent sine wave components, it is necessary to define not only the frequency of each component but also its amplitude and phase. If in the above example the relative amplitude and phase relations of the individual sine waves are altered, summation can

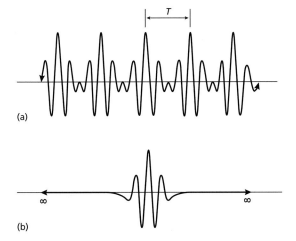

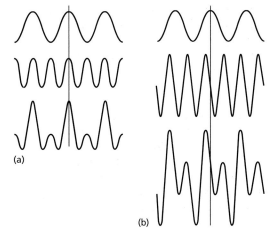

Fig. 2.4 (a) Periodic and (b) transient waveforms.

Fig. 2.5 Complex waveforms resulting from the summation of two sine wave components of frequency f and $2f$. (a) The two sine wave components are of equal amplitude and in phase. (b) The higher frequency component has twice the amplitude of the lower frequency component and is $\pi/2$ out of phase. (After Anstey 1965.)

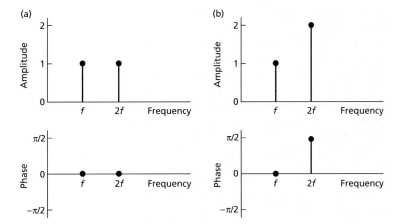

Fig. 2.6 Representation in the frequency domain of the waveforms illustrated in Fig. 2.5, showing their amplitude and phase spectra.

produce the quite different waveform illustrated in Fig. 2.5(b).

From the above it follows that a periodic waveform can be expressed in two different ways: in the familiar *time domain*, expressing wave amplitude as a function of time, or in the *frequency domain*, expressing the amplitude and phase of its constituent sine waves as a function of frequency. The waveforms shown in Fig. 2.5(a) and (b) are represented in Fig. 2.6(a) and (b) in terms of their amplitude and phase spectra. These spectra, known as line spectra, are composed of a series of discrete values of the amplitude and phase components of the waveform at set frequency values distributed between 0 Hz and the Nyquist frequency.

Transient waveforms do not repeat themselves; that is, they have an infinitely long period. They may be regarded, by analogy with a periodic waveform, as having an infinitesimally small fundamental frequency ($1/T \rightarrow 0$) and, consequently, harmonics that occur at infinitesimally small frequency intervals to give continuous amplitude and phase spectra rather than the line spectra of periodic waveforms. However, it is impossible to cope analytically with a spectrum containing an infinite number of sine wave components. Digitization of the waveform in the time domain (Section 2.2) provides a means of dealing with the continuous spectra of transient waveforms. A digitally sampled transient waveform has its amplitude and phase spectra subdivided into a number of

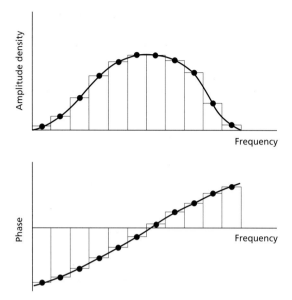

Fig. 2.7 Digital representation of the continuous amplitude and phase spectra associated with a transient waveform.

thin frequency slices, with each slice having a frequency equal to the mean frequency of the slice and an amplitude and phase proportional to the area of the slice of the appropriate spectrum (Fig. 2.7). This digital expression of a continuous spectrum in terms of a finite number of discrete frequency components provides an approximate representation in the frequency domain of a transient waveform in the time domain. Increasing the sampling frequency in the time domain not only improves the time-domain representation of the waveform, but also increases the number of frequency slices in the frequency domain and improves the accuracy of the approximation here too.

Fourier transformation may be used to convert a time function $g(t)$ into its equivalent amplitude and phase spectra $A(f)$ and $\phi(f)$, or into a complex function of frequency $G(f)$ known as the *frequency spectrum*, where

$$G(f) = A(f)e^{i\phi(f)} \qquad (2.3)$$

The time- and frequency-domain representations of a waveform, $g(t)$ and $G(f)$, are known as a *Fourier pair*, represented by the notation

$$g(t) \leftrightarrow G(f) \qquad (2.4)$$

Components of a Fourier pair are interchangeable, such that, if $G(f)$ is the Fourier transform of $g(t)$, then $g(t)$ is the Fourier transform of $G(f)$. Figure 2.8 illustrates Fourier pairs for various waveforms of geophysical significance. All the examples illustrated have *zero phase spectra*; that is, the individual sine wave components of the waveforms are in phase at zero time. In this case $\phi(f) = 0$ for all values of ϕ. Figure 2.8(a) shows a spike function (also known as a *Dirac function*), which is the shortest possible transient waveform. Fourier transformation shows that the spike function has a continuous frequency spectrum of constant amplitude from zero to infinity; thus, a spike function contains all frequencies from zero to infinity at equal amplitude. The 'DC bias' waveform of Fig. 2.8(b) has, as would be expected, a line spectrum comprising a single component at zero frequency. Note that Fig. 2.8(a) and (b) demonstrate the principle of interchangeability of Fourier pairs stated above (equation (2.4)). Figures 2.8(c) and (d) illustrate transient waveforms approximating the shape of seismic pulses, together with their amplitude spectra. Both have a band-limited amplitude spectrum, the spectrum of narrower bandwidth being associated with the longer transient waveform. In general, the shorter a time pulse the wider is its frequency bandwidth and in the limiting case a spike pulse has an infinite bandwidth.

Waveforms with zero phase spectra such as those illustrated in Fig. 2.8 are symmetrical about the time axis and, for any given amplitude spectrum, produce the maximum peak amplitude in the resultant waveform. If phase varies linearly with frequency, the waveform remains unchanged in shape but is displaced in time; if the phase variation with frequency is non-linear the shape of the waveform is altered. A particularly important case in seismic data processing is the phase spectrum associated with *minimum delay* in which there is a maximum concentration of energy at the front end of the waveform. Analysis of seismic pulses sometimes assumes that they exhibit minimum delay (see Chapter 4).

Fourier transformation of digitized waveforms is readily programmed for computers, using a '*fast Fourier transform*' (FFT) algorithm as in the Cooley–Tukey method (Brigham 1974). FFT subroutines can thus be routinely built into data processing programs in order to carry out spectral analysis of geophysical waveforms. Fourier transformation is supplied as a function to standard spreadsheets such as Microsoft Excel. Fourier transformation can be extended into two dimensions (Rayner 1971), and can thus be applied to areal distributions of data such as gravity and magnetic contour maps.

Time domain   Frequency domain

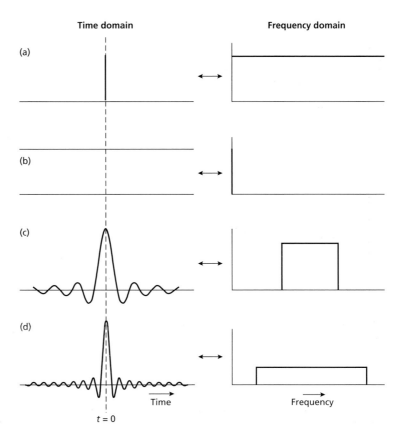

Fig. 2.8 Fourier transform pairs for various waveforms. (a) A spike function. (b) A 'DC bias'. (c) and (d) Transient waveforms approximating seismic pulses.

In this case, the time variable is replaced by horizontal distance and the frequency variable by wavenumber (number of waveform cycles per unit distance). The application of two-dimensional Fourier techniques to the interpretation of potential field data is discussed in Chapters 6 and 7.

2.4 Waveform processing

The principles of convolution, deconvolution and correlation form the common basis for many methods of geophysical data processing, especially in the field of seismic reflection surveying. They are introduced here in general terms and are referred to extensively in later chapters. Their importance is that they quantitatively describe how a waveform is affected by a filter. Filtering modifies a waveform by discriminating between its constituent sine wave components to alter their relative amplitudes or phase relations, or both. Most audio systems are provided with simple filters to cut down on high-

frequency 'hiss', or to emphasize the low-frequency 'bass'. Filtering is an inherent characteristic of any system through which a signal is transmitted.

2.4.1 Convolution

Convolution (Kanasewich 1981) is a mathematical operation defining the change of shape of a waveform resulting from its passage through a filter. Thus, for example, a seismic pulse generated by an explosion is altered in shape by filtering effects, both in the ground and in the recording system, so that the seismogram (the filtered output) differs significantly from the initial seismic pulse (the input).

As a simple example of filtering, consider a weight suspended from the end of a vertical spring. If the top of the spring is perturbed by a sharp up-and-down movement (the input), the motion of the weight (the filtered output) is a series of damped oscillations out of phase with the initial perturbation (Fig. 2.9).

The effect of a filter may be categorized by its *impulse*

Amplitude

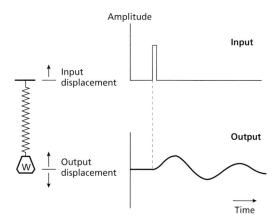

Input

↑ Input
displacement

Output

↑ Output
displacement

Time

Fig. 2.9 The principle of filtering illustrated by the perturbation of a suspended weight system.

response which is defined as the output of the filter when the input is a spike function (Fig. 2.10). The impulse response is a waveform in the time domain, but may be transformed into the frequency domain as for any other waveform. The Fourier transform of the impulse response is known as the *transfer function* and this specifies the amplitude and phase response of the filter, thus defining its operation completely. The effect of a filter is described mathematically by a *convolution* operation such that, if the input signal $g(t)$ to the filter is convolved with the impulse response $f(t)$ of the filter, known as the convolution operator, the filtered output $y(t)$ is obtained:

$$y(t) = g(t) \star f(t) \qquad (2.5)$$

where the asterisk denotes the convolution operation.

Figure 2.11(a) shows a spike function input to a filter whose impulse response is given in Fig. 2.11(b). Clearly the latter is also the filtered output since, by definition, the impulse response represents the output for a spike input. Figure 2.11(c) shows an input comprising two separate spike functions and the filtered output (Fig. 2.11(d)) is now the superposition of the two impulse response functions offset in time by the separation of the

input spikes and scaled according to the individual spike amplitudes. Since any transient wave can be represented as a series of spike functions (Fig. 2.11(e)), the general form of a filtered output (Fig. 2.11(f)) can be regarded as the summation of a set of impulse responses related to a succession of spikes simulating the overall shape of the input wave.

The mathematical implementation of convolution involves time inversion (or folding) of one of the functions and its progressive sliding past the other function, the individual terms in the convolved output being derived by summation of the cross-multiplication products over the overlapping parts of the two functions. In general, if $g_i (i = 1, 2, \ldots, m)$ is an input function and $f_j (j = 1, 2, \ldots, n)$ is a convolution operator, then the convolution output function y_k is given by

$$y_k = \sum_{i=1}^{m} g_i f_{k-i} \ (k = 1, 2, \ldots, m + n - 1) \qquad (2.6)$$

In Fig. 2.12 the individual steps in the convolution process are shown for two digital functions, a double spike function given by $g_i = g_1, g_2, g_3 = 2, 0, 1$ and an impulse response function given by $f_i = f_1, f_2, f_3, f_4 = 4, 3, 2, 1$, where the numbers refer to discrete amplitude values at the sampling points of the two functions. From Fig. 2.11 it can be seen that the convolved output $y_i = y_1, y_2, y_3, y_4, y_5, y_6 = 8, 6, 8, 5, 2, 1$. Note that the convolved output is longer than the input waveforms; if the functions to be convolved have lengths of m and n, the convolved output has a length of $(m + n - 1)$.

The convolution of two functions in the time domain becomes increasingly laborious as the functions become longer. Typical geophysical applications may have functions which are each from 250 to a few thousand samples long. The same mathematical result may be obtained by transforming the functions to the frequency domain, then multiplying together equivalent frequency terms of their amplitude spectra and adding terms of their phase spectra. The resulting output amplitude and phase spectra can then be transformed back to the time domain. Thus, digital filtering can be enacted in either the time

Spike input

Output = Impulse response

Filter

Fig. 2.10 The impulse response of a filter.

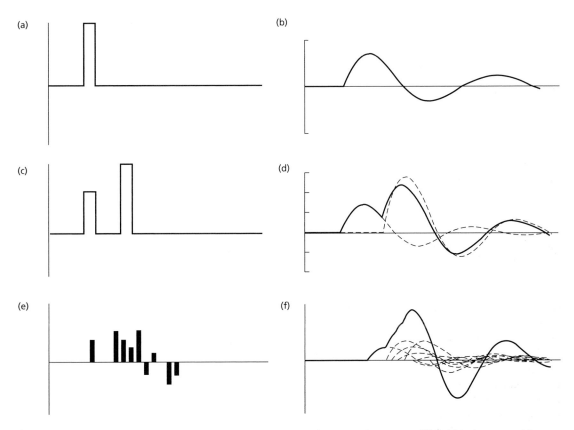

Fig. 2.11 Examples of filtering. (a) A spike input. (b) Filtered output equivalent to impulse response of filter. (c) An input comprising two spikes. (d) Filtered output given by summation of two impulse response functions offset in time. (e) A complex input represented by a series of contiguous spike functions. (f) Filtered output given by the summation of a set of impulse responses.

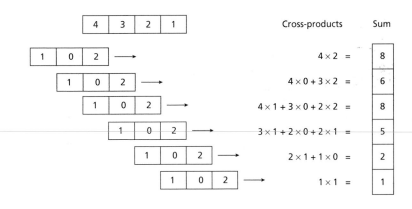

Fig. 2.12 A method of calculating the convolution of two digital functions.

domain or the frequency domain. With large data sets, filtering by computer is more efficiently carried out in the frequency domain since fewer mathematical operations are involved.

Convolution, or its equivalent in the frequency domain, finds very wide application in geophysical data processing, notably in the digital filtering of seismic and potential field data and the construction of synthetic seismograms for comparison with field seismograms (see Chapters 4 and 6).

2.4.2 Deconvolution

Deconvolution or *inverse filtering* (Kanasewich 1981) is a process that counteracts a previous convolution (or filtering) action. Consider the convolution operation given in equation (2.5)

$$y(t) = g(t) \star f(t)$$

$y(t)$ is the filtered output derived by passing the input waveform $g(t)$ through a filter of impulse response $f(t)$. Knowing $y(t)$ and $f(t)$, the recovery of $g(t)$ represents a deconvolution operation. Suppose that $f'(t)$ is the function that must be convolved with $y(t)$ to recover $g(t)$

$$g(t) = y(t) \star f'(t) \tag{2.7}$$

Substituting for $y(t)$ as given by equation (2.5)

$$g(t) = g(t) \star f(t) \star f'(t) \tag{2.8}$$

Now recall also that

$$g(t) = g(t) \star \delta(t) \tag{2.9}$$

where $\delta(t)$ is a spike function (a unit amplitude spike at zero time); that is, a time function $g(t)$ convolved with a spike function produces an unchanged convolution output function $g(t)$. From equations (2.8) and (2.9) it follows that

$$f(t) \star f'(t) = \delta(t) \tag{2.10}$$

Thus, provided the impulse response $f(t)$ is known, $f'(t)$ can be derived for application in equation (2.7) to recover the input signal $g(t)$. The function $f'(t)$ represents the deconvolution operator.

Deconvolution is an essential aspect of seismic data processing, being used to improve seismic records by removing the adverse filtering effects encountered by seismic waves during their passage through the ground. In the seismic case, referring to equation (2.5), $y(t)$ is the seismic record resulting from the passage of a seismic wave $g(t)$ through a portion of the Earth, which acts as a filter with an impulse response $f(t)$. The particular problem with deconvolving a seismic record is that the input waveform $g(t)$ and the impulse response $f(t)$ of the Earth filter are in general unknown. Thus the 'deterministic' approach to deconvolution outlined above cannot be employed and the deconvolution operator has to be designed using statistical methods. This special approach to the deconvolution of seismic records, known as predictive deconvolution, is discussed further in Chapter 4.

2.4.3 Correlation

Cross-correlation of two digital waveforms involves cross-multiplication of the individual waveform elements and summation of the cross-multiplication products over the common time interval of the waveforms. The cross-correlation function involves progressively sliding one waveform past the other and, for each time shift, or *lag*, summing the cross-multiplication products to derive the cross-correlation as a function of lag value. The cross-correlation operation is similar to convolution but does not involve folding of one of the waveforms. Given two digital waveforms of finite length, x_i and y_i ($i = 1, 2, \ldots, n$), the cross-correlation function is given by

$$\phi_{xy}(\tau) = \sum_{i=1}^{n-\tau} x_{i+\tau} y_i \quad (-m < \tau < +m) \tag{2.11}$$

where τ is the lag and m is known as the maximum lag value of the function. It can be shown that cross-correlation in the time domain is mathematically equivalent to multiplication of amplitude spectra and subtraction of phase spectra in the frequency domain.

Clearly, if two identical non-periodic waveforms are cross-correlated (Fig. 2.13) all the cross-multiplication products will sum at zero lag to give a maximum positive value. When the waveforms are displaced in time, however, the cross-multiplication products will tend to cancel out to give small values. The cross-correlation function therefore peaks at zero lag and reduces to small values at large time shifts. Two closely similar waveforms will likewise produce a cross-correlation function that is strongly peaked at zero lag. On the other hand, if two dissimilar waveforms are cross-correlated the sum of cross-multiplication products will always be near to zero due to the tendency for positive and negative products to cancel out at all values of lag. In fact, for two waveforms containing only random noise the cross-correlation function $\phi_{xy}(\tau)$ is zero for all non-zero values of τ. Thus, the cross-correlation function measures the degree of similarity of waveforms.

An important application of cross-correlation is in the detection of weak signals embedded in noise. If a waveform contains a known signal concealed in noise at unknown time, cross-correlation of the waveform with the signal function will produce a cross-correlation function

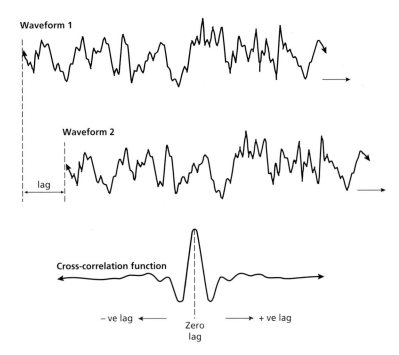

Fig. 2.13 Cross-correlation of two identical waveforms.

centred on the time value at which the signal function and its concealed equivalent in the waveform are in phase (Fig. 2.14).

A special case of correlation is that in which a waveform is cross-correlated with itself, to give the *autocorrelation* function $\phi_{xx}(\tau)$. This function is symmetrical about a zero lag position, so that

$$\phi_{xx}(\tau) = \phi_{xx}(-\tau) \qquad (2.12)$$

The autocorrelation function of a periodic waveform is also periodic, with a frequency equal to the repetition frequency of the waveform. Thus, for example, the autocorrelation function of a cosine wave is also a cosine wave. For a transient waveform, the autocorrelation function decays to small values at large values of lag. These differing properties of the autocorrelation function of periodic and transient waveforms determine one of its main uses in geophysical data processing, namely, the detection of hidden periodicities in any given waveform. Side lobes in the autocorrelation function (Fig. 2.15) are an indication of the existence of periodicities in the original waveform, and the spacing of the side lobes defines the repetition period. This property is particularly useful in the detection and suppression of multiple reflections in seismic records (see Chapter 4).

The autocorrelation function contains all the amplitude information of the original waveform but none of the phase information, the original phase relationships being replaced by a zero phase spectrum. In fact, the autocorrelation function and the square of the amplitude spectrum $A(f)$ can be shown to form a Fourier pair

$$\phi_{xx}(\tau) \leftrightarrow A(f)^2 \qquad (2.13)$$

Since the square of the amplitude represents the power term (energy contained in the frequency component) the autocorrelation function can be used to compute the *power spectrum* of a waveform.

2.5 Digital filtering

In waveforms of geophysical interest, it is standard practice to consider the waveform as a combination of *signal* and *noise*. The signal is that part of the waveform that relates to the geological structures under investigation. The noise is all other components of the waveform. The noise can be further subdivided into two components, *random* and *coherent* noise. Random noise is just that, statistically

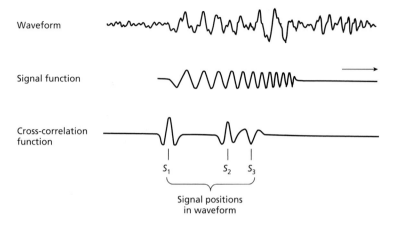

Waveform

Signal function

Cross-correlation
function

S_1 S_2 S_3

Signal positions
in waveform

Fig. 2.14 Cross-correlation to detect occurrences of a known signal concealed in noise. (After Sheriff 1973.)

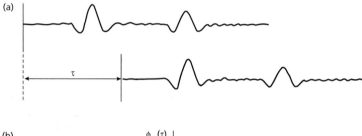

(a)

τ

(b)

$\phi_{xx}(\tau)$

τ

Fig. 2.15 Autocorrelation of the waveform exhibiting periodicity shown in (a) produces the autocorrelation function with side lobes shown in (b). The spacing of the side lobes defines the repetition period of the original waveform.

random, and usually due to effects unconnected with the geophysical survey. Coherent noise is, on the other hand, components of the waveform which are generated by the geophysical experiment, but are of no direct interest for the geological interpretation. For example, in a seismic survey the signal might be the seismic pulse arriving at a detector after being reflected by a geological boundary at depth. Random noise would be background vibration due to wind, rain or distant traffic. Coherent noise would be the surface waves generated by the seismic source, which also travel to the detector and may obscure the desired signal.

In favourable circumstances the signal-to-noise ratio (SNR) is high, so that the signal is readily identified and extracted for subsequent analysis. Often the SNR is low and special processing is necessary to enhance the information content of the waveforms. Different approaches are needed to remove the effect of different types of noise. Random noise can often be suppressed by re-peated measurement and averaging. Coherent noise may be filtered out by identifying the particular characteristics of that noise and designing a special filter to remove it. The remaining signal itself may be distorted due to the effects of the recording system, and again, if the nature of the recording system is accurately known, suitable filtering can be designed. Digital filtering is widely employed in geophysical data processing to improve SNR or otherwise improve the signal characteristics. A very wide range of digital filters is in routine use in geophysical, and especially seismic, data processing (Robinson & Treitel 2000). The two main types of digital filter are frequency filters and inverse (deconvolution) filters.

2.5.1 Frequency filters

Frequency filters discriminate against selected frequency components of an input waveform and may be low-pass (LP), high-pass (HP), band-pass (BP) or band-reject

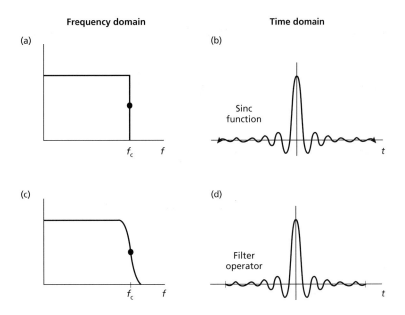

Fig. 2.16 Design of a digital low-pass filter.

(BR) in terms of their frequency response. Frequency filters are employed when the signal and noise components of a waveform have different frequency characteristics and can therefore be separated on this basis.

Analogue frequency filtering is still in widespread use and analogue antialias (LP) filters are an essential component of analogue-to-digital conversion systems (see Section 2.2). Nevertheless, digital frequency filtering by computer offers much greater flexibility of filter design and facilitates filtering of much higher performance than can be obtained with analogue filters. To illustrate the design of a digital frequency filter, consider the case of a LP filter whose cut-off frequency is f_c. The desired output characteristics of the ideal LP filter are represented by the amplitude spectrum shown in Fig. 2.16(a). The spectrum has a constant unit amplitude between 0 and f_c and zero amplitude outside this range: the filter would therefore pass all frequencies between 0 and f_c without attenuation and would totally suppress frequencies above f_c. This amplitude spectrum represents the transfer function of the ideal LP filter.

Inverse Fourier transformation of the transfer function into the time domain yields the impulse response of the ideal LP filter (see Fig. 2.16(b)). However, this impulse response (a sinc function) is infinitely long and must therefore be truncated for practical use as a convolution operator in a digital filter. Figure 2.16(c) represents the frequency response of a practically realizable LP filter operator of finite length (Fig. 2.16(d)). Convolu-

tion of the input waveform with the latter will result in LP filtering with a ramped cut-off (Fig. 2.16(c)) rather than the instantaneous cut-off of the ideal LP filter.

HP, BP and BR time-domain filters can be designed in a similar way by specifying a particular transfer function in the frequency domain and using this to design a finite-length impulse response function in the time domain. As with analogue filtering, digital frequency filtering generally alters the phase spectrum of the waveform and this effect may be undesirable. However, *zero phase filters* can be designed that facilitate digital filtering without altering the phase spectrum of the filtered signal.

2.5.2 Inverse (deconvolution) filters

The main applications of inverse filtering to remove the adverse effects of a previous filtering operation lie in the field of seismic data processing. A discussion of inverse filtering in the context of deconvolving seismic records is given in Chapter 4.

2.6 Imaging and modelling

Once the geophysical waveforms have been processed to maximize the signal content, that content must be extracted for geological interpretation. Imaging and modelling are two different strategies for this work.

As the name implies, in imaging the measured waveforms themselves are presented in a form in which they simulate an image of the subsurface structure. The most obvious examples of this are in seismic reflection (Chapter 4) and ground-penetrating radar (Chapter 9) sections, where the waveform of the variation of reflected energy with time is used to derive an image related to the occurrence of geological boundaries at depth. Often magnetic surveys for shallow engineering or archaeological investigations are processed to produce shaded, coloured, or contoured maps where the shading or colour correlates with variations of magnetic field which are expected to correlate with the structures being sought. Imaging is a very powerful tool, as it provides a way of summarizing huge volumes of data in a format which can be readily comprehended, that is, the visual image. A disadvantage of imaging is that often it can be difficult or impossible to extract quantitative information from the image.

In modelling, the geophysicist chooses a particular type of structural model of the subsurface, and uses this to predict the form of the actual waveforms recorded. The model is then adjusted to give the closest match between the predicted (modelled) and observed waveforms. The goodness of the match obtained depends on both the signal-to-noise ratio of the waveforms and the initial choice of the model used. The results of modelling are usually displayed as cross-sections through the structure under investigation. Modelling is an essential part of most geophysical methods and is well exemplified in gravity and magnetic interpretation (see Chapters 6 and 7).

Problems

1. Over the distance between two seismic recording sites at different ranges from a seismic source, seismic waves have been attenuated by 5 dB. What is the ratio of the wave amplitudes observed at the two sites?

2. In a geophysical survey, time-series data are sampled at 4 ms intervals for digital recording. (a) What is the Nyquist frequency? (b) In the absence of antialias filtering, at what frequency would noise at 200 Hz be aliased back into the Nyquist interval?

3. If a digital recording of a geophysical time series is required to have a dynamic range of 120 dB, what number of bits is required in each binary word?

4. If the digital signal $(-1, 3, -2, -1)$ is convolved with the filter operator $(2, 3, 1)$, what is the convolved output?

5. Cross-correlate the signal function $(-1, 3, -1)$ with the waveform $(-2, -4, -4, -3, 3, 1, 2, 2)$ containing signal and noise, and indicate the likely position of the signal in the waveform on the basis of the cross-correlation function.

6. A waveform is composed of two in-phase components of equal amplitude at frequencies f and $3f$. Draw graphs to represent the waveform in the time domain and the frequency domain.

Further reading

Brigham, E.O. (1974) *The Fast Fourier Transform*. Prentice-Hall, New Jersey.

Camina, A.R. & Janacek, G.J. (1984) *Mathematics for Seismic Data Processing and Interpretation*. Graham & Trotman, London.

Claerbout, J.F. (1985) *Fundamentals of Geophysical Data Processing*. McGraw-Hill, New York.

Dobrin, M.B. & Savit, C.H. (1988) *Introduction to Geophysical Prospecting* (4th edn). McGraw-Hill, New York.

Kanasewich, E.R. (1981) *Time Sequence Analysis in Geophysics* (3rd edn). University of Alberta Press.

Kulhanek, O. (1976) *Introduction to Digital Filtering in Geophysics*. Elsevier, Amsterdam.

Menke, W. (1989) *Geophysical Data Analysis: Discrete Inverse Theory*. Academic Press, London.

Rayner, J.N. (1971) *An Introduction to Spectral Analysis*. Pion, England.

Robinson, E.A. & Trietel, S. (2000) *Geophysical Signal Analysis*. Prentice-Hall, New Jersey.

Sheriff, R.E. & Geldart, L.P. (1983) *Exploration Seismology Vol 2: Data-Processing and Interpretation*. Cambridge University Press, Cambridge.

3 Elements of seismic surveying

3.1 Introduction

In seismic surveying, seismic waves are created by a controlled source and propagate through the subsurface. Some waves will return to the surface after refraction or reflection at geological boundaries within the subsurface. Instruments distributed along the surface detect the ground motion caused by these returning waves and hence measure the arrival times of the waves at different ranges from the source. These travel times may be converted into depth values and, hence, the distribution of subsurface geological interfaces may be systematically mapped.

Seismic surveying was first carried out in the early 1920s. It represented a natural development of the already long-established methods of earthquake seismology in which the travel times of earthquake waves recorded at seismological observatories are used to derive information on the internal structure of the Earth. Earthquake seismology provides information on the gross internal layering of the Earth, and measurement of the velocity of earthquake waves through the various Earth layers provides information about their physical properties and composition. In the same way, but on a smaller scale, seismic surveying can provide a clear and detailed picture of subsurface geology. It undoubtedly represents the single most important geophysical surveying method in terms of the amount of survey activity and the very wide range of its applications. Many of the principles of earthquake seismology are applicable to seismic surveying. However, the latter is concerned solely with the structure of the Earth down to tens of kilometres at most and uses artificial seismic sources, such as explosions, whose location, timing and source characteristics are, unlike earthquakes, under the direct control of the geophysicist. Seismic surveying also uses specialized recording systems and associated data processing and interpretation techniques.

Seismic methods are widely applied to exploration problems involving the detection and mapping of subsurface boundaries of, normally, simple geometry. They also identify significant physical properties of each subsurface unit. The methods are particularly well suited to the mapping of layered sedimentary sequences and are therefore widely used in the search for oil and gas. The methods are also used, on a smaller scale, for the mapping of near-surface sediment layers, the location of the water table and, in an engineering context, site investigation of foundation conditions including the determination of depth to bedrock. Seismic surveying can be carried out on land or at sea and is used extensively in offshore geological surveys and the exploration for offshore resources.

In this chapter the fundamental physical principles on which seismic methods are based are reviewed, starting with a discussion of the nature of seismic waves and going on to consider their mode of propagation through the ground, with particular reference to reflection and refraction at interfaces between different rock types. To understand the different types of seismic wave that propagate through the ground away from a seismic source, some elementary concepts of stress and strain need to be considered.

3.2 Stress and strain

When external forces are applied to a body, balanced internal forces are set up within it. *Stress* is a measure of the intensity of these balanced internal forces. The stress acting on an area of any surface within the body may be resolved into a component of normal stress perpendicular to the surface and a component of shearing stress in the plane of the surface.

At any point in a stressed body three orthogonal planes can be defined on which the components of stress are wholly normal stresses, that is, no shearing stresses act along them. These planes define three orthogonal axes

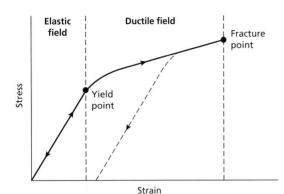

Fig. 3.1 A typical stress–strain curve for a solid body.

known as the principal axes of stress, and the normal stresses acting in these directions are known as the *principal stresses*. Each principal stress represents a balance of equal-magnitude but oppositely-directed force components. The stress is said to be compressive if the forces are directed towards each other and tensile if they are directed away from each other.

If the principal stresses are all of equal magnitude within a body the condition of stress is said to be *hydrostatic*, since this is the state of stress throughout a fluid body at rest. A fluid body cannot sustain shearing stresses (since a fluid has no shear strength), hence there cannot be shear stresses in a body under hydrostatic stress. If the principal stresses are unequal, shearing stresses exist along all surfaces within the stressed body, except for the three orthogonal planes intersecting in the principal axes.

A body subjected to stress undergoes a change of shape and/or size known as *strain*. Up to a certain limiting value of stress, known as the *yield strength* of a material, the strain is directly proportional to the applied stress (Hooke's Law). This elastic strain is reversible so that removal of stress leads to a removal of strain. If the yield strength is exceeded the strain becomes non-linear and partly irreversible (i.e. permanent strain results), and this is known as plastic or ductile strain. If the stress is increased still further the body fails by fracture. A typical stress–strain curve is illustrated in Fig. 3.1.

The linear relationship between stress and strain in the elastic field is specified for any material by its various *elastic moduli*, each of which expresses the ratio of a particular type of stress to the resultant strain. Consider a rod of original length l and cross-sectional area A which is extended by an increment Δl through the application of a stretching force F to its end faces (Fig. 3.2(a)). The relevant elastic modulus is Young's modulus E, defined by

$$E = \frac{longitudinal\ stress\ F/A}{longitudinal\ strain\ \Delta l/l}$$

Note that extension of such a rod will be accompanied by a reduction in its diameter; that is, the rod will suffer lateral as well as longitudinal strain. The ratio of the lateral to the longitudinal strain is known as *Poisson's ratio* (σ).

The *bulk modulus K* expresses the stress–strain ratio in the case of a simple hydrostatic pressure P applied to a cubic element (Fig. 3.2(b)), the resultant volume strain being the change of volume Δv divided by the original volume v

$$K = \frac{volume\ stress\ P}{volume\ strain\ \Delta v/v}$$

In a similar manner the *shear modulus* (μ) is defined as the ratio of shearing stress (τ) to the resultant shear strain $tan\ \theta$ (Fig. 3.2(c))

$$\mu = \frac{shear\ stress\ \tau}{shear\ strain\ tan\ \theta}$$

Finally, the *axial modulus* ψ defines the ratio of longitudinal stress to longitudinal strain in the case when there is no lateral strain; that is, when the material is constrained to deform uniaxially (Fig. 3.2(d))

$$\psi = \frac{longitudinal\ stress\ F/A}{longitudinal\ strain\ (uniaxial)\ \Delta l/l}$$

3.3 Seismic waves

Seismic waves are parcels of elastic strain energy that propagate outwards from a seismic source such as an earthquake or an explosion. Sources suitable for seismic surveying usually generate short-lived wave trains, known as pulses, that typically contain a wide range of frequencies, as explained in Section 2.3. Except in the immediate vicinity of the source, the strains associated with the passage of a seismic pulse are minute and may be assumed to be elastic. On this assumption the propagation velocities of seismic pulses are determined by the elastic moduli and densities of the materials through

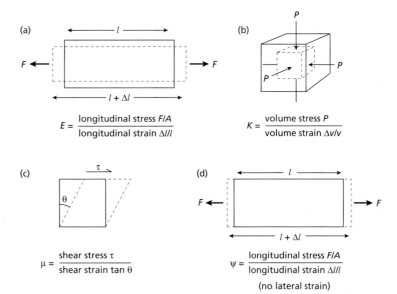

Fig. 3.2 The elastic moduli. (a) Young's modulus *E*. (b) Bulk modulus *K*. (c) Shear modulus *μ*. (d) Axial modulus *ψ*.

$$E = \frac{\text{longitudinal stress } F/A}{\text{longitudinal strain } \Delta l/l}$$

$$K = \frac{\text{volume stress } P}{\text{volume strain } \Delta v/v}$$

$$\mu = \frac{\text{shear stress } \tau}{\text{shear strain } \tan \theta}$$

$$\psi = \frac{\text{longitudinal stress } F/A}{\text{longitudinal strain } \Delta l/l}$$
(no lateral strain)

which they pass. There are two groups of seismic waves, *body waves* and *surface waves*.

3.3.1 Body waves

Body waves can propagate through the internal volume of an elastic solid and may be of two types. *Compressional waves* (the longitudinal, primary or *P-waves* of earthquake seismology) propagate by compressional and dilational uniaxial strains in the direction of wave travel. Particle motion associated with the passage of a compressional wave involves oscillation, about a fixed point, in the direction of wave propagation (Fig. 3.3(a)). *Shear waves* (the transverse, secondary or *S-waves* of earthquake seismology) propagate by a pure shear strain in a direction perpendicular to the direction of wave travel. Individual particle motions involve oscillation, about a fixed point, in a plane at right angles to the direction of wave propagation (Fig. 3.3(b)). If all the particle oscillations are confined to a plane, the shear wave is said to be plane-polarized.

The velocity of propagation of any body wave in any homogeneous, isotropic material is given by:

$$v = \left[\frac{\text{appropriate elastic modulus of material}}{\text{density of material } \rho}\right]^{1/2}$$

Hence the velocity v_p of a compressional body wave, which involves a uniaxial compressional strain, is given by

$$v_p = \left[\frac{\psi}{\rho}\right]^{1/2}$$

or, since $\psi = K + \frac{4}{3}\mu$, by

$$v_p = \left[\frac{K + \frac{4}{3}\mu}{\rho}\right]^{1/2}$$

and the velocity v_s of a shear body wave, which involves a pure shear strain, is given by

$$v_s = \left[\frac{\mu}{\rho}\right]^{1/2}$$

It will be seen from these equations that compressional waves always travel faster than shear waves in the same medium. The ratio v_p/v_s in any material is determined solely by the value of Poisson's ratio (σ) for that material

$$v_p/v_s = \left[\frac{2(1-\sigma)}{(1-2\sigma)}\right]^{1/2}$$

and since Poisson's ratio for consolidated rocks is typically about 0.25, $v_p \approx 1.7v_s$. While knowledge of the P-wave velocity is useful, it is a function of three separate properties of the rock and is only a very ambiguous indicator of rock lithology. The v_p/v_s ratio, however, is

(a) P-wave

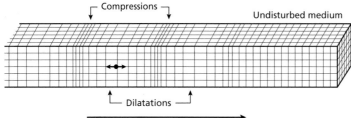

(b) S-wave

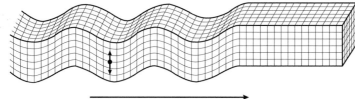

Fig. 3.3 Elastic deformations and ground particle motions associated with the passage of body waves. (a) P-wave. (b) S-wave. (From Bolt 1982.)

independent of density and can be used to derive Poisson's ratio, which is a much more diagnostic lithological indicator. If this information is required, then both v_p and v_s must be determined in the seismic survey.

These fundamental relationships between the velocity of the wave propagation and the physical properties of the materials through which the waves pass are independent of the frequency of the waves. Body waves are non-dispersive; that is, all frequency components in a wave train or pulse travel through any material at the same velocity, determined only by the elastic moduli and density of the material.

Historically, most seismic surveying has used only compressional waves, since this simplifies the survey technique in two ways. Firstly, seismic detectors which record only the vertical ground motion can be used, and these are insensitive to the horizontal motion of S-waves. Secondly, the higher velocity of P-waves ensures that they always reach a detector before any related S-waves, and hence are easier to recognize. Recording S-waves, and to a lesser extent surface waves, gives greater information about the subsurface, but at a cost of greater data acquisition (three-component recording) and consequent processing effort. As technology advances multicomponent surveys are becoming more commonplace.

One application of shear wave seismology is in engineering site investigation where the separate measurement of v_p and v_s for near-surface layers allows direct calculation of Poisson's ratio and estimation of the elas-

tic moduli, which provide valuable information on the *in situ* geotechnical properties of the ground. These may be of great practical importance, such as the value of *rippability* (see Section 5.11.1).

3.3.2 Surface waves

In a bounded elastic solid, seismic waves known as surface waves can propagate along the boundary of the solid. *Rayleigh waves* propagate along a free surface, or along the boundary between two dissimilar solid media, the associated particle motions being elliptical in a plane perpendicular to the surface and containing the direction of propagation (Fig. 3.4(a)). The orbital particle motion is in the opposite sense to the circular particle motion associated with an oscillatory water wave, and is therefore sometimes described as *retrograde*. A further major difference between Rayleigh waves and oscillatory water waves is that the former involve a shear strain and are thus restricted to solid media. The amplitude of Rayleigh waves decreases exponentially with distance below the surface. They have a propagation velocity lower than that of shear body waves and in a homogeneous half-space they would be non-dispersive. In practice, Rayleigh waves travelling round the surface of the Earth are observed to be dispersive, their waveform undergoing progressive change during propagation as a result of the different frequency components travelling at different velocities. This dispersion is directly attributable to velocity variation with depth in the Earth's

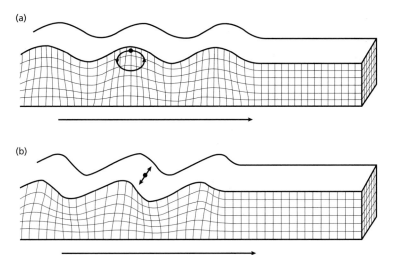

Fig. 3.4 Elastic deformations and ground particle motions associated with the passage of surface waves. (a) Rayleigh wave. (b) Love wave. (From Bolt 1982.)

interior. Analysis of the observed pattern of dispersion of earthquake waves is a powerful method of studying the velocity structure of the lithosphere and asthenosphere (Knopoff 1983). The same methodology, applied to the surface waves generated by a sledgehammer, can be used to examine the strength of near-surface materials for civil engineering investigations.

If the surface is layered and the surface layer shear wave velocity is lower than that of the underlying layer, a second set of surface waves is generated. *Love waves* are polarized shear waves with a particle motion parallel to the free surface and perpendicular to the direction of wave propagation (Fig. 3.4(b)). The velocity of Love waves is intermediate between the shear wave velocity of the surface layer and that of deeper layers, and Love waves are inherently dispersive. The observed pattern of Love wave dispersion can be used in a similar way to Rayleigh wave dispersion to study the subsurface structure.

3.3.3 Waves and rays

A seismic pulse propagates outwards from a seismic source at a velocity determined by the physical properties of the surrounding rocks. If the pulse travels through a homogeneous rock it will travel at the same velocity in all directions away from the source so that at any subsequent time the *wavefront*, defined as the locus of all points which the pulse has reached at a particular time, will be a sphere. *Seismic rays* are defined as thin pencils of seismic energy travelling along ray paths that, in isotropic media, are everywhere perpendicular to wavefronts (Fig. 3.5).

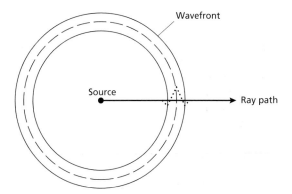

Fig. 3.5 The relationship of a ray path to the associated wavefront.

Rays have no physical significance but represent a useful concept in discussing travel paths of seismic energy through the ground.

It should be noted that the propagation velocity of a seismic wave is the velocity with which the seismic energy travels through a medium. This is completely independent of the velocity of a particle of the medium perturbed by the passage of the wave. In the case of compressional body waves, for example, their propagation velocity through rocks is typically a few thousand metres per second. The associated oscillatory ground motions involve *particle velocities* that depend on the amplitude of the wave. For the weak seismic events routinely recorded in seismic surveys, particle velocities may be as small as $10^{-8}\,\mathrm{m\,s^{-1}}$ and involve ground displacements of only about 10^{-10} m. The detection of seismic waves involves measuring these very small particle velocities.

3.4 Seismic wave velocities of rocks

By virtue of their various compositions, textures (e.g. grain shape and degree of sorting), porosities and contained pore fluids, rocks differ in their elastic moduli and densities and, hence, in their seismic velocities. Information on the compressional and shear wave velocities, v_p and v_s, of rock layers encountered by seismic surveys is important for two main reasons: firstly, it is necessary for the conversion of seismic wave travel times into depths; secondly, it provides an indication of the lithology of a rock or, in some cases, the nature of the pore fluids contained within it.

To relate rock velocities to lithology, the assumption that rocks are uniform and isotropic in structure must be reviewed. A typical rock texture can be regarded as having mineral grains making up most of the rock (the *matrix*), with the remaining volume being occupied by void space (the *pores*). The fractional volume of pore space is the *porosity* (ϕ) of the rock. For simplicity it may be assumed that all the matrix grains have the same physical properties. This is a surprisingly good approximation since the major rock-forming minerals, quartz, feldspar and calcite, have quite similar physical properties. In this case, the properties of the bulk rock will be an average of the properties of the matrix minerals and the pore fluid, weighted according to the porosity. The simplest case is for the density of a rock, where the bulk density ρ_b can be related to the matrix and pore fluid densities (ρ_m, ρ_f):

$$\rho_b = \rho_f \phi + (1 - \phi) \rho_m$$

For P-wave velocity a similar relationship exists, but the velocity weighting is proportional to the percentage of travel-time spent in each component of the system, which is inversely proportional to velocity, giving the relationship:

$$\frac{1}{v_b} = \frac{\phi}{v_f} + \frac{(1 - \phi)}{v_m}$$

From the above equations it is possible to produce cross-plot graphs (Fig. 3.6) which allow the estimation of the matrix grain type and the porosity of a rock, purely from the seismic P-wave velocity and density.

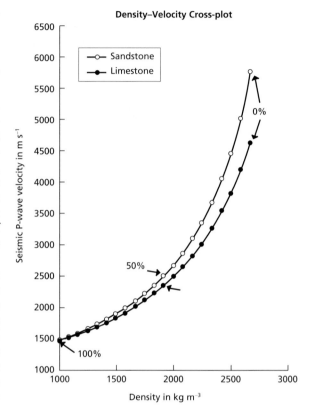

Fig. 3.6 The relationship of seismic velocity and density to porosity, calculated for mono-mineralic granular solids: open circles – sandstone, calculated for a quartz matrix; solid circles – limestone, calculated for a calcite matrix. Points annotated with the corresponding porosity value 0–100%. Such relationships are useful in borehole log interpretation (see Chapter 11).

For S-wave velocity, the derivation of bulk velocity is more complex since S-waves will not travel through pore spaces at all. This is an interesting point, since it suggests that the S-wave velocity depends only on the matrix grain properties and their texture, while the P-wave velocity is also influenced by the pore fluids. In principle it is then possible, if both the P-wave and S-wave velocity of a formation are known, to detect variations in pore fluid. This technique is used in the hydrocarbon industry to detect gas-filled pore spaces in underground hydrocarbon reservoirs.

Rock velocities may be measured *in situ* by field measurement, or in the laboratory using suitably prepared rock samples. In the field, seismic surveys yield estimates of velocity for rock layers delineated by reflecting or refracting interfaces, as discussed in detail in Chapters 4 and 5. If boreholes exist in the vicinity of a seismic survey, it may be possible to correlate velocity values so derived with individual rock units encountered within borehole sequences. As discussed in Chapter 11, velocity may also be measured directly in boreholes using a sonic probe, which emits high-frequency pulses and measures the travel time of the pulses through a small vertical interval of wall rock. Drawing the probe up through the borehole yields a *sonic log*, or continuous velocity log (CVL), which is a record of velocity variation through the borehole section (see Section 11.8, Fig. 11.14).

In the laboratory, velocities are determined by measuring the travel-time of high-frequency (about 1 MHz) acoustic pulses transmitted through cylindrical rock specimens. By this means, the effect on velocity of varying temperature, confining pressure, pore fluid pressure or composition may be quantitatively assessed. It is important to note that laboratory measurements at low confining pressures are of doubtful validity. The intrinsic velocity of a rock is not normally attained in the laboratory below a confining pressure of about 100 MPa (megapascals), or 1 kbar, at which pressure the original solid contact between grains characteristic of the pristine rock is re-established.

The following empirical findings of velocity studies are noteworthy:

1. Compressional wave velocity increases with confining pressure (very rapidly over the first 100 MPa).

2. Sandstone and shale velocities show a systematic increase with depth of burial and with age, due to the combined effects of progressive compaction and cementation.

3. For a wide range of sedimentary rocks the compressional wave velocity is related to density, and well-established velocity–density curves have been published (Sheriff & Geldart 1983; see Section 6.9, Fig. 6.16). Hence, the densities of inaccessible subsurface layers may be predicted if their velocity is known from seismic surveys.

4. The presence of gas in sedimentary rocks reduces the elastic moduli, Poisson's ratio and the v_p/v_s ratio. v_p/v_s ratios greater than 2.0 are characteristic of unconsolidated sand, whilst values less than 2.0 may indicate either a consolidated sandstone or a gas-filled unconsolidated sand. The potential value of v_s in detecting gas-filled sediments accounts for the current interest in shear wave seismic surveying.

Typical compressional wave velocity values and ranges for a wide variety of Earth materials are given in Table 3.1.

Table 3.1 Compressional wave velocities in Earth materials.

	v_p (km s^{-1})
Unconsolidated materials	
Sand (dry)	0.2–1.0
Sand (water-saturated)	1.5–2.0
Clay	1.0–2.5
Glacial till (water-saturated)	1.5–2.5
Permafrost	3.5–4.0
Sedimentary rocks	
Sandstones	2.0–6.0
Tertiary sandstone	2.0–2.5
Pennant sandstone (Carboniferous)	4.0–4.5
Cambrian quartzite	5.5–6.0
Limestones	2.0–6.0
Cretaceous chalk	2.0–2.5
Jurassic oolites and bioclastic limestones	3.0–4.0
Carboniferous limestone	5.0–5.5
Dolomites	2.5–6.5
Salt	4.5–5.0
Anhydrite	4.5–6.5
Gypsum	2.0–3.5
Igneous/Metamorphic rocks	
Granite	5.5–6.0
Gabbro	6.5–7.0
Ultramafic rocks	7.5–8.5
Serpentinite	5.5–6.5
Pore fluids	
Air	0.3
Water	1.4–1.5
Ice	3.4
Petroleum	1.3–1.4
Other materials	
Steel	6.1
Iron	5.8
Aluminium	6.6
Concrete	3.6

3.5 Attenuation of seismic energy along ray paths

As a seismic pulse propagates in a homogeneous material, the original energy E transmitted outwards from the source becomes distributed over a spherical shell, the wavefront, of expanding radius. If the radius of the wavefront is r, the amount of energy contained within a unit area of the shell is $E/4\pi r^2$. With increasing distance along a ray path, the energy contained in the ray falls off as r^{-2} due to the effect of the *geometrical spreading* of the energy.

Wave amplitude, which is proportional to the square root of the wave energy, therefore falls off as r^{-1}.

A further cause of energy loss along a ray path arises because, even at the low strains involved, the ground is imperfectly elastic in its response to the passage of seismic waves. Elastic energy is gradually absorbed into the medium by internal frictional losses, leading eventually to the total disappearance of the seismic disturbance. The mechanisms for the absorption of energy are complex, but the loss of energy is usually regarded as being a fixed proportion of the total energy, for each oscillation of the rock particles involved, during which time the wavefront will have moved forward one wavelength. The *absorption coefficient* α expresses the proportion of energy lost during transmission through a distance equivalent to a complete wavelength λ. Values of α for common Earth materials range from 0.25 to 0.75 dB λ^{-1} (for a definition of decibels, dB, see Section 2.2).

Over the range of frequencies used in seismic surveying the absorption coefficient is normally assumed to be independent of frequency. If the amount of absorption per wavelength is constant, it follows that higher frequency waves attenuate more rapidly than lower frequency waves as a function of time or distance. To illustrate this point, consider two waves with frequencies of 10 Hz and 100 Hz to propagate through a rock in which $v_p = 2.0\,\mathrm{km\,s^{-1}}$ and $\alpha = 0.5\,\mathrm{dB}\,\lambda^{-1}$. The 100 Hz wave ($\lambda = 20\,\mathrm{m}$) will be attenuated due to absorption by 5 dB over a distance of 200 m, whereas the 10 Hz wave ($\lambda = 200\,\mathrm{m}$) will be attenuated by only 0.5 dB over the same distance. The shape of a seismic pulse with a broad frequency content therefore changes continuously during propagation due to the progressive loss of the higher frequencies. In general, the effect of absorption is to produce a progressive lengthening of the seismic pulse (Fig. 3.7). This effect of absorption is a familiar experience as it applies to P-waves in air, sound. The sharp crack of a nearby lightning flash is heard far away as the distant 'rumble' of thunder.

3.6 Ray paths in layered media

At an interface between two rock layers there is generally a change of propagation velocity resulting from the difference in physical properties of the two layers. At such an interface, the energy within an incident seismic pulse is partitioned into transmitted and reflected pulses. The relative amplitudes of the transmitted and reflected pulses depend on the velocities and densities

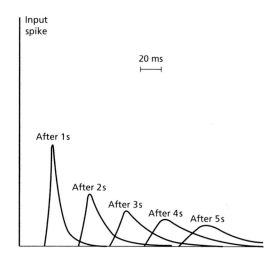

Fig. 3.7 The progressive change of shape of an original spike pulse during its propagation through the ground due to the effects of absorption. (After Anstey 1977.)

of the two layers, and the angle of incidence on the interface.

3.6.1 Reflection and transmission of normally incident seismic rays

Consider a compressional ray of amplitude A_0 normally incident on an interface between two media of differing velocity and density (Fig. 3.8). A transmitted ray of amplitude A_2 travels on through the interface in the same direction as the incident ray and a reflected ray of amplitude A_1 returns back along the path of the incident ray.

The total energy of the transmitted and reflected rays must equal the energy of the incident ray. The relative proportions of energy transmitted and reflected are determined by the contrast in *acoustic impedance* Z across the interface. The acoustic impedance of a rock is the product of its density (ρ) and its wave velocity (v); that is,

$$Z = \rho v$$

It is difficult to relate acoustic impedance to a tangible rock property but, in general, the harder a rock, the higher is its acoustic impedance. Intuitively, the smaller the contrast in acoustic impedance across a rock interface the greater is the proportion of energy transmitted through the interface. Obviously all the energy is transmitted if the rock material is the same on both sides of the

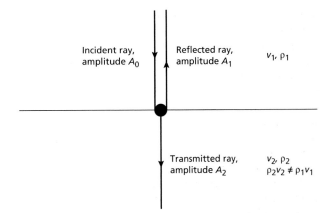

Fig. 3.8 Reflected and transmitted rays associated with a ray normally incident on an interface of acoustic impedance contrast.

interface, and more energy is reflected the greater the contrast. From common experience with sound, the best echoes come from rock or brick walls. In terms of physical theory, acoustic impedance is closely analogous to electrical impedance and, just as the maximum transmission of electrical energy requires a matching of electrical impedances, so the maximum transmission of seismic energy requires a matching of acoustic impedances.

The *reflection coefficient* R is a numerical measure of the effect of an interface on wave propagation, and is calculated as the ratio of the amplitude A_1 of the reflected ray to the amplitude A_0 of the incident ray

$$R = A_1 / A_0$$

To relate this simple measure to the physical properties of the materials at the interface is a complex problem. As we have already seen, the propagation of a P-wave depends on the bulk and shear elastic moduli, as well as the density of the material. At the boundary the stress and strain in the two materials must be considered. Since the materials are different, the relations between stress and strain in each will be different. The orientation of stress and strain to the interface also becomes important. The formal solution of this physical problem was derived early in the 20th century, and the resulting equations are named the Zoeppritz equations (Zoeppritz 1919; and for explanation of derivations see Sheriff & Geldart 1982). Here, the solutions of these equations will be accepted. For a normally incident ray the relationships are fairly simple, giving:

$$R = \frac{\rho_2 v_2 - \rho_1 v_1}{\rho_2 v_2 + \rho_1 v_1} = \frac{Z_2 - Z_1}{Z_2 + Z_1}$$

where ρ_1, v_1, Z_1 and ρ_2, v_2, Z_2 are the density, P-wave velocity and acoustic impedance values in the first and second layers, respectively. From this equation it follows that $-1 \leq R \leq +1$. A negative value of R signifies a phase change of π (180°) in the reflected ray.

The *transmission coefficient* T is the ratio of the amplitude A_2 of the transmitted ray to the amplitude A_0 of the incident ray

$$T = A_2 / A_0$$

For a normally incident ray this is given, from solution of Zoeppritz's equations, by

$$T = \frac{2Z_1}{Z_2 + Z_1}$$

Reflection and transmission coefficients are sometimes expressed in terms of energy rather than wave amplitude. If energy intensity I is defined as the amount of energy flowing through a unit area normal to the direction of wave propagation in unit time, so that I_0, I_1 and I_2 are the intensities of the incident, reflected and transmitted rays respectively, then

$$R' = \frac{I_1}{I_0} = \left[\frac{Z_2 - Z_1}{Z_2 + Z_1} \right]^2$$

and

$$T' = \frac{I_2}{I_1} = \frac{4 Z_1 Z_2}{(Z_2 + Z_1)^2}$$

where R' and T' are the reflection and transmission coefficients expressed in terms of energy.

If R or $R' = 0$, all the incident energy is transmitted. This is the case when there is no contrast of acoustic impedance across an interface, even if the density and velocity values are different in the two layers (i.e. $Z_1 = Z_2$). If R or $R' = +1$ or -1, all the incident energy is reflected. A good approximation to this situation occurs at the free surface of a water layer: rays travelling upwards from an explosion in a water layer are almost totally reflected back from the water surface with a phase change of π ($R = -0.9995$).

Values of reflection coefficient R for interfaces between different rock types rarely exceed ± 0.5 and are typically much less than ± 0.2. Thus, normally the bulk of seismic energy incident on a rock interface is transmitted and only a small proportion is reflected. By use of an empirical relationship between velocity and density (see also Section 6.9), it is possible to estimate the reflection coefficient from velocity information alone (Gardner *et al.* 1974, Meckel & Nath 1977):

$$R = 0.625 \ln(v_1/v_2)$$

Such relationships can be useful, but must be applied with caution since rock lithologies are highly variable and laterally heterogeneous as pointed out in Section 3.4.

3.6.2 Reflection and refraction of obliquely incident rays

When a *P-wave ray* is obliquely incident on an interface of acoustic impedance contrast, reflected and transmitted P-wave rays are generated as in the case of normal incidence. Additionally, some of the incident compressional energy is converted into reflected and transmitted S-wave rays (Fig. 3.9) that are polarized in a vertical plane. Zoeppritz's equations show that the amplitudes of the four phases are a function of the angle of incidence θ. The converted rays may attain a significant magnitude at large angles of incidence. Detection and identification of converted waves can be difficult in seismic surveys, but they do have potential to provide more constraints on the physical properties of the media at the interface. Here consideration will be confined to the P-waves.

In the case of oblique incidence, the transmitted P-wave ray travels through the lower layer with a changed direction of propagation (Fig. 3.10) and is referred to as a *refracted ray*. The situation is directly analogous to the be-

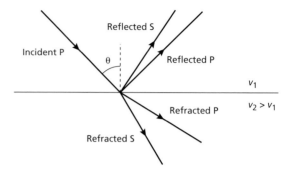

Fig. 3.9 Reflected and refracted P- and S-wave rays generated by a P-wave ray obliquely incident on an interface of acoustic impedance contrast.

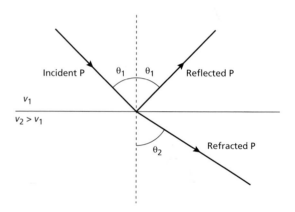

Fig. 3.10 Reflected and refracted P-wave rays associated with a P-wave rays obliquely incident on an interface of acoustic impedance contrast.

haviour of a light ray obliquely incident on the boundary between, say, air and water and *Snell's Law of Refraction* applies equally to the optical and seismic cases. Snell defined the *ray parameter* $p = \sin i/v$, where i is the angle of inclination of the ray in a layer in which it is travelling with a velocity v. The generalized form of Snell's Law states that, along any one ray, the ray parameter remains a constant.

For the refracted P-wave ray shown in Fig. 3.10, therefore

$$\frac{\sin \theta_1}{v_1} = \frac{\sin \theta_2}{v_2}$$

or

$$\frac{\sin \theta_1}{\sin \theta_2} = \frac{v_1}{v_2}$$

Note that if $v_2 > v_1$ the ray is refracted away from the normal to the interface; that is, $\theta_2 > \theta_1$. Snell's Law also applies to the reflected ray, from which it follows that the angle of reflection equals the angle of incidence (Fig. 3.10).

3.6.3 Critical refraction

When the velocity is higher in the underlying layer there is a particular angle of incidence, known as the *critical angle* θ_c, for which the angle of refraction is 90°. This gives rise to a critically refracted ray that travels along the interface at the higher velocity v_2. At any greater angle of incidence there is total internal reflection of the incident energy (apart from converted S-wave rays over a further range of angles). The critical angle is given by

$$\frac{\sin \theta_c}{v_1} = \frac{\sin 90°}{v_2} = \frac{1}{v_2}$$

so that

$$\theta_c = \sin^{-1}(v_1/v_2)$$

The passage of the critically refracted ray along the top of the lower layer causes a perturbation in the upper layer that travels forward at the velocity v_2, which is greater than the seismic velocity v_1 of that upper layer. The situation is analogous to that of a projectile travelling through air at a velocity greater than the velocity of sound in air and the result is the same, the generation of a shock wave. This wave is known as a *head wave* in the seis-

mic case, and it passes up obliquely through the upper layer towards the surface (Fig. 3.11). Any ray associated with the head wave is inclined at the critical angle i_c. By means of the head wave, seismic energy is returned to the surface after critical refraction in an underlying layer of higher velocity.

3.6.4 Diffraction

In the above discussion of the reflection and transmission of seismic energy at interfaces of acoustic impedance contrast it was implicitly assumed that the interfaces were continuous and approximately planar. At abrupt discontinuities in interfaces, or structures whose radius of curvature is shorter than the wavelength of incident waves, the laws of reflection and refraction no longer apply. Such phenomena give rise to a radial scattering of incident seismic energy known as *diffraction*. Common sources of diffraction in the ground include the edges of faulted layers (Fig. 3.12) and small isolated objects, such as boulders, in an otherwise homogeneous layer.

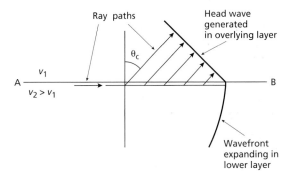

Fig. 3.11 Generation of a head wave in the upper layer by a wave propagating through the lower layer.

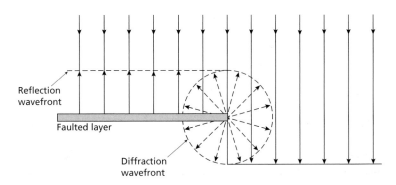

Fig. 3.12 Diffraction caused by the truncated end of a faulted layer.

(a)

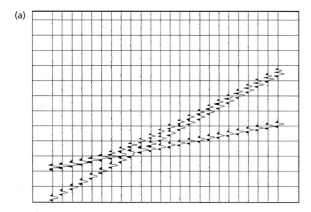

(b)

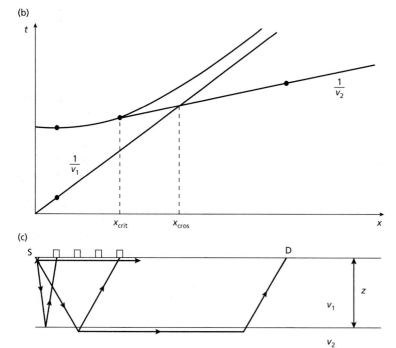

(c)

Fig. 3.13 (a) Seismogram showing the output traces of 24 geophones distributed along the Earth's surface as a function of time. (b) Travel-time curves for direct, reflected and refracted rays in the case of a simple two-layer model. (c) Direct, reflected and refracted ray paths from a near surface source to a surface detector in the case of a simple two-layer model.

Diffracted phases are commonly observed in seismic recordings and are sometimes difficult to discriminate from reflected and refracted phases, as discussed in Chapter 4.

3.7 Reflection and refraction surveying

Consider the simple geological section shown in Fig. 3.13 involving two homogeneous layers of seismic velocities v_1 and v_2 separated by a horizontal interface at a depth z, the compressional wave velocity being higher in the underlying layer (i.e. $v_2 > v_1$).

From a near-surface seismic source S there are three types of ray path by which energy reaches the surface at a distance from the source, where it may be recorded by a suitable detector as at D, a horizontal distance x from S. The *direct ray* travels along a straight line through the top layer from source to detector at velocity v_1. The *reflected ray* is obliquely incident on the interface and is reflected back through the top layer to the detector, travelling along its entire path at the top layer velocity v_1. The *refracted ray* travels obliquely down to the interface at velocity v_1, along a segment of the interface at the higher velocity v_2, and back up through the upper layer at v_1.

The travel time of a direct ray is given simply by

$$t_{dir} = x/v_1$$

which defines a straight line of slope $1/v_1$ passing through the time–distance origin.

The travel time of a reflected ray is given by

$$t_{refl} = \frac{(x^2 + 4z^2)^{1/2}}{v_1}$$

which, as discussed in Chapter 4, is the equation of an hyperbola.

The travel time of a refracted ray (for derivation see Chapter 5) is given by

$$t_{refr} = \frac{x}{v_2} + \frac{2z \cos \theta_c}{v_1}$$

which is the equation of a straight line having a slope of $1/v_2$ and an intercept on the time axis of

$$\frac{2z \cos \theta_c}{v_1}$$

Travel-time curves, or time–distance curves, for direct, refracted and reflected rays are illustrated in Fig. 3.13. By suitable analysis of the travel-time curve for reflected or refracted rays it is possible to compute the depth to the underlying layer. This provides two independent seismic surveying methods for locating and mapping subsurface interfaces, *reflection surveying* and *refraction surveying*. These have their own distinctive methodologies and fields of application and they are discussed separately in detail in Chapters 4 and 5. However, some general remarks about the two methods may be made here with reference to the travel-time curves and seismogram of Fig. 3.13. The curves are more complicated in the case of a multilayered model, but the following remarks still apply.

The first arrival of seismic energy at a surface detector offset from a surface source is always a direct ray or a refracted ray. The direct ray is overtaken by a refracted ray at the *crossover distance* x_{cros}. Beyond this offset distance the first arrival is always a refracted ray. Since critically re-

fracted rays travel down to the interface at the critical angle there is a certain distance, known as the *critical distance* x_{crit}, within which refracted energy will not be returned to the surface. At the critical distance, the travel times of reflected rays and refracted rays coincide because they follow effectively the same path. Reflected rays are never first arrivals; they are always preceded by direct rays and, beyond the critical distance, by refracted rays also.

The above characteristics of the travel-time curves determine the methodology of refraction and reflection surveying. In refraction surveying, recording ranges are chosen to be sufficiently large to ensure that the crossover distance is well exceeded in order that refracted rays may be detected as first arrivals of seismic energy. Indeed, some types of refraction survey consider only these first arrivals, which can be detected with unsophisticated field recording systems. In general, this approach means that the deeper a refractor, the greater is the range over which recordings of refracted arrivals need to be taken.

In reflection surveying, by contrast, reflected phases are sought that are never first arrivals and are normally of very low amplitude because geological reflectors tend to have small reflection coefficients. Consequently, reflections are normally concealed in seismic records by higher amplitude events such as direct or refracted body waves, and surface waves.

Reflection surveying methods therefore have to be capable of discriminating between reflected energy and many types of synchronous noise. Recordings are normally restricted to small offset distances, well within the critical distance for the reflecting interfaces of main interest. However, in multichannel reflection surveying recordings are conventionally taken over a significant range of offset distances, for reasons that are discussed fully in Chapter 4.

3.8 Seismic data acquisition systems

The fundamental purpose of seismic surveys is accurately to record the ground motion caused by a known source in a known location. The record of ground motion with time constitutes a *seismogram* and is the basic information used for interpretation through either modelling or imaging (see Chapter 2). The essential instrumental requirements are to

• generate a seismic pulse with a suitable *source*
• detect the seismic waves in the ground with a suitable *transducer*

• record and display the seismic waveforms on a suitable *seismograph*.

The general methodology of examining hidden structures by studying their effects on artificially generated acoustic or seismic waves has an enormously wide range of applications covering a wide range of spatial scales. Perhaps the smallest scale is ultrasound imaging in medicine, which can also be applied industrially to examining engineering structures. Within the more geophysical applications, the scales range from depths of a metre or less in engineering, environmental or archaeological surveys to tens of kilometres for crustal and upper mantle studies.

For each application there is a limit to the smallest structures that can be detected, known as the *resolution* of the survey. The resolution is basically determined by the pulse length: for a pulse of any particular length there is a minimum separation below which the pulses will overlap in time in the seismic recording. Although the pulse length may be shortened at the processing stage by deconvolution (see Section 4.8.2), this is only possible if the data are of good quality, and is a complement to, not a substitute for, good survey design. The pulse width is determined by both the maximum frequency and the frequency bandwidth of the recorded signal. Since Earth materials absorb seismic energy in a frequency-selective way (Section 3.5), the optimum waveform will be specific to each survey. It is an important characteristic of all geophysical surveys, and particularly seismic ones, that they must be designed individually for each specific case. The general aspects of the equipment used for seismic surveys are reviewed here; specific variations for reflection and refraction surveying are described in Chapters 4 and 5.

3.8.1 Seismic sources and the seismic/acoustic spectrum

A seismic source is a localized region within which the sudden release of energy leads to a rapid stressing of the surrounding medium. The archetypal seismic source is an explosion. While explosives are still used, there is an increasing number of more sophisticated and efficient (and safe!) ways to collect seismic data.

The main requirements of the seismic source are:
• Sufficient energy across the broadest possible frequency range, extending up to the highest recordable frequencies.
• Energy should be concentrated in the type of wave energy which is required for a specific survey, either P-

wave or S-wave, and generate minimum energy of other wave types. Such other unwanted energy would degrade the recorded data and be classed as *coherent noise*.
• The source waveform must be repeatable. Seismic surveys almost always involve comparing the seismograms generated by a series of sources at different locations. Variations on the seismograms should be diagnostic of the ground structure, not due to random variations of the source.
• The source must be safe, efficient, and environmentally acceptable. Most seismic surveys are commercial operations which are governed by safety and environmental legislation. They must be as cost-effective as possible. Sometimes the requirements for efficiency lead to higher safety and environmental standards than legally enforced. Whether involving personal injury or not, accidents are referred to as 'lost-time incidents'. Safety aids efficiency as well as being desirable from many other viewpoints.

The complete seismic/acoustic spectrum is shown in Fig. 3.14. There is a very wide variety of seismic sources, characterized by differing energy levels and frequency characteristics. In general, a seismic source contains a wide range of frequency components within the range from 1 Hz to a few hundred hertz, though the energy is often concentrated in a narrower frequency band.

Source characteristics can be modified by the use of several similar sources in an array designed, for example, to improve the frequency spectrum of the transmitted pulse. This matter is taken up in Chapter 4 when discussing the design parameters of seismic reflection surveys.

Explosive sources

On land, explosives are normally detonated in shallow shot holes to improve the coupling of the energy source with the ground and to minimize surface damage. Explosives offer a reasonably cheap and highly efficient seismic source with a wide frequency spectrum, but their use normally requires special permission and presents logistical difficulties of storage and transportation. They are slow to use on land because of the need to drill shot holes. Their main shortcoming, however, is that they do not provide the type of precisely repeatable source signature required by modern processing techniques, nor can the detonation of explosives be repeated at fixed and precise time intervals as required for efficient reflection profiling at sea carried out by survey vessels underway. Since explosive sources thus fail at least two,

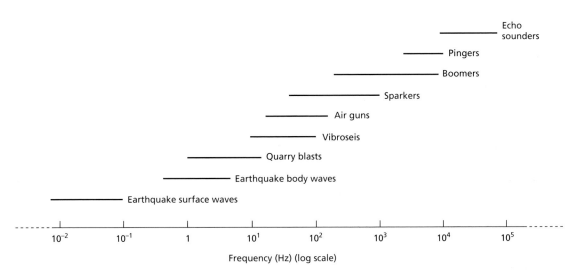

Fig. 3.14 The seismic/acoustic spectrum.

and usually three, of the basic requirements for modern surveys, their use is steadily declining and limited to locations where alternative sources cannot be used.

Non-explosive land sources

Vibroseis® is the most common non-explosive source used for reflection surveying. It uses truck-mounted vibrators to pass into the ground an extended vibration of low amplitude and continuously varying frequency, known as a *sweep signal*. A typical sweep signal lasts from several seconds up to a few tens of seconds and varies progressively in frequency between limits of about 10 and 80 Hz. The field recordings consist of overlapping reflected wave trains of very low amplitude concealed in the ambient seismic noise. In order both to increase the signal-to-noise ratio (SNR) and to shorten the pulse length, each recorded seismogram is cross-correlated (see Section 2.4.3) with the known sweep signal to produce a correlated seismogram or *correlogram*. The correlogram has a similar appearance to the type of seismogram that would be obtained with a high-energy impulsive source such as an explosion, but the seismic arrivals appear as symmetrical (zero phase) wavelets known as *Klauder wavelets* (Fig. 3.15).

The Vibroseis® source is quick and convenient to use and produces a precisely known and repeatable signal. The vibrator unit needs a firm base on which to operate, such as a tarmac road, and it will not work well on soft ground. The peak force of a vibrator is only about 10^5 N

and, to increase the transmitted energy for deep penetration surveys, vibrators are typically employed in groups with a phase-locked response. Multiple sweeps are commonly employed, the recordings from individual sweeps being added together (stacked) to increase the SNR. A particular advantage of vibrators is that they can be used in towns since they cause no damage or significant disturbance to the environment. The cross-correlation method of extracting the signal is also capable of coping with the inherently high noise levels of urban areas. Some Vibroseis® trucks are adapted so that the vibration direction can be horizontal rather than vertical. In this case the truck can also be used as an S-wave source. A principal disadvantage of the Vibroseis® method is that each fully configured truck costs of the order of half a million dollars. While the method is effective for major hydrocarbon surveys, the costs are prohibitive for small surveys. Small electro-mechanical vibrators have been developed for shallow geophysical surveys, and these are gaining increasing acceptance as seismographs capable of receiving and correlating the signals are developed.

Mini-Sosie adapts the principle of using a precisely known source signature of long duration to cheaper, lower energy applications. A pneumatic hammer delivers a random sequence of impacts to a base plate, thus transmitting a pulse-encoded signal of low amplitude into the ground. The source signal is recorded by a detector on the base plate and used to cross-correlate with the field recordings of reflected arrivals of the

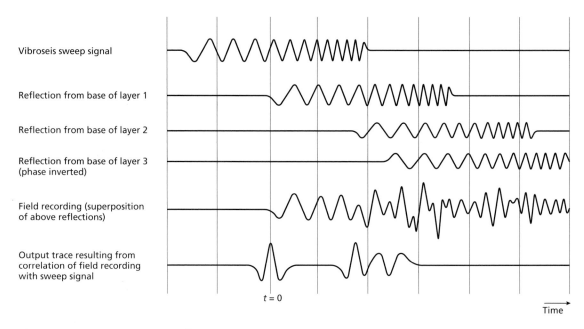

Vibroseis sweep signal

Reflection from base of layer 1

Reflection from base of layer 2

Reflection from base of layer 3
(phase inverted)

Field recording (superposition
of above reflections)

Output trace resulting from
correlation of field recording
with sweep signal

$t = 0$

Time

Fig. 3.15 Cross-correlation of a Vibroseis® seismogram with the input sweep signal to locate the positions of occurrence of reflected arrivals.

pulse-encoded signal from buried interfaces. Peaks in the cross-correlation function reveal the positions of reflected signals in the recordings.

Weight drops and hammers. Perhaps the simplest land seismic source is a large mass dropped on to the ground surface. Weight drops have been manufactured in a wide variety of forms from eight-wheel trucks dropping a weight of several tonnes, to a single person with a sledgehammer. If the source energy required is relatively low, these types of sources can be fast and efficient. The horizontal impact of a weight or hammer on to one side of a vertical plate partially embedded in the ground can be used as a source for shear wave seismology.

Shotguns, buffalo guns and rifles. One solution to gaining additional energy for small-scale surveys is to use the compact chemical energy in small-arms ammunition. Rifles have been used as seismic sources by firing the bullet into the ground. While effective as a very high-frequency source, this is banned by legislation in many countries. An alternative is to fire a blank shotgun cartridge in a hole using a suitable device, commonly termed a buffalo gun (Fig 3.16). The blank shotgun cartridge offers an impulsive source giving considerably more energy than a sledgehammer, with few of the safety problems of explosives.

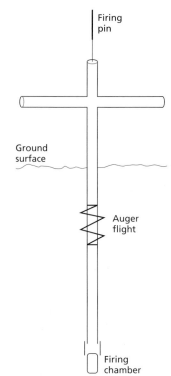

Firing
pin

Ground
surface

Auger
flight

Firing
chamber

Fig. 3.16 Schematic cross-section of a typical buffalo gun. The cartridge is fired by dropping a simple firing pin on to the cartridge.

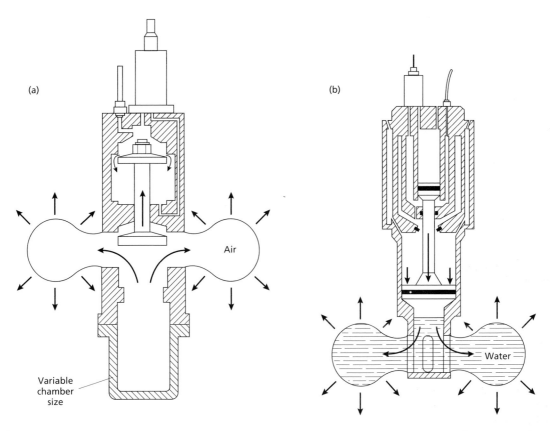

Fig. 3.17 Schematic cross-sections through (a) a Bolt air gun and (b) a Sodera water gun to illustrate the principles of operation. (Redrawn with permission of Bolt Associates and Sodera Ltd.)

Marine sources

Air guns (Fig. 3.17(a)) are pneumatic sources in which a chamber is charged with very high-pressure (typically 10–15 MPa) compressed air fed through a hose from a shipboard compressor. The air is released, by electrical triggering, through vents into the water in the form of a high-pressure bubble. A wide range of chamber volumes are available, leading to different energy outputs and frequency characteristics. The primary pulse generated by an air gun is followed by a train of *bubble pulses* that increase the overall length of the pulse. Bubble pulses are caused by the oscillatory expansion and collapse of secondary gas bubbles following collapse of the initial bubble. They have the effect of unduly lengthening the seismic pulse. Steps can, however, be taken to suppress the effect of the bubble pulse by detonating near to the water surface so that the gas bubble escapes into the air. While this does remove the bubble pulse

effect, much energy is wasted and the downgoing seismic pulse is weakened. More sophisticated methods can be used to overcome the bubble pulse problem while preserving seismic efficiency. Arrays of guns of differing dimensions and, therefore, different bubble pulse periods can be combined to produce a high-energy source in which primary pulses interfere constructively whilst bubble pulses interfere destructively (Fig. 3.18). For deep penetration surveys the total energy transmitted may be increased by the use of multiple arrays of air guns mounted on a frame that is towed behind the survey vessel. Air guns are mechanically simple and can operate with great reliability and repeatability. They have become the standard marine seismic source.

Water guns (Fig. 3.17(b)) are an adaptation of air guns to avoid the bubble pulse problem. The compressed air, rather than being released into the water layer, is used to drive a piston that ejects a water jet into the surrounding

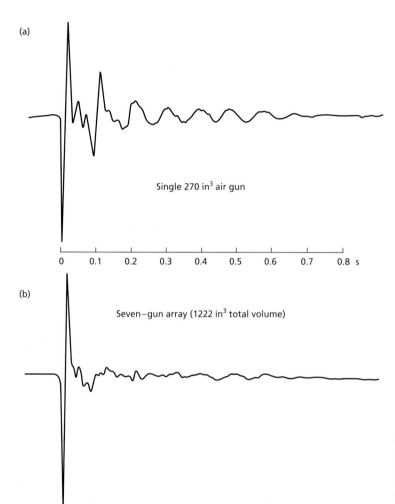

(a)

Single 270 in³ air gun

0   0.1   0.2   0.3   0.4   0.5   0.6   0.7   0.8 s

(b)

Seven–gun array (1222 in³ total volume)

Fig. 3.18 Comparison of the source signatures of (a) a single air gun (peak pressure: 4.6 bar metres) and (b) a seven-gun array (peak pressure: 19.9 bar metres). Note the effective suppression of bubble pulses in the latter case. (Redrawn with permission of Bolt Associates.)

water. When the piston stops, a vacuum cavity is created behind the advancing water jet and this implodes under the influence of the ambient hydrostatic pressure, generating a strong acoustic pulse free of bubble oscillations. Since the implosion represents collapse into a vacuum, no gaseous material is compressed to 'bounce back' as a bubble pulse. The resulting short pulse length offers a potentially higher resolution than is achieved with air guns but at the expense of a more complex initial source pulse due to the piston motion.

Several marine sources utilize explosive mixtures of gases, but these have not achieved the same safety and reliability, and hence industry acceptance, as air guns. In *sleeve exploders*, propane and oxygen are piped into a submerged flexible rubber sleeve where the gaseous mix-

ture is fired by means of a spark plug. The products of the resultant explosion cause the sleeve to expand rapidly, generating a shock wave in the surrounding water. The exhaust gases are vented to surface through a valve that opens after the explosion, thus attenuating the growth of bubble pulses.

Marine Vibroseis®. Whilst vibrators were developed for land surveys, it is of interest to note that experiments have been carried out using marine vibrator units, with special baseplates, deployed in fixtures attached to a survey vessel (Baeten *et al.* 1988).

Sparkers are devices for converting electrical energy into acoustic energy. The sparker pulse is generated by the discharge of a large capacitor bank directly into the sea water through an array of electrodes towed in a frame behind the survey vessel. Operating voltages are typically 3.5–4.0 kV and peak currents may exceed 200 A. This electrical discharge leads to the formation and rapid growth of a plasma bubble and the consequent generation of an acoustic pulse. For safety reasons, sparkers are increasingly being replaced by other sources.

Boomers comprise a rigid aluminium plate attached below a heavy-duty electrical coil by a spring-loaded mounting. A capacitor bank is discharged through the coil and the electromagnetic induction thus generated forces the aluminium plate rapidly downwards, setting up a compressional wave in the water. The device is typically towed behind the survey vessel in a catamaran mounting.

Sparkers and boomers generate broad-band acoustic pulses and can be operated over a wide range of energy levels so that the source characteristics can to some extent be tailored to the needs of a particular survey. In general, boomers offer better resolution (down to 0.5 m) but more restricted depth penetration (a few hundred metres maximum).

Pingers consist of small ceramic piezoelectric transducers, mounted in a towing fish, which, when activated by an electrical impulse, emit a very short, high-frequency acoustic pulse of low energy. They offer a very high resolving power (down to 0.1 m) but limited penetration (a few tens of metres in mud, much less in sand or rock). They are useful in offshore engineering applications such as surveys of proposed routes for submarine pipelines.

Chirp systems are electro-mechanical transducers that produce an extended, repeatable, source waveform which allows greater energy output. This longer signal can be compressed in processing to give greater resolution and/or better signal-to-noise ratio.

Further discussion of the use of air guns, sparkers, boomers and pingers in single-channel seismic reflection profiling systems is given in Section 4.15.

3.8.2 Seismic transducers

Conversion of the ground motion to an electrical signal requires a transducer which is sensitive to some component of the ground motion, and can record the required range of frequencies and amplitudes without distortion. The first issue is which component of the motion to measure. As the ground oscillates, it is possible to measure either the displacement, velocity or acceleration of the ground particles as the wave passes. The ground motion also takes place in three dimensions. To record it

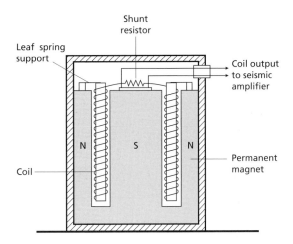

Fig. 3.19 Schematic cross-section through a moving-coil geophone.

faithfully requires knowledge of the components of motion in the vertical, east–west and north–south directions. Historically, the solution has been to measure the vertical component of velocity only. The vertical component has maximum sensitivity to P-waves relative to S-waves and surface waves. Velocity is technically easier to detect and record than either displacement or acceleration. Recording only one component minimizes the technical problems of data storage.

Devices used on land to detect seismic ground motions are known as *seismometers* or *geophones*. In water, the passage of a compressional seismic wave is marked by transient pressure changes and these are detected by *hydrophones* towed or suspended in the water column or, in very shallow water, laid on the sea bed. Hydrophones may also be used in the water-saturated ground conditions encountered in swamps or marshland. Detectors may comprise individual geophones or hydrophones, or arrays of these devices connected together in series or parallel to provide a summed output.

Geophones are made to several designs, but the most common is the *moving-coil* geophone (Fig. 3.19). A cylindrical coil is suspended from a spring support in the field of a permanent magnet which is attached to the instrument casing. The magnet has a cylindrical pole piece inside the coil and an annular pole piece surrounding the coil. The suspended coil represents an oscillatory system with a resonant frequency determined by the mass of the coil and the stiffness of its spring suspension.

The geophone is fixed by a spike base into soft ground or mounted firmly on hard ground. It moves in

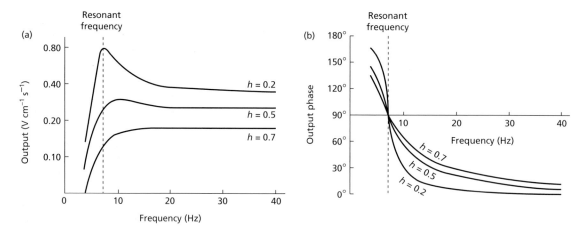

Fig. 3.20 Amplitude and phase responses of a geophone with a resonant frequency of 7 Hz, for different damping factors *h*. Output phase is expressed relative to input phase. (After Telford *et al.* 1976.)

sympathy with the ground surface during the passage of a seismic wave, causing relative motion between the suspended coil and the fixed magnet. Movement of the coil in the magnetic field generates a voltage across the terminals of the coil. The oscillatory motion of the coil is inherently damped because the current flowing in the coil induces a magnetic field that interacts with the field of the magnet to oppose the motion of the coil. The amount of this damping can be altered by connecting a shunt resistance across the coil terminals to control the amount of current flowing in the coil.

Ideally, the output waveform of a geophone closely mirrors the ground motion and this is arranged by careful selection of the amount of damping. Too little damping results in an oscillatory output at the resonant frequency, whilst overdamping leads to a reduction of sensitivity. Damping is typically arranged to be about 0.7 of the critical value at which oscillation would just fail to occur for an impulsive mechanical input such as a sharp tap. With this amount of damping the frequency response of the geophone is effectively flat above the resonant frequency. The effect of differing amounts of damping on the frequency and phase response of a geophone is shown in Fig. 3.20.

To preserve the shape of the seismic waveform, geophones should have a flat frequency response and minimal phase distortion within the frequency range of interest. Consequently, geophones should be arranged to have a resonant frequency well below the main frequency band of the seismic signal to be recorded. Most commercial seismic reflection surveys employ

geophones with a resonant frequency between 4 and 15 Hz.

Above the resonant frequency, the output of a moving-coil geophone is proportional to the velocity of the coil. Note that the coil velocity is related to the very low particle velocity associated with a seismic ground motion and not to the much higher propagation velocity of the seismic energy (see Section 3.4). The sensitivity of a geophone, measured in output volts per unit of velocity, is determined by the number of windings in the coil and the strength of the magnetic field; hence, instruments of larger and heavier construction are required for higher sensitivity. The miniature geophones used in commercial reflection surveying typically have a sensitivity of about 10 V per m s^{-1}.

Moving-coil geophones are sensitive only to the component of ground motion along the axis of the coil. Vertically travelling compressional waves from subsurface reflectors cause vertical ground motions and are therefore best detected by geophones with an upright coil as illustrated in Fig. 3.19. The optimal recording of seismic phases that involve mainly horizontal ground motions, such as horizontally-polarized shear waves, requires geophones in which the coil is mounted and constrained to move horizontally.

Hydrophones are composed of ceramic piezoelectric elements which produce an output voltage proportional to the pressure variations associated with the passage of a compressional seismic wave through water. The sensitivity is typically 0.1 mV Pa^{-1}. For multichannel seismic surveying at sea, large numbers of individual hydrophones are made up into hydrophone streamers by distributing them along an oil-filled plastic tube. The tube is arranged to have neutral buoyancy and is manufactured from materials with an acoustic impedance close to that of water to ensure good transmission of seismic energy to the hydrophone elements. Since piezoelectric elements are also sensitive to accelerations, hydrophones are often composed of two elements mounted back to back and connected in series so that the effects of accelerations of the streamer as it is towed through the water are cancelled out in the hydrophone outputs. The response of each element to pressure change is, however, unaffected and the seismic signal is fully preserved.

Arrays of geophones or hydrophones may be connected together into linear or areal arrays containing tens or even hundreds of transducers whose individual outputs are summed. Such arrays provide detectors with a directional response that facilitates the enhancement of signal and the suppression of certain types of noise as discussed further in Chapter 4.

3.8.3 Seismic recording systems

Recording a seismogram is a very difficult technical operation from at least three key aspects:
1. The recording must be timed accurately relative to the seismic source.
2. Seismograms must be recorded with multiple transducers simultaneously, so that the speed and direction of travel of seismic waves can be interpreted.
3. The electrical signals must be stored for future use.

The least difficult of these problems is the timing. For nearly all seismic surveys, times need to be accurate to better than one thousandth of a second (one millisecond). For very small-scale surveys the requirement may be for better than 0.1 ms. In fact, with modern electronics, measuring such short time intervals is not difficult. Usually the biggest uncertainty is in deciding how to measure the instant when the seismic source started the wave. Even in a simple case, as for a sledge-hammer hitting the ground, is the correct instant when the hammer first hits the ground, or when it stops compressing the ground and a seismic wave radiates outward? The first is easy to measure, the second is probably more important, and they are usually separated by more than 1 ms.

In order to determine the subsurface path of the seismic energy, the direction from which the wave arrives at the surface must be determined. This is achieved by having many transducers on the surface, and measuring the small changes of arrival time as the waves move across them. Typically, this number might be 24 for a small engineering survey, to several thousand for a large hydrocarbon exploration survey.

The electrical signals from the transducers must be recorded in real time. Before the availability of portable, powerful, computer systems this was a fundamental problem. Before the 1960s the majority of seismograms were recorded as wiggly traces written directly to paper or photographic film charts. The seismic computor was a human geophysicist with a slide rule. While direct paper recording is still used for some very specialist applications, virtually all seismic data are now recorded by digitizing the analogue transducer output, and storing the series of digital samples in some computer format. It is a little surprising to realize that recording a seismic source is technically more demanding than recording a classical orchestra. The dynamic range of signals and the required accuracy of amplitude recording are both more stringent in the seismic case. An amplitude ratio of one million is equivalent to a dynamic range of 120 dB. A maximum dynamic range for geophones of about 140 dB and an inherent minimum noise level in seismic amplifiers of about 1 μV effectively limits the maximum dynamic range of a seismic recording to 120 dB.

Seismic signals from the transducer must be amplified, filtered if necessary, digitized then stored with appropriate index information. International standards produced by the Society of Exploration Geophysicists (SEG 1997) are used for the format of seismic data storage. Virtually all seismic data, from small engineering surveys to lithospheric studies, are now recorded on computer systems in these formats. The physical nature of the computer media used is continuously being upgraded from magnetic tape, to magnetic cartridge and CDROM. The data volumes produced can be impressive. A seismic acquisition ship working on the continental shelf can easily record around 40 gigabytes of data per 24-hour day. This generates a data storage problem (about 60 CDROMs), but more importantly a huge data cataloguing and data processing task.

The high capacity of modern computer systems for data recording and processing has allowed experimentation with more data-intensive survey methods. It is becoming commonplace to record three-component surveys, with three geophones at each survey station recording the east–west and north–south components of motion as well as the vertical. This triples the data volume, but does allow investigation of S-waves and potentially reveals more information about the physical properties of each subsurface layer.

Distributed systems

In seismic surveying the outputs of several detectors are fed to a multichannel recording system mounted in a recording vehicle. The individual detector outputs may be fed along a multicore cable. The weight and complexity of multicore cables becomes prohibitive as the number of channels of data rises into the hundreds. Modern systems distribute the task of amplification, digitization and recording of data from groups of detectors to individual computer units left unattended in the field. These are connected together to make a field computer network using lightweight fibre-optic cables or telemetry links. The separate units can then be controlled by a central recording station, and upload their digital seismograms to it on command.

Problems

1. How does the progressive loss of higher frequencies in a propagating seismic pulse lead to an increase in pulse length?

2. A 10 Hz seismic wave travelling at 5 km s^{-1} propagates for 1000 m through a medium with an absorption coefficient of 0.2 dB λ^{-1}. What is the wave attenuation in decibels due solely to absorption?

3. A wave component with a wavelength of 100 m propagates through a homogeneous medium from a seismic source at the bottom of a borehole. Between two detectors, located in boreholes at radial distances of 1 km and 2 km from the source, the wave amplitude is found to be attenuated by 10 dB. Calculate the contribution of geometrical spreading to this value of attenuation and, thus, determine the absorption coefficient of the medium.

4. What is the crossover distance for direct and critically refracted rays in the case of a horizontal interface at a depth of 200 m separating a top layer of velocity 3.0 km s^{-1} from a lower layer of velocity 5.0 km s^{-1}?

5. A seismic pulse generated by a surface source is returned to the surface after reflection at the tenth of a series of horizontal interfaces, each of which has a reflection coefficient R of 0.1. What is the attenuation in amplitude of the pulse caused by energy partitioning at all interfaces encountered along its path?

6. At what frequency would a 150 Hz signal be recorded by a digital recording system with a sampling rate of 100 Hz?

Further reading

Al-Sadi, H.N. (1980) *Seismic Exploration*. Birkhauser Verlag, Basel.

Anstey, N.A. (1977) *Seismic Interpretation: The Physical Aspects*. IHRDC, Boston.

Anstey, N.A. (1981) *Seismic Prospecting Instruments. Vol 1: Signal Characteristics and Instrument Specifications*. Gebruder Borntraeger, Berlin.

Dobrin, M.B. & Savit, C.H. (1988) *Introduction to Geophysical Prospecting* (4th edn). McGraw-Hill, New York.

Gregory, A.R. (1977) Aspects of rock physics from laboratory and log data that are important to seismic interpretation. *In*: Payton, C.E. (ed.), *Seismic Stratigraphy—Applications to Hydrocarbon Exploration*. Memoir 26, American Association of Petroleum Geologists, Tulsa.

Lavergne, M. (1989) *Seismic Methods*. Editions Technip, Paris.

SEG (1997) Digital Tape Standards (SEG-A, SEG-B, SEG-C, SEG-Y and SEG-D formats plus SEG-D rev 1&2). Compiled by SEG Technical Standards Committee. Society of Exploration Geophysicists, Tulsa, USA.

Sheriff, R.E. & Geldart. L.P. (1982) *Exploration Seismology Vol 1: History. Theory and Data Acquisition*. Cambridge University Press, Cambridge.

Sheriff, R.E. & Geldart, L.P. (1983) *Exploration Seismology Vol 2: Data-Processing and Interpretation*. Cambridge University Press, Cambridge.

Waters, K.H. (1978) *Reflection Seismology—A Tool For Energy Resource Exploration*. Wiley, New York.

Zoeppritz, K, 1919. Uber reflexion und durchgang seismischer wellen durch Unstetigkerlsflaschen. Berlin, Uber Erdbebenwellen VII B, Nachrichten der Koniglichen Gesellschaft der Wissensschaften zu Gottingen, math-phys. Kl. pp. 57–84.

4 Seismic reflection surveying

4.1 Introduction

Seismic reflection surveying is the most widely used and well-known geophysical technique. The current state of sophistication of the technique is largely a result of the enormous investment in its development made by the hydrocarbon industry, coupled with the development of advanced electronic and computing technology. Seismic sections can now be produced to reveal details of geological structures on scales from the top tens of metres of drift to the whole lithosphere. Part of the spectacular success of the method lies in the fact that the raw data are processed to produce a seismic section which is an image of the subsurface structure. This also provides a trap for the unwary, since the seismic section is similar to, but fundamentally different from, a depth section of the geology. Only by understanding how the reflection method is used and seismic sections are created, can the geologist make informed interpretations. This chapter provides the essential knowledge and understanding to support interpretation of seismic reflection data. It builds up systematically from the basics of seismic wave reflection from rock layers, and refers back to relevant material in Chapters 2 and 3.

4.2 Geometry of reflected ray paths

In seismic reflection surveys seismic energy pulses are reflected from subsurface interfaces and recorded at near-normal incidence at the surface. The travel times are measured and can be converted into estimates of depths to the interfaces. Reflection surveys are most commonly carried out in areas of shallowly dipping sedimentary sequences. In such situations, velocity varies as a function of depth, due to the differing physical properties of the individual layers. Velocity may also vary horizontally, due to lateral lithological changes within the individual layers. As a first approxi-

mation, the horizontal variations of velocity may be ignored.

Figure 4.1 shows a simple physical model of horizontally-layered ground with vertical reflected ray paths from the various layer boundaries. This model assumes each layer to be characterized by an *interval velocity* v_i, which may correspond to the uniform velocity within a homogeneous geological unit or the average velocity over a depth interval containing more than one unit. If z_i is the thickness of such an interval and τ_i is the one-way travel time of a ray through it, the interval velocity is given by

$$v_i = \frac{z_i}{\tau_i}$$

The interval velocity may be averaged over several depth intervals to yield a *time-average velocity* or, simply, *average velocity* $\overline{V}$. Thus the average velocity of the top n layers in Fig. 4.1 is given by

$$\overline{V} = \frac{\sum_{i=1}^{n} z_i}{\sum_{i=1}^{n} \tau_i} = \frac{\sum_{i=1}^{n} v_i \tau_i}{\sum_{i=1}^{n} \tau_i}$$

or, if Z_n is the total thickness of the top n layers and T_n is the total one-way travel time through the n layers,

$$\overline{V} = \frac{Z_n}{T_n}$$

4.2.1 Single horizontal reflector

The basic geometry of the reflected ray path is shown in Fig. 4.2(a) for the simple case of a single horizontal re-

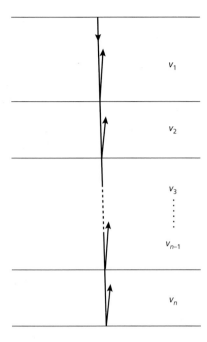

Fig. 4.1 Vertical reflected ray paths in a horizontally-layered ground.

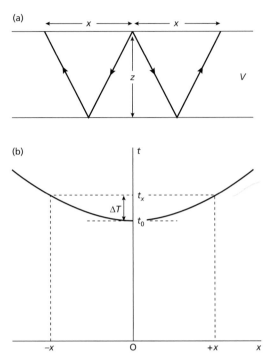

Fig. 4.2 (a) Section through a single horizontal layer showing the geometry of reflected ray paths and (b) time–distance curve for reflected rays from a horizontal reflector. ΔT = normal moveout (NMO).

flector lying at a depth z beneath a homogeneous top layer of velocity V. The equation for the travel time t of the reflected ray from a shot point to a detector at a horizontal offset, or shot–detector separation, x is given by the ratio of the travel path length to the velocity

$$t = \left(x^2 + 4z^2\right)^{1/2} \big/ V \tag{4.1}$$

In a reflection survey, reflection time t is measured at an offset distance x. These values can be applied to equation (4.1), but still leave two unknown values which are related to the subsurface structure, z and V. If many reflection times t are measured at different offsets x, there will be enough information to solve equation (4.1) for both these unknown values. The graph of travel time of reflected rays plotted against offset distance (the *time–distance curve*) is a hyperbola whose axis of symmetry is the time axis (Fig. 4.2(b)).

Substituting $x = 0$ in equation (4.1), the travel time t_0 of a vertically reflected ray is obtained:

$$t_0 = \frac{2z}{V} \tag{4.2}$$

This is the intercept on the time axis of the time–distance curve (see Fig. 4.2(b)). Equation (4.1) can be written

$$t^2 = \frac{4z^2}{V^2} + \frac{x^2}{V^2} \tag{4.3}$$

Thus

$$t^2 = t_0^2 + \frac{x^2}{V^2} \tag{4.4}$$

This form of the travel-time equation (4.4) suggests the simplest way of determining the velocity V. If t^2 is plotted against x^2, the graph will produce a straight line of slope $1/V^2$. The intercept on the time axis will also give the vertical two-way time, t_0, from which the depth to the reflector can be found. In practice, however, this method is unsatisfactory since the range of values of x is restricted, and the slope of the best-fit straight line has large uncertainty. A much better method of determining velocity is by considering the increase of reflected travel time with offset distance, the *moveout*, as discussed below.

Equation (4.3) can also be rearranged

$$t = \frac{2z}{V}\left[1 + \left(\frac{x}{2z}\right)^2\right]^{1/2} = t_0\left[1 + \left(\frac{x}{Vt_0}\right)^2\right]^{1/2} \quad (4.5)$$

This form of the equation is useful since it indicates clearly that the travel time at any offset x will be the vertical travel time plus an additional amount which increases as x increases, V and t_0 being constants. This relationship can be reduced to an even simpler form with a little more rearrangement. Using the standard binomial expansion of equation (4.5) gives

$$t = t_0\left[1 + \frac{1}{2}\left(\frac{x}{Vt_0}\right)^2 - \frac{1}{8}\left(\frac{x}{Vt_0}\right)^4 + \dots\right]$$

Remembering that $t_0 = 2z/V$, the term x/Vt_0 can be written as $x/2z$. If $x = z$, the second term in this series becomes $1/8$ of $(1/2)^4$, i.e. 0.0078, which is less than a 1% change in the value of t. For small offset/depth ratios (i.e. $x/z \ll 1$), the normal case in reflection surveying, this equation may be truncated after the first term to obtain the approximation

$$t \approx t_0\left[1 + \frac{1}{2}\left(\frac{x}{Vt_0}\right)^2\right] \approx t_0 + \frac{x^2}{2V^2t_0} \quad (4.6)$$

This is the most convenient form of the time–distance equation for reflected rays and it is used extensively in the processing and interpretation of reflection data.

Moveout is defined as the difference between the travel times t_1 and t_2 of reflected-ray arrivals recorded at two offset distances x_1 and x_2. Substituting t_1, x_1 and t_2, x_2 in equation (4.6), and subtracting the resulting equations gives

$$t_2 - t_1 \approx \frac{x_2^2 - x_1^2}{2V^2t_0}$$

Normal moveout (NMO) at an offset distance x is the difference in travel time ΔT between reflected arrivals at x and at zero offset (see Fig. 4.2)

$$\Delta T = t_x - t_0 \approx \frac{x^2}{2V^2t_0} \quad (4.7)$$

Note that NMO is a function of offset, velocity and reflector depth z (since $z = Vt_0/2$). The concept of moveout is fundamental to the recognition, correlation and enhancement of reflection events, and to the calculation of velocities using reflection data. It is used explicitly or implicitly at many stages in the processing and interpretation of reflection data.

As an important example of its use, consider the $T-\Delta T$ method of velocity analysis. Rearranging the terms of equation (4.7) yields

$$V \approx \frac{x}{(2t_0\Delta T)^{1/2}} \quad (4.8)$$

Using this relationship, the velocity V above the reflector can be computed from knowledge of the zero-offset reflection time (t_0) and the NMO (ΔT) at a particular offset x. In practice, such velocity values are obtained by computer analysis which produces a statistical estimate based upon many such calculations using large numbers of reflected ray paths (see Section 4.7). Once the velocity has been derived, it can be used in conjunction with t_0 to compute the depth z to the reflector using $z = Vt_0/2$.

4.2.2 Sequence of horizontal reflectors

In a multilayered ground, inclined rays reflected from the nth interface undergo refraction at all higher interfaces to produce a complex travel path (Fig. 4.3(a)). At offset distances that are small compared to reflector depths, the travel-time curve is still essentially hyperbolic but the homogeneous top layer velocity V in equations (4.1) and (4.7) is replaced by the *average velocity* $\bar{V}$ or, to a closer approximation (Dix 1955), the *root-mean-square velocity* V_{rms} of the layers overlying the reflector. As the offset increases, the departure of the actual travel-time curve from a hyperbola becomes more marked (Fig. 4.3(b)).

The root-mean-square velocity of the section of ground down to the nth interface is given by

$$V_{rms,n} = \left[\sum_{i=1}^{n} v_i^2 \tau_i \bigg/ \sum_{i=1}^{n} \tau_i\right]^{1/2}$$

where v_i is the interval velocity of the ith layer and τ_i is the one-way travel time of the reflected ray through the ith layer.

(a)

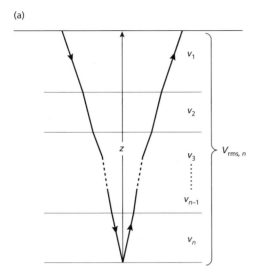

(b)

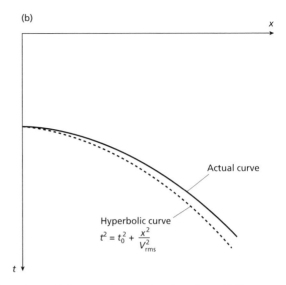

Fig. 4.3 (a) The complex travel path of a reflected ray through a multilayered ground, showing refraction at layer boundaries. (b) The time–distance curve for reflected rays following such a travel path. Note that the divergence from the hyperbolic travel-time curve for a homogeneous overburden of velocity V_{rms} increases with offset.

Thus at small offsets x ($x \ll z$), the total travel time t_n of the ray reflected from the nth interface at depth z is given to a close approximation by

$$t_n = (x^2 + 4z^2)^{1/2} \big/ V_{rms} \qquad \text{cf. equation (4.1)}$$

and the NMO for the nth reflector is given by

$$\Delta T_n \approx \frac{x^2}{2V_{rms,n}^2 t_0} \qquad \text{cf. equation (4.7)}$$

The individual NMO value associated with each reflection event may therefore be used to derive a root-mean-square velocity value for the layers above the reflector. Values of V_{rms} down to different reflectors can then be used to compute interval velocities using the *Dix formula*. To compute the interval velocity v_n for the nth interval

$$v_n = \left[\frac{V_{rms,n}^2 \, t_n - V_{rms,n-1}^2 \, t_{n-1}}{t_n - t_{n-1}} \right]^{1/2}$$

where $V_{rms,n-1}^2$, t_{n-1} and $V_{rms,n}$, t_n are, respectively, the root-mean-square velocity and reflected ray travel times to the $(n-1)$th and nth reflectors (Dix 1955).

4.2.3 Dipping reflector

In the case of a dipping reflector (Fig. 4.4(a)) the value of dip θ enters the time–distance equation as an additional unknown. The equation is derived similarly to that for horizontal layers by considering the ray path length divided by the velocity:

$$t = \frac{(x^2 + 4z^2 + 4xz \sin \theta)^{1/2}}{V} \qquad \text{cf. equation (4.1)}$$

The equation still has the form of a hyperbola, as for the horizontal reflector, but the axis of symmetry of the hyperbola is now no longer the time axis (Fig. 4.4(b)). Proceeding as in the case of a horizontal reflector, using a truncated binomial expansion, the following expression is obtained:

$$t \approx t_0 + \frac{(x^2 + 4xz \sin \theta)}{2V^2 t_0} \qquad (4.9)$$

Consider two receivers at equal offsets x updip and downdip from a central shot point (Fig. 4.4). Because of the dip of the reflector, the reflected ray paths are of different length and the two rays will therefore have different travel times. *Dip moveout* ΔT_d is defined as the

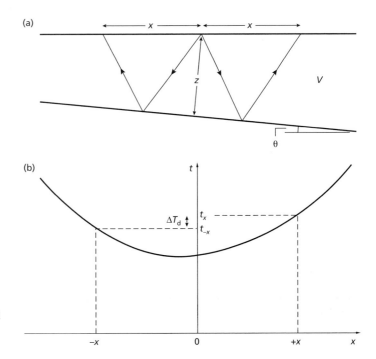

Fig. 4.4 (a) Geometry of reflected ray paths and (b) time–distance curve for reflected rays from a dipping reflector. ΔT_{d} = dip moveout.

difference in travel times t_x and t_{-x} of rays reflected from the dipping interface to receivers at equal and opposite offsets x and $-x$

$$\Delta T_{\mathrm{d}} = t_x - t_{-x}$$

Using the individual travel times defined by equation (4.9)

$$\Delta T_{\mathrm{d}} = 2x \sin \theta / V$$

Rearranging terms, and for small angles of dip (when $\sin \theta \approx \theta$)

$$\theta \approx V \Delta T_{\mathrm{d}} / 2x$$

Hence the *dip moveout* ΔT_{d} may be used to compute the reflector dip θ if V is known. V can be derived via equation (4.8) using the NMO ΔT which, for small dips, may be obtained with sufficient accuracy by averaging the updip and downdip moveouts:

$$\Delta T \approx (t_x + t_{-x} - 2t_0)/2$$

4.2.4 Ray paths of multiple reflections

In addition to rays that return to the surface after reflection at a single interface, known as *primary reflections*, there are many paths in a layered subsurface by which rays may return to the surface after reflection at more than one interface. Such rays are called *reverberations*, *multiple reflections* or simply *multiples*. A variety of possible ray paths involving multiple reflection is shown in Fig. 4.5(a).

Generally, multiple reflections tend to have lower amplitudes than primary reflections because of the loss of energy at each reflection. However, there are two types of multiple that are reflected at interfaces of high reflection coefficient and therefore tend to have amplitudes comparable with primary reflections:

1. *Ghost reflections*, where rays from a buried explosion on land are reflected back from the ground surface or the base of the weathered layer (see Section 4.6) to produce a reflection event, known as a ghost reflection, that arrives a short time after the primary.

2. *Water layer reverberations*, where rays from a marine source are repeatedly reflected at the sea bed and sea surface.

Multiple reflections that involve only a short additional path length arrive so soon after the primary event that they merely extend the overall length of the

(a)

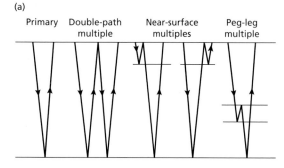

(b)

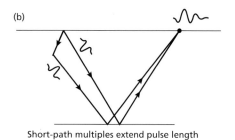

Short-path multiples extend pulse length

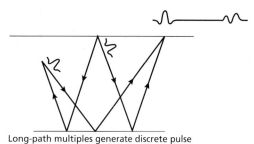

Long-path multiples generate discrete pulse

Fig. 4.5 (a) Various types of multiple reflection in a layered ground. (b) The difference between short-path and long-path multiples.

recorded pulse. Such multiples are known as *short-path multiples* (or short-period reverberations) and these may be contrasted with *long-path multiples* whose additional path length is sufficiently long that the multiple reflection is a distinct and separate event in the seismic record (Fig. 4.5(b)).

The correct recognition of multiples is essential. Misidentification of a long-path multiple as a primary event, for example, would lead to serious interpretation error. The arrival times of multiple reflections are predictable, however, from the corresponding primary reflection times. Multiples can therefore be suppressed by suitable data processing techniques to be described later (Section 4.8).

4.3 The reflection seismogram

The graphical plot of the output of a single detector in a reflection spread is a visual representation of the local pattern of vertical ground motion (on land) or pressure variation (at sea) over a short interval of time following the triggering of a nearby seismic source. This *seismic trace* represents the combined response of the layered ground and the recording system to a seismic pulse. Any display of a collection of one or more seismic traces is termed a *seismogram*. A collection of such traces representing the responses of a series of detectors to the energy from one shot is termed a *shot gather*. A collection of the traces relating to the seismic response at one surface mid-point is termed a *common mid-point gather* (*CMP gather*). The collection of the seismic traces for each CMP and their transformation to a component of the image presented as a seismic section is the main task of seismic reflection processing.

4.3.1 The seismic trace

At each layer boundary a proportion of the incident energy in the pulse is reflected back towards the detector. The proportion is determined by the contrast in acoustic impedances of the two layers, and for a vertically travelling ray, the reflection coefficient can be simply calculated (see Section 3.6). Figure 4.6 shows the relationship of the geological layering, the variation in acoustic impedance and the reflection coefficients as a function of depth. The detector receives a series of reflected pulses, scaled in amplitude according to the distance travelled and the reflection coefficients of the various layer boundaries. The pulses arrive at times determined by the depths to the boundaries and the velocities of propagation between them.

Assuming that the pulse shape remains unchanged as it propagates through such a layered ground, the resultant seismic trace may be regarded as the *convolution* of the input pulse with a time series known as a *reflectivity function* composed of a series of spikes. Each spike has an amplitude related to the reflection coefficient of a boundary and a travel time equivalent to the two-way reflection time for that boundary. This time series represents the *impulse response* of the layered ground (i.e. the output for a spike input). The convolution model is illustrated schematically in Fig. 4.6. Since the pulse has a finite length, individual reflections from closely-spaced boundaries are seen to overlap in time on the resultant seismogram.

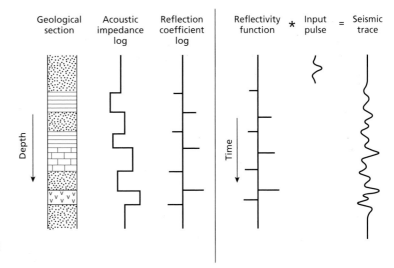

Fig. 4.6 The convolutional model of the reflection seismic trace, showing the trace as the convolved output of a reflectivity function with an input pulse, and the relationship of the reflectivity function to the physical properties of the geological layers.

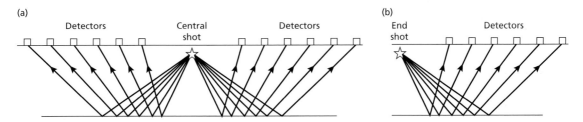

Fig. 4.7 Shot–detector configurations used in multichannel seismic reflection profiling. (a) Split spread, or straddle spread. (b) Single-ended or on-end spread.

In practice, as the pulse propagates it lengthens due to the progressive loss of its higher frequency components by absorption. The basic reflection seismic trace may then be regarded as the convolution of the reflectivity function with a *time-varying* seismic pulse. The trace will be further complicated by the superposition of various types of noise such as multiple reflections, direct and refracted body waves, surface waves (ground roll), air waves and coherent and incoherent noise unconnected with the seismic source. In consequence of these several effects, seismic traces generally have a complex appearance and reflection events are often not recognizable without the application of suitable processing techniques.

In seismic reflection surveying, the seismic traces are recorded, and the purpose of seismic processing can be viewed as an attempt to reconstruct the various columns of Fig. 4.6, moving from right to left. This will involve:

- removing noise
- determining the input pulse and removing that to give the reflectivity function
- determining the velocity function to allow conversion from time to depth axis
- determination of the acoustic impedances (or related properties) of the formations.

4.3.2 The shot gather

The initial display of seismic profile data is normally in groups of seismic traces recorded from a common shot, known as *common shot point gathers* or, simply, *shot gathers*. The seismic detectors (e.g. geophones) may be distributed on either side of the shot, or only on one side as illustrated in Fig. 4.7. The display of shot gathers at the time of field recording provides a means of checking that a satisfactory recording has been achieved from any particular shot. In shot gathers, the seismic traces are plotted side by side in their correct relative positions and the records are commonly displayed with their time axes arranged vertically in a draped fashion. In these seismic records, recognition of reflection events and their correlation from trace to trace is much assisted if one half of the normal 'wiggly-trace' waveform is blocked out. Figure 4.8 shows a draped section with this mode of display, derived from a split-spread multichannel survey. A short time after the shot instant the first arrival of seismic energy reaches the innermost geophones (the central traces) and this energy passes out symmetrically through the two arms of the split spread. The first arrivals are followed by a series of reflection events revealed by their hyperbolic moveout.

4.3.3   The CMP gather

Each seismic trace has three primary geometrical factors which determine its nature. Two of these are the shot position and the receiver position. The third, and perhaps most critical, is the position of the subsurface reflection point. Before seismic processing this position is unknown, but a good approximation can be made by assuming this reflection point lies vertically under the position on the surface mid-way between the shot and receiver for that trace. This point is termed the *mid-point*. Older terminology is to call this point the *depth point*, but the former term is a description of what the position is, rather than what it is wished to represent, and is hence preferred. Collecting all the traces with a common mid-point forms a *common mid-point (CMP) gather* (Fig. 4.9). The seismic industry and the literature use the older term *common depth point (CDP)* interchangeably for CMP.

The CMP gather lies at the heart of seismic processing for two main reasons:

1. The simple equations derived in Section 4.2 assume horizontal uniform layers. They can be applied with less error to a set of traces that have passed through the same geological structure. The simplest approximation to such a set of traces is the CMP gather. In the case of horizontal layers, reflection events on each CMP gather are reflected from a common depth point (CDP – see Fig. 4.9(a)). For these traces, the variation of travel time with offset, the moveout, will depend only on the velocity of the subsurface layers, and hence the subsurface velocity can be derived.

2. The reflected seismic energy is usually very weak. It is imperative to increase the signal-to-noise ratio of most

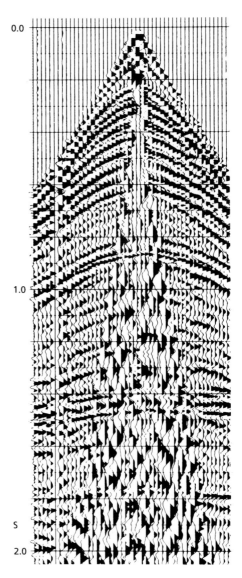

Fig. 4.8 A draped seismic record of a shot gather from a split spread (courtesy Prakla-Seismos GmbH). Sets of reflected arrivals from individual interfaces are recognizable by the characteristic hyperbolic alignment of seismic pulses. The late-arriving, high-amplitude, low-frequency events, defining a triangular-shaped central zone within which reflected arrivals are masked, represent surface waves (ground roll). These latter waves are a typical type of coherent noise.

data. Once the velocity is known, the traces in a CMP can be corrected for NMO to correct each trace to the equivalent of a *zero-offset trace*. These will all have the same reflected pulses at the same times, but different

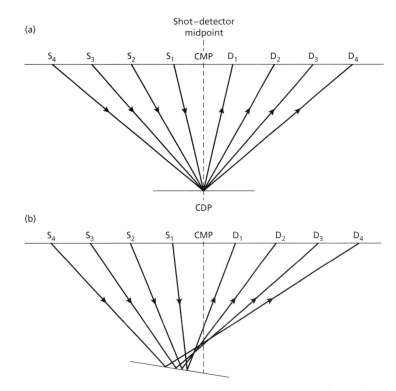

Fig. 4.9 Common mid-point (CMP) reflection profiling. (a) A set of rays from different shots to detectors reflected off a common depth point (CDP) on a horizontal reflector. (b) The common depth point is not achieved in the case of a dipping reflector.

random and coherent noise. Combining all the traces in a CMP together will average out the noise, and increase the signal-to-noise ratio (SNR). This process is termed *stacking*.

Strictly, the common mid-point principle breaks down in the presence of dip because the common depth point then no longer directly underlies the shot–detector mid-point and the reflection point differs for rays travelling to different offsets (see Fig. 4.9(b)). Nevertheless, the method is sufficiently robust that CMP stacks almost invariably result in marked improvements in SNR compared to single traces.

In two-dimensional CMP surveying, known as *CMP profiling*, the reflection points are all assumed to lie within the vertical section containing the survey line; in three-dimensional surveying, the reflection points are distributed across an area of any subsurface reflector, and the CMP is defined as a limited area on the surface.

4.4 Multichannel reflection survey design

The basic requirement of a multichannel reflection survey is to obtain recordings of reflected pulses at several offset distances from a shot point. As discussed in Chapter 3, this requirement is complicated in practice by the fact that the reflected pulses are never the first arrivals of seismic energy, and they are generally of very low amplitude. Owing to this and other problems to be discussed later, each individual reflection survey is designed specifically to optimize the data for the required purpose. It is essential that the geologists and interpreting geophysicists commissioning a survey understand this, and communicate their requirements to the geophysical contractor performing the survey.

In *two-dimensional surveys* (*reflection profiling*), data are collected along survey lines that nominally contain all shot points and receivers. For the purpose of data processing, reflected ray paths are assumed to lie in the vertical plane containing the survey line. Thus, in the presence of cross-dip the resultant seismic sections do not provide a true representation of the subsurface structure, since actual reflection points then lie outside the vertical plane. Two-dimensional survey methods are adequate for the mapping of structures (such as cylindrical folds, or faults) which maintain uniform geometry along strike. They may also be used to investigate three-dimensional structures by mapping lateral changes across a series of closely-spaced survey lines or around a grid of lines. However, as discussed later in Section 4.10, *three-dimensional surveys* provide a much better means of mapping three-dimensional structures and, in areas of structural complexity, they may provide the only means of obtaining reliable structural interpretations.

Reflection profiling is normally carried out along profile lines with the shot point and its associated spread of detectors being moved progressively along the line to build up lateral coverage of the underlying geological section. This progression is carried out in a stepwise fashion on land but continuously, by a ship under way, at sea.

The two most common shot–detector configurations in multichannel reflection profiling surveys are the *split spread* (or *straddle spread*) and the *single-ended spread* (Fig. 4.7), where the number of detectors in a spread may be several hundred. In split spreads, the detectors are distributed on either side of a central shot point; in single-ended spreads, the shot point is located at one end of the detector spread. Surveys on land are commonly carried out with a split-spread geometry, but in marine reflection surveys single-ended spreads are the normal configuration due to the constraint of having to tow equipment behind a ship. The marine source is towed close behind the ship, with the hydrophone streamer (which may be several kilometers long) trailing behind.

4.4.1 Vertical and horizontal resolution

Reflection surveys are normally designed to provide a specified depth of penetration and a particular degree of resolution of the subsurface geology in both the vertical and horizontal dimensions. The vertical resolution is a measure of the ability to recognize individual, closely-spaced reflectors and is determined by the pulse length on the recorded seismic section. For a reflected pulse represented by a simple wavelet, the maximum resolution possible is between one-quarter and one-eighth of the dominant wavelength of the pulse (Sheriff & Geldart 1983). Thus, for a reflection survey involving a signal with a dominant frequency of 50 Hz propagating in sedimentary strata with a velocity of 2.0 km s^{-1}, the dominant wavelength would be 40 m and the vertical resolution may therefore be no better than about 10 m. This figure is worth noting since it serves as a reminder that the smallest geological structures imaged on seismic sections tend to be an order of magnitude larger than the structures usually seen by geologists at rock exposures. Since deeper-travelling seismic waves tend to have a

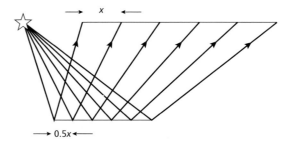

Fig. 4.10 The horizontal sampling of a seismic reflection survey is half the detector spacing.

lower dominant frequency due to the progressive loss of higher frequencies by absorption (Section 3.5) and higher velocity due to the effects of sediment compaction, vertical resolution decreases as a function of depth. It should be noted that the vertical resolution of a seismic survey may be improved at the data processing stage by a shortening of the recorded pulse length using inverse filtering (deconvolution) (Section 4.8).

There are two main controls on the horizontal resolution of a reflection survey, one being intrinsic to the physical process of reflection and the other being determined by the detector spacing. To deal with the latter point first, the horizontal resolution is clearly determined by the spacing of the individual depth estimates from which the reflector geometry is reconstructed. From Fig. 4.10 it can be seen that, for a flat-lying reflector, the horizontal sampling is equal to half the detector spacing. Note, also, that the length of reflector sampled by any detector spread is half the spread length. The spacing of detectors must be kept small to ensure that reflections from the same interface can be correlated reliably from trace to trace in areas of complex geology.

Notwithstanding the above, there is an absolute limit to the achievable horizontal resolution in consequence of the actual process of reflection. The path by which energy from a source is reflected back to a detector may be expressed geometrically by a simple ray path. However, such a ray path is only a geometrical abstraction. The actual reflection process is best described by considering any reflecting interface to be composed of an infinite number of point scatterers, each of which contributes energy to the reflected signal (Fig. 4.11). The actual reflected pulse then results from interference of an infinite number of backscattered rays.

Energy that is returned to a detector within half a wavelength of the initial reflected arrival interferes con-

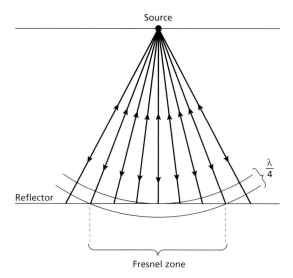

Source

$\frac{\lambda}{4}$

Reflector

Fresnel zone

Fig. 4.11 Energy is returned to source from all points of a reflector. The part of the reflector from which energy is returned within half a wavelength of the initial reflected arrival is known as the Fresnel zone.

structively to build up the reflected signal, and the part of the interface from which this energy is returned is known as the first *Fresnel zone* (Fig. 4.11) or, simply, the Fresnel zone. Around the first Fresnel zone are a series of annular zones from which the overall reflected energy tends to interfere destructively and cancel out. The width of the Fresnel zone represents an absolute limit on the horizontal resolution of a reflection survey since reflectors separated by a distance smaller than this cannot be individually distinguished. The width w of the Fresnel zone is related to the dominant wavelength λ of the source and the reflector depth z by

$$w = (2z\lambda)^{1/2} \quad \text{(for } z \gg \lambda)$$

The size of the first Fresnel zone increases as a function of reflector depth. Also, as noted in Section 3.5, deeper-travelling reflected energy tends to have a lower dominant frequency due to the effects of absorption. The lower dominant frequency is coupled with an increase in interval velocity, and both lead to an increase in the wavelength. For both these reasons the horizontal resolution, like the vertical resolution, reduces with increasing reflector depth.

As a practical rule of thumb, the Fresnel zone width for the target horizons should be estimated, then the geophone spacing fixed at no more than one-quarter of that width. In this case the horizontal resolution will be limited only by the physics of the seismic wave, not by the survey design.

4.4.2 Design of detector arrays

Each detector in a conventional reflection spread consists of an *array* (or *group*) of several geophones or hydrophones arranged in a specific pattern and connected together in series or parallel to produce a single channel of output. The effective offset of an array is taken to be the distance from the shot to the centre of the array. Arrays of geophones provide a directional response and are used to enhance the near-vertically travelling reflected pulses and to suppress several types of horizontally travelling *coherent* noise. Coherent noise is that which can be correlated from trace to trace as opposed to random noise (Fig. 4.12). To exemplify this, consider a Rayleigh surface wave (a vertically polarized wave travelling along the surface) and a vertically travelling compressional wave reflected from a deep interface to pass simultaneously through two geophones connected in series and spaced at half the wavelength of the Rayleigh wave. At any given instant, ground motions associated with the Rayleigh wave will be in opposite directions at the two geophones and the individual outputs of the geophones at any instant will therefore be equal and opposite and be cancelled by summing. However, ground motions associated with the reflected compressional wave will be in phase at the two geophones and the summed outputs of the geophones will therefore be twice their individual outputs.

The directional response of any linear array is governed by the relationship between the apparent wavelength λ_a of a wave in the direction of the array, the number of elements n in the array and their spacing Δx. The response is given by a response function R

$$R = \frac{\sin n\beta}{\sin \beta}$$

where

$$\beta = \pi\Delta x / \lambda_a$$

R is a periodic function that is fully defined in the interval $0 \le \Delta x / \lambda_a \le 1$ and is symmetrical about $\Delta x / \lambda_a = 0.5$. Typical array response curves are shown in Fig. 4.13.

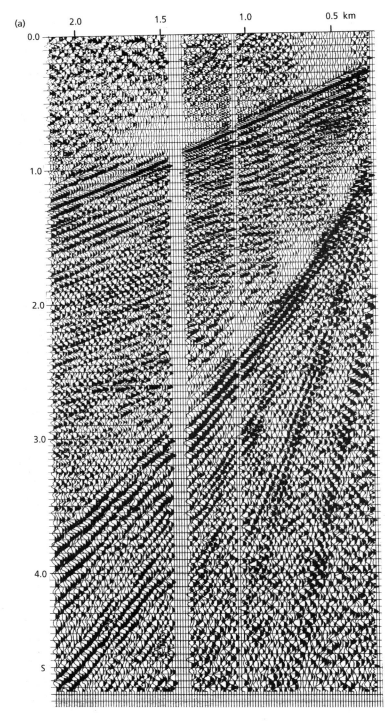

Fig. 4.12 Noise test to determine the appropriate detector array for a seismic reflection survey. (a) Draped seismic record obtained with a noise spread composed of clustered (or 'bunched') geophones. (b) Seismic record obtained over the same ground with a spread composed of 140 m long geophone arrays. (From Waters 1978.)

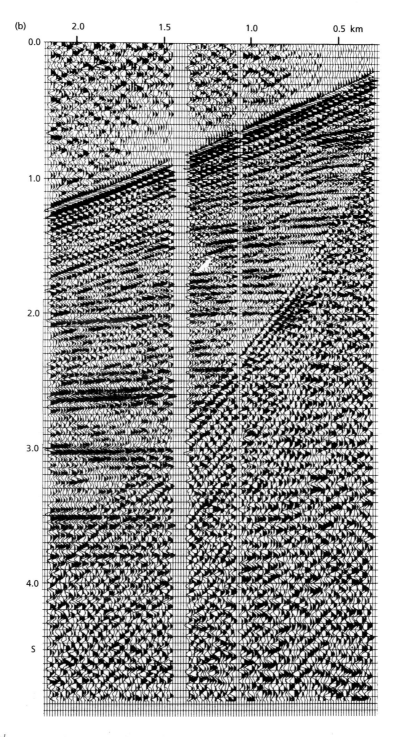

Fig. 4.12 *Continued*

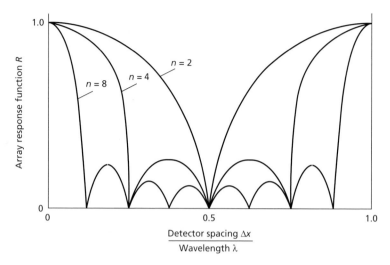

Fig. 4.13 Response functions for different detector arrays. (After Al-Sadi 1980.)

Arrays comprising areal rather than linear patterns of geophones may be used to suppress horizontal noise travelling along different azimuths.

The initial stage of a reflection survey involves field trials in the survey area to determine the most suitable combination of source, offset recording range, array geometry and detector spacing (the horizontal distance between the centres of adjacent geophone arrays, often referred to as the group interval) to produce good seismic data in the prevailing conditions.

Source trials involve tests of the effect of varying, for example, the shot depth and charge size of an explosive source, or the number, chamber sizes and trigger delay times of individual guns in an air gun array. The detector array geometry needs to be designed to suppress the prevalent coherent noise events (mostly source-generated). On land, the local noise is investigated by means of a *noise test* in which shots are fired into a spread of closely-spaced detectors (*noise spread*) consisting of individual geophones, or arrays of geophones clustered together to eliminate their directional response. A series of shots is fired with the noise spread being moved progressively out to large offset distances. For this reason such a test is sometimes called a *walk-away spread*. The purpose of the noise test is to determine the characteristics of the coherent noise, in particular, the velocity across the spread and dominant frequency of the air waves (shot noise travelling through the air), surface waves (ground roll), direct and shallow refracted arrivals, that together tend to conceal the low-amplitude reflections. A typical *noise section* derived from such a test is shown in Fig.

4.12(a). This clearly reveals a number of coherent noise events that need to be suppressed to enhance the SNR of reflected arrivals. Such noise sections provide the necessary information for the optimal design of detector phone arrays. Figure 4.12(b) shows a time section obtained with suitable array geometry designed to suppress the local noise events and reveals the presence of reflection events that were totally concealed in the noise section.

It is apparent from the above account that the use of suitably designed arrays can markedly improve the SNR of reflection events on field seismic recordings. Further improvements in SNR and survey resolution are achievable by various types of data processing discussed later in the chapter. Unfortunately, the noise characteristics tend to vary along any seismic line, due to near-surface geological variations and cultural effects. With the technical ability of modern instrumentation to record many hundreds of separate channels of data, there is an increasing tendency to use smaller arrays in the field, record more separate channels of data, then have the ability to experiment with different array types by combining recorded traces during processing. This allows more sophisticated noise cancellation, at the cost of some increase in processing time.

4.4.3 Common mid-point (CMP) surveying

If the shot–detector spread in a multichannel reflection survey is moved forward in such a way that no two reflected ray paths sample the same point on a subsurface reflector, the survey coverage is said to be *single-fold*. Each seismic trace then represents a unique sampling of some point on the reflector. In common mid-point (CMP) profiling, which has become the standard method of two-dimensional multichannel seismic surveying, it is arranged that a set of traces recorded at different offsets contains reflections from a common depth point (CDP) on the reflector (Fig. 4.14).

The *fold* of the stacking refers to the number of traces in the CMP gather and may conventionally be 24, 30, 60 or, exceptionally, over 1000. The fold is alternatively expressed as a percentage: single-fold = 100% coverage, six-fold = 600% coverage and so on. The fold of a CMP profile is determined by the quantity $N/2n$, where N is the number of geophone arrays along a spread and n is the number of geophone array spacings by which the spread is moved forward between shots (the *move-up rate*). Thus with a 96-channel spread ($N = 96$) and a move-up rate of 8 array spacings per shot interval ($n = 8$), the coverage would be $96/16 = 6$-fold. A field procedure for the routine collection of six-fold CMP coverage using a single-ended 12-channel spread configuration progressively moved forward along a profile line is shown in Fig. 4.14.

The theoretical improvement in SNR brought about by stacking n traces containing a mixture of coherent in-phase signals and random (incoherent) noise is $\sqrt{n}$. Stacking also attenuates long-path multiples. They have travelled in nearer-surface, lower velocity layers and have a significantly different moveout from the primary reflections. When the traces are stacked with the correct velocity function, the multiples are not in phase and do not sum. The stacked trace is the equivalent of a trace recorded with a vertical ray path, and is often referred to as a *zero-offset* trace.

4.4.4 Display of seismic reflection data

Profiling data from two-dimensional surveys are conventionally displayed as seismic sections in which the individual stacked *zero-offset* traces are plotted side by side, in close proximity, with their time axes arranged vertically. Reflection events may then be traced across the section by correlating pulses from trace to trace and in this way the distribution of subsurface reflectors beneath the survey line may be mapped. However, whilst it is tempting to envisage seismic sections as straightforward images of geological cross-sections it must not be forgotten that the vertical dimension of the sections is time, not depth.

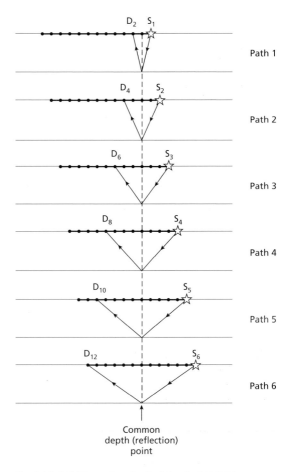

Fig. 4.14 A field procedure for obtaining six-fold CDP coverage with a single-ended 12-channel detector spread moved progressively along the survey line.

4.5 Time corrections applied to seismic traces

Two main types of correction need to be applied to reflection times on individual seismic traces in order that the resultant seismic sections give a true representation of geological structure. These are the *static* and *dynamic* corrections, so-called because the former is a fixed time correction applied to an entire trace whereas the latter varies as a function of reflection time.

4.6 Static correction

All previous consideration in this chapter on reflected seismic traces has assumed that the source and detector are placed on the planar horizontal surface of a uniform velocity layer. This is clearly not true for field data where the surface elevations vary, and the near-surface geology is usually highly variable, primarily due to variable weathering of bedrock, drift deposits and variable depth of the water table.

Reflection times on seismic traces have to be corrected for time differences introduced by these near-surface irregularities, which have the effect of shifting reflection events on adjacent traces out of their true time relationships. If the *static corrections* are not performed accurately, the traces in a CMP gather will not stack correctly. Furthermore, the near-surface static effects may be interpreted as spurious structures on deeper reflectors.

Accurate determination of static corrections is one of the most important problems which must be overcome in seismic processing (Cox 2001). In order to have enough information to arrive at a satisfactory correction, the data collection must be carefully designed to include information on the weathered layer. Two reflected pulses on traces within a CMP gather will only add together provided that the time offset between them is less than one-quarter of a period of the pulse. For typical deep seismic data with a dominant frequency of 50 Hz, this implies that static errors will be less than 5 ms. A 2 m thick layer of sand, soil or peat under one geophone station is sufficient to produce a local delay in a vertically travelling ray of about 5 ms.

Separate components of these *static corrections* are caused by the near-surface structure under each shot and each geophone for each trace. The corrections for each survey station occupied by either a shot and/or geophone comprise two components:

1. *Elevation static corrections*, which correct for the surface heights of the shot and geophone above a standard height datum (usually taken at sea-level).
2. *Weathering static corrections*, which correct for the heterogeneous surface layer, a few metres to several tens of metres thick, of abnormally low seismic velocity. The weathered layer is mainly caused by the presence within the surface zone of open joints and micro-fractures and by the unsaturated state of the zone. Although it may be only a few metres thick, its abnormally low velocity causes large time delays to rays passing through it. Thus variations in thickness of the weathered layer may, if not corrected for, lead to false structural relief on underlying reflectors shown on resulting seismic sections.

In marine surveys there is no elevation difference between individual shots and detectors but the water layer represents a surface layer of anomalously low velocity in some ways analogous to the weathered layer on land.

Static corrections are calculated on the assumption that the reflected ray path is effectively vertical immediately beneath any shot or detector. The travel time of the ray is then corrected for the time taken to travel the vertical distance between the shot or detector elevation and the survey datum (Fig. 4.15). Survey datum may lie above the local base of the weathered layer, or even above the local land surface. In adjusting travel times to datum, the height interval between the base of the weathered layer and datum is effectively replaced by material with the velocity of the main top layer, the *subweathering velocity*.

The *elevation static* correction is normally applied first. The *global positioning system (GPS)* satellite location system is now almost universally used for determination of the precise heights of all survey stations. Using *differential GPS* systems (DGPS), positions and heights can be determined in real time to an accuracy of better than 1 m, which is quite adequate for most surveys. Providing the subweathering velocity is also known, the corrections to datum can be computed very easily.

Calculation of the *weathering static* correction requires knowledge of the variable velocity and thickness of the weathered layer. The first arrivals of energy at detectors in a reflection spread are normally rays that have been refracted along the top of the subweathering layer. These arrivals can be used in a seismic refraction interpretation to determine the thickness and velocity of the various units within the weathered layer using methods discussed in Chapter 5. This procedure is termed a *refraction statics analysis* and is a routine part of seismic reflection processing. If the normal reflection spread does not contain recordings at sufficiently small offsets to detect these shallow refracted rays and the direct rays defining the weathered layer velocity v_w, special short refraction surveys may be carried out for this purpose. It is quite common for a seismic reflection recording crew to include a separate 'weathering' team, who conduct small-scale refraction surveys along the survey lines specifically to determine the structure of the weathered layer.

Direct measurements of the weathered layer velocity may also be obtained by *uphole surveys* in which small shots are fired at various depths down boreholes penetrating through the weathered layer and the velocities of rays travelling from the shots to a surface detector are calculated. Conversely, a surface shot may be recorded by downhole detectors. In reflection surveys using buried shots, a geophone is routinely located at the surface close

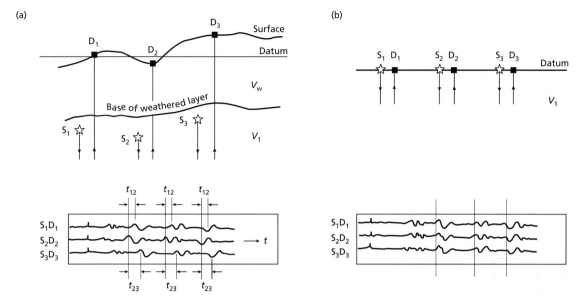

Fig. 4.15 Static corrections. (a) Seismograms showing time differences between reflection events on adjacent seismograms due to the different elevations of shots and detectors and the presence of a weathered layer. (b) The same seismograms after the application of elevation and weathering corrections, showing good alignment of the reflection events. (After O'Brien, 1974.)

to the shot hole to measure the *vertical time* (*VT*) or *uphole time*, from which the velocity of the surface layer above the shot may be calculated.

The complex variations in velocity and thickness within the weathered layer can never be precisely defined. The best estimate of the static correction derived from the field data is usually referred to as the *field static*. It always contains errors, or residuals, which have the effect of diminishing the SNR of CMP stacks and reducing the coherence of reflection events on time sections. These residuals can be investigated using sophisticated statistical analysis in a *residual static analysis*. This purely empirical approach assumes that the weathered layer and surface relief are the only cause of irregularities in the travel times of rays reflected from a shallow interface. It then operates by searching through all the data traces for systematic residual effects associated with individual shot and detector locations and applying these as corrections to the individual traces before the CMP stack. Figure 4.16 shows the marked improvement in SNR and reflection coherence achievable by the application of these automatically computed residual static corrections.

In marine reflection surveys the situation is much simpler since the shot and receivers are situated in a medium with a level surface and a constant velocity. The static correction is commonly restricted to a conversion of travel times to mean sea-level datum, without removing the overall effect of the water layer. Travel times are increased by $(d_s + d_h)v_w$, where d_s and d_h are the depths below mean sea-level of the source and hydrophone array and v_w is the seismic velocity of sea water. The effect of marine tidal height is often significant, especially in coastal waters, and demands a time-variant static correction. Tidal height data are usually readily available and the only complexity to the correction is their time-variant nature.

4.7 Velocity analysis

The *dynamic correction* is applied to reflection times to remove the effect of normal moveout. The correction is therefore numerically equal to the NMO and, as such, is a function of offset, velocity and reflector depth. Consequently, the correction has to be calculated separately for each time increment of a seismic trace.

Adequate correction for normal moveout is dependent on the use of accurate velocities. In common midpoint surveys the appropriate velocity is derived by

(a)

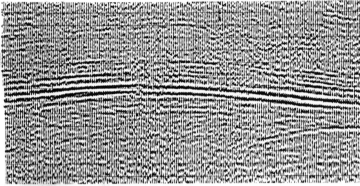

(b)

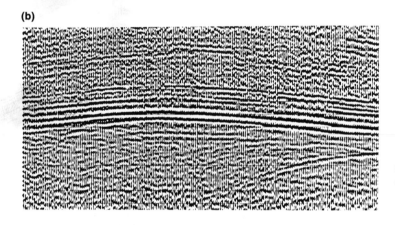

Fig. 4.16 Major improvement to a seismic section resulting from residual static analysis. (a) Field statics only. (b) After residual static correction. (Courtesy Prakla Seismos GmbH.)

computer analysis of moveout in the groups of traces from a common mid-point (*CMP gathers*). Prior to this *velocity analysis*, static corrections must be applied to the individual traces to remove the effect of the low-velocity surface layer and to reduce travel times to a common height datum. The method is exemplified with reference to Fig. 4.17 which illustrates a set of statically corrected traces containing a reflection event with a zero-offset travel time of t_0. Dynamic corrections are calculated for a range of velocity values and the dynamically corrected traces are stacked. The *stacking velocity* V_{st} is defined as that velocity value which produces the maximum amplitude of the reflection event in the stack of traces. This clearly represents the condition of successful removal of NMO. Since the stacking velocity is that which removes NMO, it is given by the equation

$$t^2 = t_0^2 + \frac{x^2}{V_{st}^2}$$ (cf. equation (4.4))

As previously noted, the travel-time curve for reflected rays in a multilayered ground is not a hyperbola (see Fig. 4.3(b)). However, if the maximum offset value x is small compared with reflector depth, the stacking velocity closely approximates the root-mean-square velocity V_{rms}, though it is obviously also affected by any reflector dip. Values of V_{st} for different reflectors can therefore be used in a similar way to derive interval velocities using the Dix formula (see Section 4.2.2). In practice, NMO corrections are computed for narrow time windows down the entire trace, and for a range of velocities, to produce a *velocity spectrum* (Fig. 4.18). The suitability of each velocity value is assessed by calculating a form of multitrace correlation, the *semblance*, between the corrected traces of the CMP gather. This assesses the power of the stacked reflected wavelet. The semblance values are contoured, such that contour peaks occur at times corresponding to reflected wavelets, and at velocities which produce an optimum

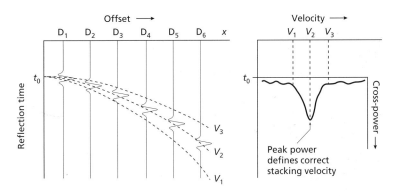

Fig. 4.17 A set of reflection events in a CMP gather is corrected for NMO using a range of velocity values. The stacking velocity is that which produces peak cross-power from the stacked events; that is, the velocity that most successfully removes the NMO. In the case illustrated, V_2 represents the stacking velocity. (After Taner & Koehler 1969.)

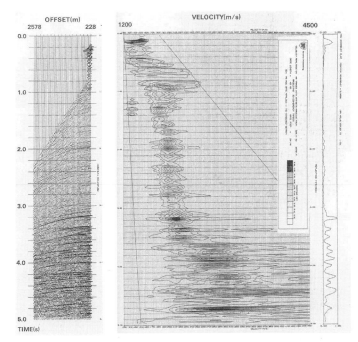

Fig. 4.18 The velocity spectrum is used to determine the stacking velocity as a function of reflection time. The cross-power function (semblance) is calculated over a large number of narrow time windows down the seismic trace, and for a range of possible velocities for each time window. The velocity spectrum is typically displayed alongside the relevant CMP gather as shown. Peaks in the contoured semblance values correspond to appropriate velocities for that travel time, where a reflection phase occurs in the CMP gather.

stacked wavelet. A velocity function defining the increase of velocity with depth for that CMP is derived by picking the location of the peaks on the velocity spectrum plot.

Velocity functions are derived at regular intervals along a CMP profile to provide stacking velocity values for use in the dynamic correction of each individual trace.

4.8 Filtering of seismic data

Several digital data processing techniques are available for the enhancement of seismic sections. In general, the aim of reflection data processing is to increase further the SNR and improve the vertical resolution of the individual seismic traces. As a broad generalization, these dual objectives have to be pursued independently. The two

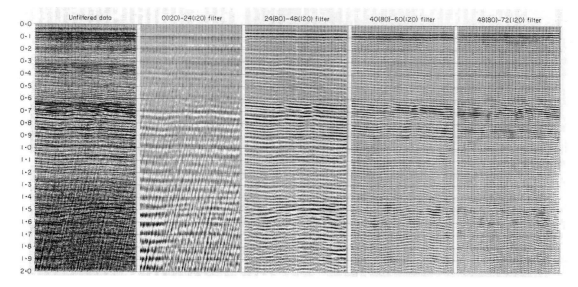

Fig. 4.19 Filter panels showing the frequency content of a panel of reflection records by passing them through a series of narrow-band frequencies. This plot allows the geophysicist to assess the frequency band that maximises the signal-to-noise ratio. Note that this may vary down the traces due to frequency-dependent absorption. (From Hatton *et al.* 1986, p. 88)

main types of waveform manipulation are frequency filtering and inverse filtering (deconvolution). Frequency filtering can improve the SNR but potentially damages the vertical resolution, while deconvolution improves the resolution, but at the expense of a decrease in the SNR. As with many aspects of seismic processing, compromises must be struck in each process to produce the optimum overall result.

4.8.1 Frequency filtering

Any coherent or incoherent noise event whose dominant frequency is different from that of reflected arrivals may be suppressed by frequency filtering (see Chapter 2). Thus, for example, ground roll in land surveys and several types of ship-generated noise in marine seismic surveying can often be significantly attenuated by low-cut filtering. Similarly, wind noise may be reduced by high-cut filtering. Frequency filtering may be carried out at several stages in the processing sequence. Normally, shot records would be filtered at a very early stage in the processing to remove obvious noise. Later applications of filters are used to remove artefacts produced by other processing stages. The final application of filters is to produce the sections to be used by the seismic interpreters, and here the choice of filters is made to produce the optimum visual display.

Since the dominant frequency of reflected arrivals decreases with increasing length of travel path, due to the selective absorption of the higher frequencies, the characteristics of frequency filters are normally varied as a function of reflection time. For example, the first second of a 3 s seismic trace might typically be band-pass filtered between limits of 15 and 75 Hz, whereas the frequency limits for the third second might be 10 and 45 Hz. The choice of frequency bands is made by inspection of filter panels (Fig. 4.19). As the frequency characteristics of reflected arrivals are also influenced by the prevailing geology, the appropriate time-variant frequency filtering may also vary as a function of distance along a seismic profile. The filtering may be carried out by computer in the time domain or the frequency domain (see Chapter 2).

4.8.2 Inverse filtering (deconvolution)

Many components of seismic noise lie within the frequency spectrum of a reflected pulse and therefore cannot be removed by frequency filtering. Inverse filters discriminate against noise and improve signal character using criteria other than simply frequency. They are thus able to suppress types of noise that have the same frequency characteristics as the reflected signal. A wide range of inverse filters is available for reflection data processing, each designed to remove some specific adverse effect of filtering in the ground along the transmission path, such as absorption or multiple reflection.

Deconvolution is the analytical process of removing the effect of some previous filtering operation (convolution). Inverse filters are designed to deconvolve seismic traces by removing the adverse filtering effects associated with the propagation of seismic pulses through a layered ground or through a recording system. In general, such effects lengthen the seismic pulse; for example, by the generation of multiple wave trains and by progressive absorption of the higher frequencies. Mutual interference of extended reflection wave trains from individual interfaces seriously degrades seismic records since onsets of reflections from deeper interfaces are totally or partially concealed by the wave trains of reflections from shallower interfaces.

Examples of inverse filtering to remove particular filtering effects include:
• *dereverberation* to remove ringing associated with multiple reflections in a water layer;
• *deghosting* to remove the short-path multiple associated with energy travelling upwards from the source and reflected back from the base of the weathered layer or the surface; and
• *whitening* to equalize the amplitude of all frequency components within the recorded frequency band (see below).

All these deconvolution operations have the effect of shortening the pulse length on processed seismic sections and, thus, improve the vertical resolution.

Consider a composite waveform w_k resulting from an initial spike source extended by the presence of short-path multiples near source such as, especially, water layer reverberations. The resultant seismic trace x_k will be given by the convolution of the reflectivity function r_k with the composite input waveform w_k as shown schematically in Fig. 4.6 (neglecting the effects of attenuation and absorption)

$$x_k = r_k \star w_k \quad \text{(plus noise)}$$

Reflected waveforms from closely-spaced reflectors will overlap in time on the seismic trace and, hence, will interfere. Deeper reflections may thus be concealed by the reverberation wave train associated with reflections from shallower interfaces, so that only by the elimination of the multiples will all the primary reflections be revealed. Note that short-path multiples have effectively the same normal moveout as the related primary reflection and are

therefore not suppressed by CDP stacking, and they have similar frequency content to the primary reflection so that they cannot be removed by frequency filtering.

Deconvolution has the general aim, not fully realizable, of compressing every occurrence of a composite waveform w_k on a seismic trace into a spike output, in order to reproduce the reflectivity function r_k that would fully define the subsurface layering. This is equivalent to the elimination of the multiple wave train. The required deconvolution operator is an inverse filter i_k which, when convolved with the composite waveform w_k, yields a spike function d_k

$$i_k \star w_k = d_k$$

Convolution of the same operator with the entire seismic trace yields the reflectivity function

$$i_k \star x_k = r_k$$

Where w_k is known, deconvolution can be achieved by the use of *matched filters* which effectively cross-correlate the output with the known input signal (as in the initial processing of *Vibroseis*® seismic records to compress the long source wave train; see Section 3.8.1). *Wiener filters* may also be used when the input signal is known. A Wiener filter (Fig. 4.20) converts the known input signal into an output signal that comes closest, in a least-squares sense, to a desired output signal. The filter optimizes the output signal by arranging that the sum of squares of differences between the actual output and the desired output is a minimum.

Although special attempts are sometimes made in marine surveys to measure the source signature directly, by suspending hydrophones in the vicinity of the source, both w_k and r_k are generally unknown in reflection surveying. The reflectivity function r_k is, of course, the main target of reflection surveying. Since, normally, only the seismic time series x_k is known, a special approach is required to design suitable inverse filters. This approach uses statistical analysis of the seismic time series, as in *predictive deconvolution* which attempts to remove the effect of multiples by predicting their arrival times from knowledge of the arrival times of the relevant primary events. Two important assumptions underlying predictive deconvolution (see e.g. Robinson & Treitel 2000) are:
1. that the reflectivity function represents a random series (i.e. that there is no systematic pattern to the distribution of reflecting interfaces in the ground); and

2. that the composite waveform w_k for an impulsive source is minimum delay (i.e. that its contained energy is concentrated at the front end of the pulse; see Chapter 2).

From assumption (1) it follows that the autocorrelation function of the seismic trace represents the autocorrelation function of the composite waveform w_k. From assumption (2) it follows that the autocorrelation function can be used to define the shape of the waveform, the

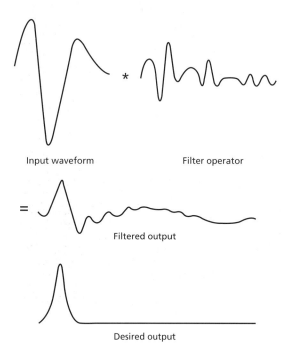

Input waveform   Filter operator

Filtered output

Desired output

Fig. 4.20 The principle of Wiener filtering.

necessary phase information coming from the minimum-delay assumption.

Such an approach allows prediction of the shape of the composite waveform for use in Wiener filtering. A particular case of Wiener filtering in seismic deconvolution is that for which the desired output is a spike function. This is the basis of *spiking deconvolution*, also known as *whitening deconvolution* because a spike has the amplitude spectrum of *white noise* (i.e. all frequency components have the same amplitude).

A wide variety of deconvolution operators can be designed for inverse filtering of real seismic data, facilitating the suppression of multiples (dereverberation and deghosting) and the compression of reflected pulses. The presence of short-period reverberation in a seismogram is revealed by an autocorrelation function with a series of decaying waveforms (Fig. 4.21(a)). Long-period reverberations appear in the autocorrelation function as a series of separate side lobes (Fig. 4.21(b)), the lobes occurring at lag values for which the primary reflection aligns with a multiple reflection. Thus the spacing of the side lobes represents the periodicity of the reverberation pattern. The first multiple is phase-reversed with respect to the primary reflection, due to reflection at the ground surface or the base of the weathered layer. Thus the first side lobe has a negative peak resulting from cross-correlation of the out-of-phase signals. The second multiple undergoes a further phase reversal so that it is in phase with the primary reflection and therefore gives rise to a second side lobe with a positive peak (see Fig. 4.21(b)). Autocorrelation functions such as those shown in Fig. 4.21 form the basis of predictive deconvolution operators for removing reverberation events from seismograms.

(a)

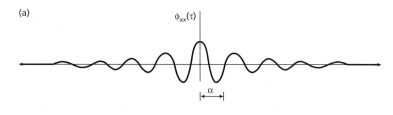

$\phi_{xx}(\tau)$

(b)

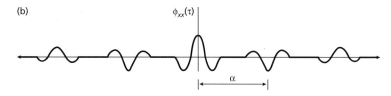

$\phi_{xx}(\tau)$

Fig. 4.21 Autocorrelation functions of seismic traces containing reverberations. (a) A gradually decaying function indicative of short-period reverberation. (b) A function with separate side lobes indicative of long-period reverberation.

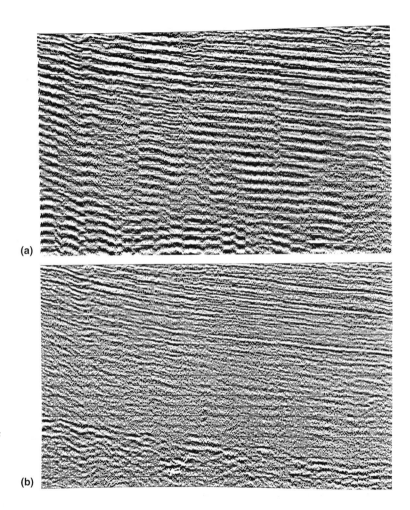

Fig. 4.22 Removal of reverberations by predictive deconvolution. (a) Seismic record dominated by strong reverberations. (b) Same section after spiking deconvolution. (Courtesy Prakla Seismos GmbH.)

(a)

(b)

Practically achievable inverse filters are always approximations to the ideal filter that would produce a reflectivity function from a seismic trace: firstly, the ideal filter operator would have to be infinitely long; secondly, predictive deconvolution makes assumptions about the statistical nature of the seismic time series that are only approximately true. Nevertheless, dramatic improvements to seismic sections, in the way of multiple suppression and associated enhancement of vertical resolution, are routinely achieved by predictive deconvolution. An example of the effectiveness of predictive deconvolution in improving the quality of a seismic section is shown in Fig. 4.22. Deconvolution may be carried out on individual seismic traces before stacking (*deconvolution before stacking*: DBS) or on CMP stacked traces (*deconvolution after stacking*: DAS), and is commonly employed at both these stages of data processing.

4.8.3 Velocity filtering

The use of *velocity filtering* (also known as *fan filtering* or *pie slice filtering*) is to remove coherent noise events from seismic records on the basis of the particular angles at which the events dip (March & Bailey 1983). The angle of dip of an event is determined from the apparent velocity with which it propagates across a spread of detectors.

A seismic pulse travelling with velocity v at an angle α to the vertical will propagate across the spread with an apparent velocity $v_a = v/\sin \alpha$ (Fig. 4.23). Along the spread direction, each individual sinusoidal component of the pulse will have an apparent wavenumber k_a related to its individual frequency f, where

$$f = v_a k_a$$

Hence, a plot of frequency f against apparent wavenumber k_a for the pulse will yield a straight-line curve with a gradient of v_a (Fig. 4.24). Any seismic event propagating across a surface spread will be characterized by an f–k curve radiating from the origin at a particular gradient determined by the apparent velocity with which the event passes across the spread. The overall set of curves for a typical shot gather containing reflected and surface propagating seismic events is shown in Fig. 4.25. Events that appear to travel across the spread away from the source will plot in the positive wavenumber field; events travelling towards the source, such as backscattered rays, will plot in the negative wavenumber field.

It is apparent that different types of seismic event fall within different zones of the f–k plot and this fact provides a means of filtering to suppress unwanted events on the basis of their apparent velocity. The normal means

by which this is achieved, known as f–k filtering, is to enact a two-dimensional Fourier transformation of the seismic data from the t–x domain to the f–k domain, then to filter the f–k plot by removing a wedge-shaped zone or zones containing the unwanted noise events (March & Bailey 1983), and finally to transform back into the t–x domain.

An important application of velocity filtering is

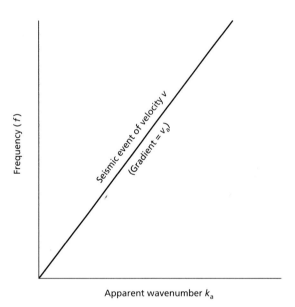

Fig. 4.24 An f–k plot for a seismic pulse passing across a surface spread of detectors.

Fig. 4.23 A wave travelling at an angle α to the vertical will pass across an in-line spread of surface detectors at a velocity of $v/\sin \alpha$.

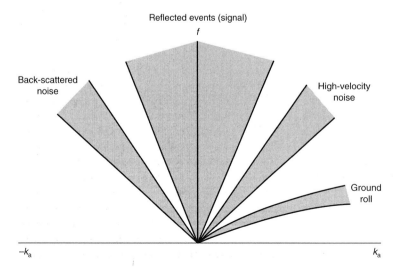

Fig. 4.25 An f–k plot for a typical shot gather (such as that illustrated in Fig. 4.8) containing reflection events and different types of noise.

the removal of ground roll from shot gathers. This leads to marked improvement in the subsequent stacking process, facilitating better estimation of stacking velocities and better suppression of multiples. Velocity filtering can also be applied to portions of seismic record sections, rather than individual shot gathers, in order to suppress coherent noise events evident because of their anomalous dip, such as diffraction patterns. An example of such velocity filtering is shown in Fig. 4.26.

It may be noted that individual detector arrays operate selectively on seismic arrivals according to their apparent velocity across the array (Section 4.4.2), and therefore function as simple velocity filters at the data acquisition stage.

4.9 Migration of reflection data

On seismic sections such as that illustrated in Fig. 4.22 each reflection event is mapped directly beneath the mid-point of the appropriate CMP gather. However, the reflection point is located beneath the mid-point only if the reflector is horizontal. In the presence of a component of dip along the survey line the actual reflection point is displaced in the up-dip direction; in the presence of a component of dip across the survey line (cross-dip) the reflection point is displaced out of the plane of the section. *Migration* is the process of reconstructing a seismic section so that reflection events are repositioned under their correct surface location and at a corrected vertical reflection time. Migration also improves the resolution of seismic sections by focusing energy spread over a Fresnel zone and by collapsing diffraction patterns produced by point reflectors and faulted beds. In *time migration*, the migrated seismic sections still have time as the vertical dimension. In *depth migration*, the migrated reflection times are converted into reflector depths using appropriate velocity information.

Two-dimensional survey data provide no information on cross-dip and, hence, in the migration of two-dimensional data the migrated reflection points are constrained to lie within the plane of the section. In the presence of cross-dip, this *two-dimensional migration* is clearly an imperfect process. Its inability to deal with effects of cross-dip mean that, even when the seismic line is along the geological strike, migration will be imperfect since the true reflection points are themselves out of the vertical section.

The conversion of reflection times recorded on non-

migrated sections into reflector depths, using one-way reflection times multiplied by the appropriate velocity, yields a reflector geometry known as the *record surface*. This coincides with the actual *reflector surface* only when the latter is horizontal. In the case of dipping reflectors the record surface departs from the reflector surface; that is, it gives a distorted picture of the reflector geometry. Migration removes the distorting effects of dipping reflectors from seismic sections and their associated record surfaces. Migration also removes the diffracted arrivals resulting from point sources since every diffracted arrival is migrated back to the position of the point source. A variety of geological structures and sources of diffraction are illustrated in Fig. 4.27(a) and the resultant non-migrated seismic section is shown in Fig. 4.27(b). Structural distortion in the non-migrated section (and record surfaces derived from it) includes a broadening of anticlines and a narrowing of synclines. The edges of fault blocks act as point sources and typically give rise to strong diffracted phases, represented by hyperbolic patterns of events in the seismic section. Synclines within which the reflector curvature exceeds the curvature of the incident wavefront are represented on non-migrated seismic sections by a 'bow-tie' event resulting from the existence of three discrete reflection points for any surface location (see Fig. 4.28).

Various aspects of migration are discussed below using the simplifying assumption that the source and detector have a common surface position (i.e. the detector has a zero offset, which is approximately the situation involved in CMP stacks). In such a case, the incident and reflected rays follow the same path and the rays are normally incident on the reflector surface. Consider a source–detector on the surface of a medium of constant seismic velocity (Fig. 4.29). Any reflection event is conventionally mapped to lie directly beneath the source–detector but in fact it may lie anywhere on the locus of equal two-way reflection times, which is a semi-circle centred on the source–detector position.

Now consider a series of source–detector positions overlying a planar dipping reflector beneath a medium of uniform velocity (Fig. 4.30). The reflection events are mapped to lie below each source–detector location but the actual reflection points are offset in the updip direction. The construction of arcs of circles (wavefront segments) through all the mapped reflection points enables the actual reflector geometry to be mapped. This represents a simple example of migration. The migrated section indicates a steeper reflector dip than the record surface derived from the non-migrated section. In gen-

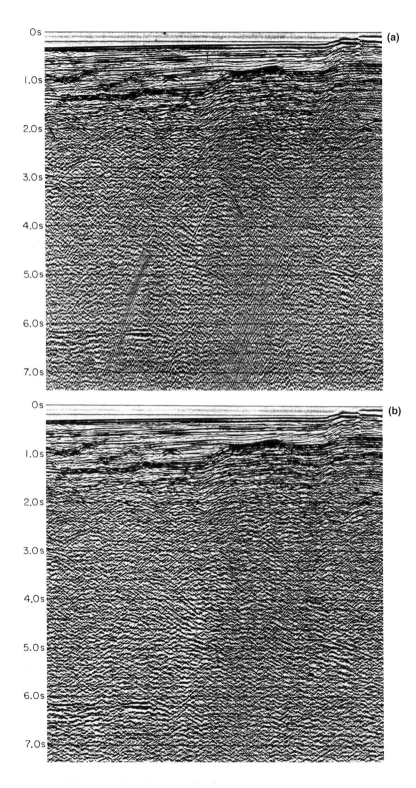

Fig. 4.26 The effect of *f–k* filtering of a seismic section. (a) Stacked section showing steeply-dipping coherent noise events, especially below 4.5 s two-way reflection time. (b) The same section after rejection of noise by *f–k* filtering (Courtesy Prakla-Seismos GmbH).

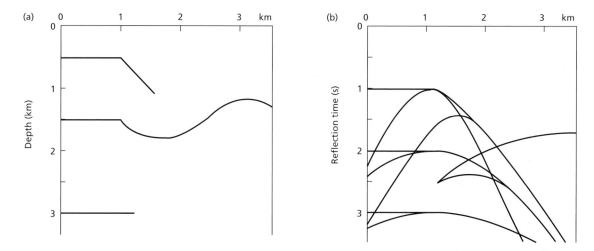

Fig. 4.27 (a) A structural model of the subsurface and (b) the resultant reflection events that would be observed in a non-migrated seismic section, containing numerous diffraction events. (After Sheriff 1978.)

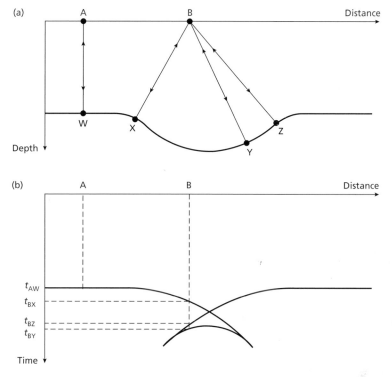

Fig. 4.28 (a) A sharp synclinal feature in a reflecting interface, and (b) the resultant 'bow-tie' shape of the reflection event on the non-migrated seismic section.

eral, if α_s is the dip of the record surface and α_t is the true dip of the reflector, $\sin \alpha_t = \tan \alpha_s$. Hence the maximum dip of a record surface is 45° and represents the case of horizontal reflection paths from a vertical reflector. This *wavefront common-envelope* method of migration can be extended to deal with reflectors of irregular geometry. If there is a variable velocity above the reflecting surface to be migrated, the reflected ray paths are not straight and the associated wavefronts are not circular. In such a case, a *wavefront chart* is constructed for the prevailing velocity–depth relationship and this is used to construct the wavefront segments passing through each reflection event to be migrated.

An alternative approach to migration is to assume that any continuous reflector is composed of a series of closely-spaced point reflectors, each of which is a source of diffractions, and that the continuity of any reflection event results from the constructive and destructive interference of these individual diffraction events. A set of diffracted arrivals from a single point reflector embedded in a uniform-velocity medium is shown in Fig. 4.31. The two-way reflection times to different surface locations define a hyperbola. If arcs of circles (wavefront segments) are drawn through each reflection event, they intersect at the actual point of diffraction (Fig. 4.31). In the case of a variable velocity above the point reflector the diffraction event will not be a hyperbola but a curve of similar convex shape. No reflection event on a seismic section can have a greater convexity than a diffraction event, hence the latter is referred to as a *curve of maximum convexity*. In *diffraction migration* all dipping reflection events are assumed to be tangential to some curve of maximum con-

vexity. By the use of a wavefront chart appropriate to the prevailing velocity–depth relationship, wavefront segments can be drawn through dipping reflection events on seismic sections and the events migrated back to their diffraction points (Fig. 4.31). Events so migrated will, overall, map the prevailing reflector geometry.

All modern approaches to migration use the *seismic wave equation* which is a partial differential equation describing the motion of waves within a medium that have been generated by a wave source. The migration problem can be considered in terms of wave propagation through the ground in the following way. For any reflection event, the form of the seismic wavefield at the surface can be reconstructed from the travel times of reflected arrivals to different source–detector locations. For the purpose of migration it is required to reconstruct the form of the wavefield within the ground, in the vicinity of a reflecting interface. This reconstruction can be achieved by solution of the wave equation, effectively tracing the propagation of the wave backwards in time. Propagation of the wavefield of a reflection event halfway back to its origin time should place the wave on the reflecting interface, hence, the form of the wavefield at that time should define the reflector geometry.

Migration using the wave equation is known as *wave equation migration* (Robinson & Treitel 2000). There are several approaches to the problem of solving the wave equation and these give rise to specific types of wave equation migration such as *finite difference migration*, in which the wave equation is approximated by a finite difference equation suitable for solution by computer, and *frequency-domain migration*, in which the wave equation is solved by means of Fourier transformations, the necessary spatial transformations to achieve migration being enacted in the frequency domain and recovered by an inverse Fourier transformation.

Migration by computer can also be carried out by

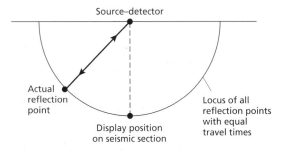

Fig. 4.29 For a given reflection time, the reflection point may be anywhere on the arc of a circle centred on the source–detector position. On a non-migrated seismic section the point is mapped to be immediately below the source–detector.

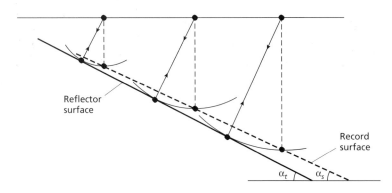

Fig. 4.30 A planar-dipping reflector surface and its associated record surface derived from a non-migrated seismic section.

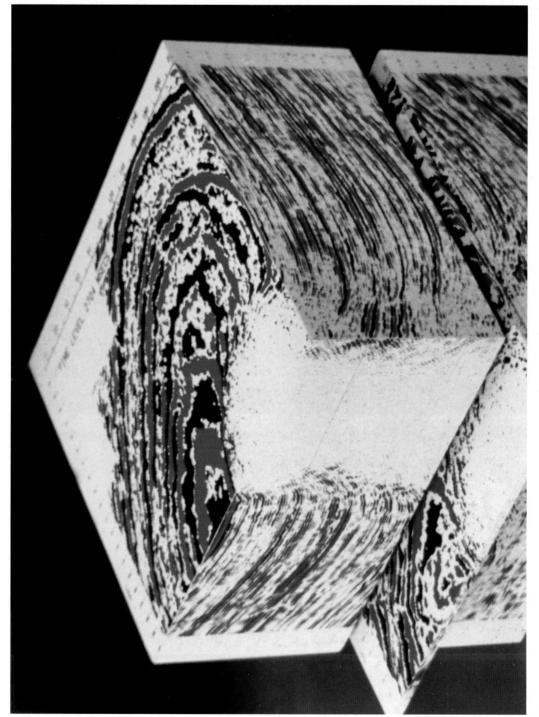

Plate 4.1 Three-dimensional data volume showing a Gulf of Mexico salt dome with an associated rim syncline. (Reproduced from *AAPG Memoir* No. 42, with the permission of the publishers.)

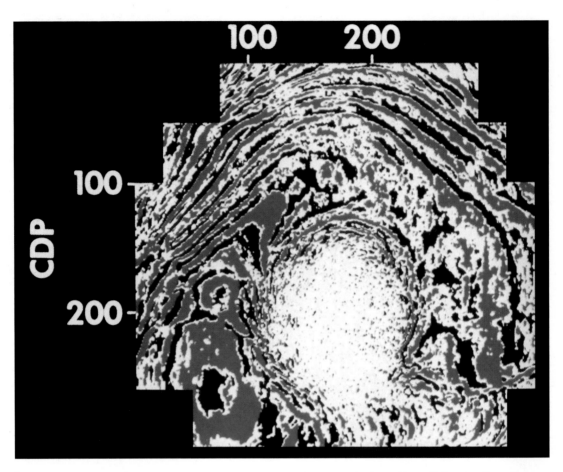

Plate 4.2 Seiscrop section at 3760 ms from a three-dimensional survey in the Eugene island area of the Gulf of Mexico. (Reproduced from *AAPG Memoir* No. 42, with the permission of the publishers.)

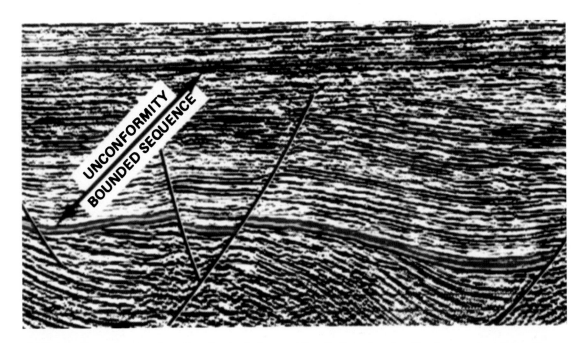

Plate 4.3 A seismic section from the northern Amadeus basin, central Australia, illustrating a depositional sequence bounded by major unconformities. (Reproduced from *AAPG Memoir* No. 39, with the permission of the publishers.)

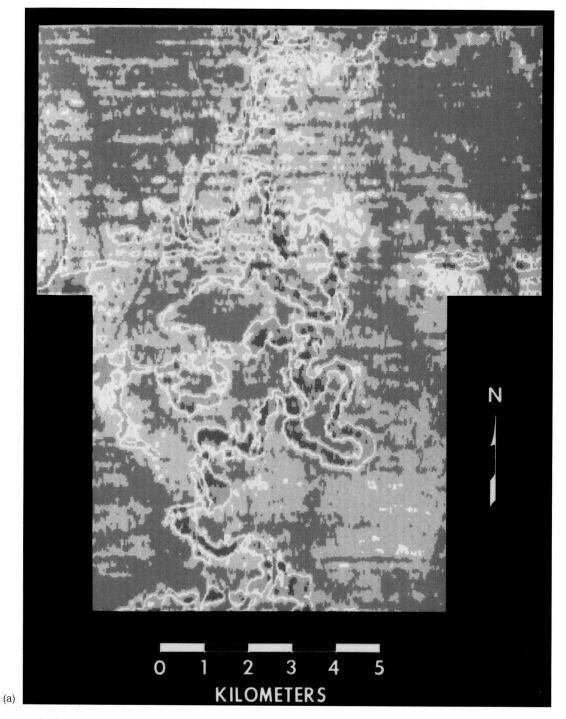

N

0 1 2 3 4 5
KILOMETERS

(a)

Plate 4.4 (a) Seiscrop section at 196 ms from a three-dimensional survey in the Gulf of Thailand area, showing a meandering stream channel.

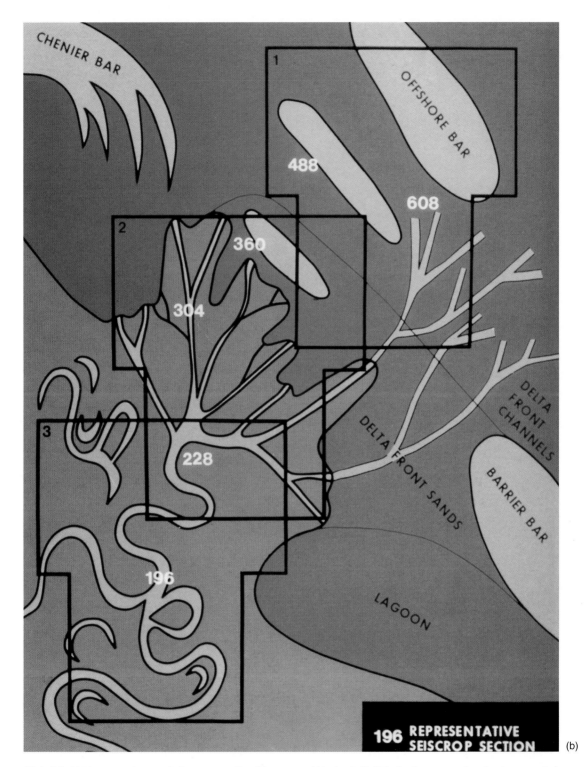

Plate 4.4 (b) Diagrammatic map of a former prograding delta system within the Gulf of Thailand survey area, based on interpretation of seiscrop sections 1, 2 and 3 shown on map. (Both illustrations reproduced from *AAPG Memoir* No. 42, with the permission of the publishers.)

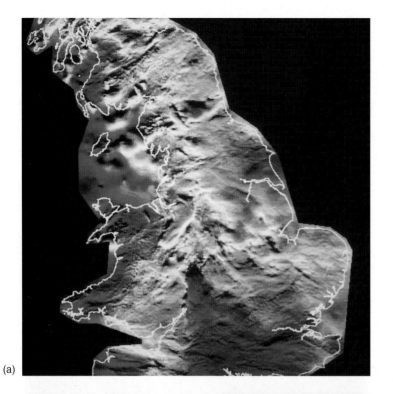

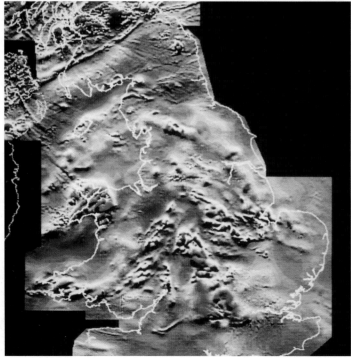

Plate 5.1 (a) Colour shaded–relief image of the gravity field of Central Britain illuminated from the north. Blue represents low values, red high values. (b) Colour shaded–relief image of the magnetic field of Central Britain illuminated from the north. Blue represents low values, red high values. (Both illustrations reproduced from Lee *et al.* 1990, with permission.)

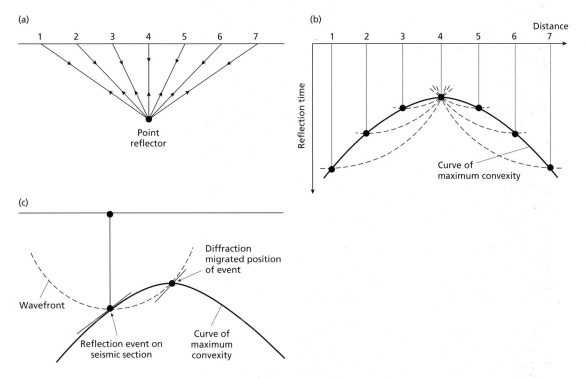

Fig. 4.31 Principles of diffraction migration. (a) Reflection paths from a point reflector. (b) Migration of individual reflection events back to position of point reflector. (c) Use of wavefront chart and curve of maximum convexity to migrate a specific reflection event; the event is tangential to the appropriate curve of maximum convexity, and the migrated position of the event is at the intersection of the wavefront with the apex of the curve.

direct modelling of ray paths through hypothetical models of the ground, the geometry of the reflecting interfaces being adjusted iteratively to remove discrepancies between observed and calculated reflection times. Particularly in the case of seismic surveys over highly complex subsurface structures, for example those encountered in the vicinity of salt domes and salt walls, this *ray trace migration* method may be the only method capable of successfully migrating the seismic sections.

In order to migrate a seismic section accurately it would be necessary to define fully the velocity field of the ground; that is, to specify the value of velocity at all points. In practice, for the purposes of migration, an estimate of the velocity field is made from prior analysis of the non-migrated seismic section, together with information from borehole logs where available. In spite of this approximation, migration almost invariably leads to major improvement in the seismic imaging of reflector geometry.

Migration of seismic profile data is normally carried out on CMP stacks, thus reducing the number of traces to be migrated by a factor equal to the fold of the survey and thereby reducing the computing time and associated costs. Migration of stacked traces is based on the assumption that the stacks closely resemble the form of individual traces recorded at zero offset and containing only normal-incidence reflection events. This assumption is clearly invalid in the case of recordings over a wide range of offsets in areas of structural complexity. A better approach is to migrate the individual seismic traces (assembled into a series of profiles containing all traces with a common offset), then to assemble the migrated traces into CMP gathers and stacks. Such an approach is not necessarily cost-effective in the case of high-fold CMP surveys, and a compromise is to migrate subsets of CMP stacks recorded over a narrow range of offset distances, and then produce a full CMP stack by summing the migrated partial stacks after correction for normal moveout. Procedures involving migration before final stacking involve extra cost but can lead to significant improvements in the migrated sections and to more reliable stacking velocities.

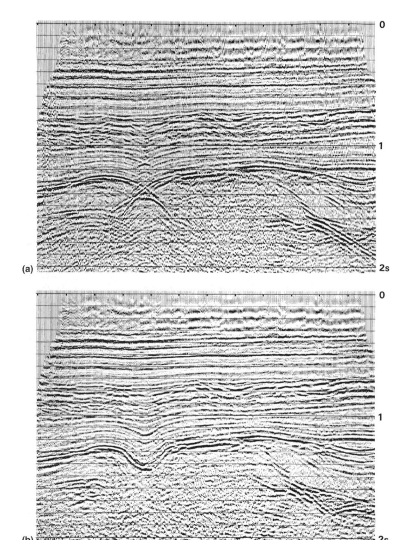

Fig. 4.32 (a) A non-migrated seismic section. (b) The same seismic section after wave equation migration. (Courtesy Prakla-Seismos GmbH.)

Any system of migration represents an approximate solution to the problem of mapping reflecting surfaces into their correct spatial positions and the various methods have different performances with real data. For example, the diffraction method performs well in the presence of steep reflector dips but is poor in the presence of a low SNR. The best all round performance is given by frequency-domain migration. Examples of the migration of seismic sections are illustrated in Figs 4.32 and 4.33. Note in particular the clarification of structural detail, including the removal of bow-tie effects, and the repositioning of structural features in the migrated sections. Clearly, when planning to test hydrocarbon

prospects in areas of structural complexity (as on the flank of a salt dome) it is important that drilling locations are based on interpretation of migrated rather than non-migrated seismic sections.

4.10 3D seismic reflection surveys

The general aim of three-dimensional surveys is to achieve a higher degree of resolution of the subsurface geology than is achievable by two-dimensional surveys. Three-dimensional survey methods involve collecting field data in such a way that recorded arrivals are not re-

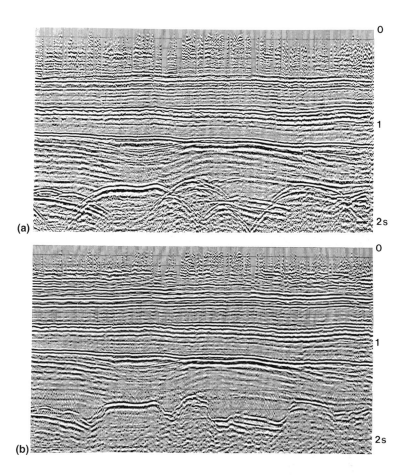

Fig. 4.33 (a) A non-migrated seismic section. (b) The same seismic section after diffraction migration. (Courtesy Prakla-Seismos GmbH.)

stricted to rays that have travelled in a single vertical plane. In a three-dimensional survey, the disposition of shots and receivers is such that groups of recorded arrivals can be assembled that represent rays reflected from an area of each reflecting interface. Three-dimensional surveying therefore samples a volume of the subsurface rather than an area contained in a vertical plane, as in two-dimensional surveying.

In three-dimensional surveying the common midpoint principle applies similarly, but each CMP gather involves an areal rather than a linear distribution of shot points and detector locations (Fig. 4.34). Thus, for example, a 20-fold coverage is obtained in a crossed-array three-dimensional survey if reflected ray paths from five shots along different shot lines to four detectors along different recording lines all have a common reflection point.

On land, three-dimensional data are normally collected using the *crossed-array method* in which shots and detectors are distributed along orthogonal sets of lines (in-lines and cross-lines) to establish a grid of recording points. For a single pair of lines, the areal coverage of a subsurface reflector is illustrated in Fig. 4.35.

At sea, three-dimensional data may be collected along closely-spaced parallel tracks with the hydrophone streamer feathered to tow obliquely to the ship's track such that it sweeps across a swathe of the sea floor as the vessel proceeds along its track. By ensuring that the swathes associated with adjacent tracks overlap, data may be assembled to provide areal coverage of subsurface reflectors. In the alternative *dual source array method*, sources are deployed on side gantries to port and starboard of the hydrophone streamer and fired alternately (Fig. 4.36). Multiple streamers may similarly be deployed to obtain both a wider swath and a denser fold of three-dimensional data.

High-quality position fixing is a prerequisite of three-dimensional marine surveys in order that the locations of

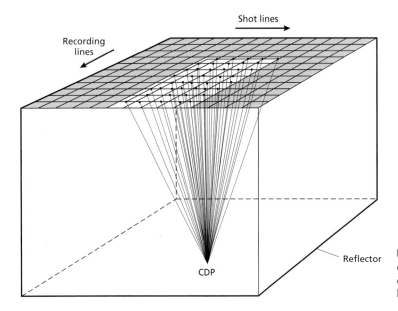

Fig. 4.34 Reflected ray paths defining a common depth point from an areal distribution of shot points and detector locations in a three-dimensional survey.

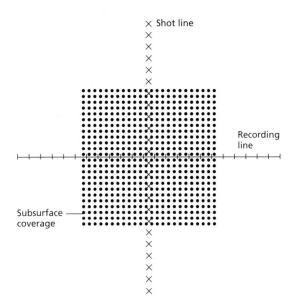

Fig. 4.35 The areal coverage derived from a single pair of crossing lines in a three-dimensional survey. Each dot represents the mid-point between a shot and a detector.

induced by radio wave distortion in the atmosphere. This can be corrected using a reference ground station of known position. In this case the system is termed differential GPS (DGPS) and can give real-time accuracy to within a few metres. In near-shore areas use may be made of radio navigation systems, in which a location is determined by calculation of range from onshore radio transmitters. Doppler sonar may also be employed to determine the velocity of the vessel along the survey track for comparison with GPS satellite fixes (Lavergne 1989).

The areal reflector coverage obtained in three-dimensional surveying provides the additional information necessary to permit full *three-dimensional migration* in which reflection points can be migrated in any azimuthal direction. This ability fully to migrate three-dimensional survey data further enhances the value of such surveys over two-dimensional surveys in areas of complex structure.

The essential difference between two-dimensional and three-dimensional migration may be illustrated with reference to a point reflector embedded in a homogeneous medium. On a seismic section derived from a two-dimensional survey the point reflector is imaged as a diffraction hyperbola, and migration involves summing amplitudes along the hyperbolic curve and plotting the resultant event at the apex of the hyperbola (see Fig. 4.31). The actual three-dimensional pattern associated with a point reflector is a hyperboloid of rotation,

all shot–detector mid-points are accurately determined. Position fixing is normally achieved using the global positioning system (GPS). The standard form of the system, as now widely available in personal handsets, is not accurate enough for seismic surveying due to the errors

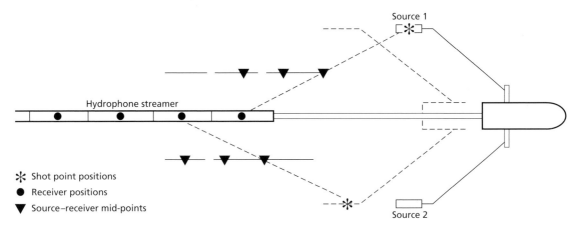

Fig. 4.36 The dual source array method of collecting three-dimensional seismic data at sea. Alternate firing of sources 1 and 2 into the hydrophone streamer produces two parallel sets of source–detector mid-points.

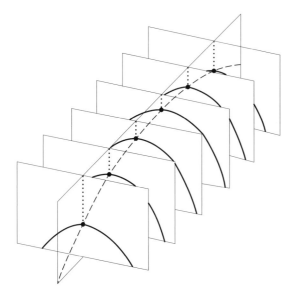

Fig. 4.37 The two-pass method of three-dimensional migration for the case of a point reflector. The apices of diffraction hyperbolas in one line direction may be used to construct a diffraction hyperbola in the orthogonal line direction. The apex of the latter hyperbola defines the position of the point reflector.

the diffraction hyperbola recorded in a two-dimensional survey representing a vertical slice through this hyperboloid. In a three-dimensional survey, reflections are recorded from a surface area of the hyperboloid and three-dimensional migration involves summing amplitudes over the surface area to define the apex of the hyperboloid.

A practical way of achieving this aim with crossed-array data from a three-dimensional land survey is the *two-pass* method (Fig. 4.37). The first pass involves collapsing diffraction hyperbolas recorded in vertical sections along one of the orthogonal line directions. The series of local apices in these sections together define a hyperbola in a vertical section along the perpendicular direction. This hyperbola can then be collapsed to define the apex of the hyperboloid.

The product of three-dimensional seismic surveying is a volume of data (Fig. 4.38, Plate 4.1) representing reflection coverage from an area of each subsurface reflector. From this reflection data volume, conventional two-dimensional seismic sections may be constructed not only along the actual shot lines and recording lines employed but also along any other vertical slice through the data volume. Hence, seismic sections may be simulated for any azimuth across the survey area by taking a vertical slice through the data volume, and this enables optimal two-dimensional representation of any recorded structural features.

More importantly, horizontal slices may be taken through the data volume to display the pattern of reflections intersected by any time plane. Such a representation of the three-dimensional data is known as a *time slice* or *seiscrop*, and analysis of reflection patterns displayed in time slices provides a powerful means of mapping three-dimensional structures (see Plates 4.1 & 4.2). In particular, structures may be traced laterally through the data volume, rather than having to be interpolated between adjacent lines as is the case in two-dimensional surveys. The manipulation of data volumes obtained from three-

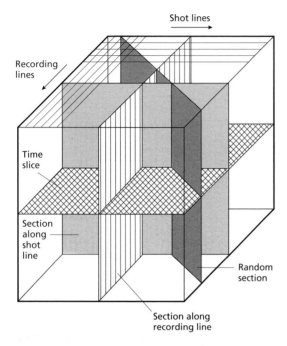

Fig. 4.38 The reflection data volume obtained from a three-dimensional seismic survey. By taking vertical slices through this data volume, it is possible to generate seismic sections in any azimuthal direction; by taking horizontal slices (time slices), the areal distribution of reflection events can be studied at any two-way reflection time.

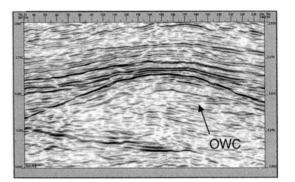

Fig. 4.39 Seismic section from a 3D data volume showing the horizontal reflector produced by the oil–water contact. This is clearly distinguishable from the reflections from geological formations due to its strictly horizontal nature. Example from the Fulmar field, UK North Sea. (From Jack 1997.)

dimensional surveys is carried out at computer work stations using software routines that enable seismic sections and time slices to be displayed as required. Automatic event picking and contouring are also facilitated (Brown 1986).

On high-quality modern seismic data it is quite common to image the oil–water contact within a hydrocarbon reservoir (Fig. 4.39), or the bright spot, a particularly strong reflection, caused by the high reflection coefficient at the top of a gas-filled formation.

4.11 Three component (3C) seismic reflection surveys

All the previous discussion has only considered seismic recording using vertical geophones. These only record one component of the total seismic wave motion. Vertical geophones are chosen in preference since they are most sensitive to vertically travelling P-waves. The actual ground motion consists of movement in all direc-

tions. This can be measured fully by having three geophones at each location, oriented mutually at right angles, and each recording one component. Thus three components of the ground motion are recorded, giving the method its name. Often these are labelled as having their sensitive axes oriented to vertical, north–south and east–west, though any set of orthogonal components is sufficient. In this case the true ground motion is fully recorded, and can be analysed in detail.

The three component (3C) technique requires three times as many recording sensors, and more stages of data analysis than vertical component recording. With developing technology, the additional sophistication of the field equipment (Fig. 4.40) and the availability of large computing power for the data analysis have made 3C recording practicable. In fact, 3C data recording is becoming increasingly common, and is now a routine operation in the exploration for hydrocarbons.

The analysis of 3C data provides two major benefits. These are the ability to identify S-waves in addition to P-waves in the same data, and the ability to perform more sophisticated filtering to identify and remove unwanted wave energy, whether from surface waves, or noise sources. Naturally the improved filtering contributes substantially to the ability to detect the separate P- and S-waves. S-waves are generated at any interface where a P-wave is obliquely incident (see Section 3.6.2). Thus, any seismic data will always contain energy from both P- and S-waves. With appropriate processing, principally exploiting the different particle motions and velocities of the two waves, the P-wave and S-wave energy can be separated and analysed.

Fig. 4.40 A three-component geophone.

Knowledge of the behaviour of both body waves provides important additional information. In a lithified rock formation, such as an oil reservoir, the P-wave is transmitted through both the rock matrix and the fluids in the pore spaces. The behaviour of the P-wave is thus determined by the average of the rock matrix and pore fluid properties, weighted with respect to the porosity of the rock.

The S-wave on the other hand is only transmitted through the rock matrix, since the shear wave cannot propagate through a fluid. Comparison of the P-wave and S-wave velocities of the same formation thus can give information about the porosity of the formation and the nature of the fluids filling the pore spaces. The relationships can be complex, but the presence of hydrocarbons, especially if accompanied by gas, can be identified directly from the seismic data in favourable circumstances. Derivation of measures which reliably predict the presence of hydrocarbons, *direct hydrocarbon indicators* (DHIs), is an important part of modern seismic processing (Yilmaz 1987, 2001), though the details of this are beyond the scope of this book.

The ability to detect these features is an enormous advantage to the hydrocarbon industry and has had a marked effect on the success rate of exploration boreholes in locating oil or gas reservoirs. Since the cost of drilling a borehole can often reach or exceed $10 m, the additional effort in seismic data acquisition and processing is very cost-effective.

4.12 4D seismic surveys

Once an oilfield is in production, the oil and/or gas is extracted and its place in the pore spaces of the reservoir rock is taken by inflowing groundwater. Since the pore fluids are changing, the seismic response of the formations also changes. Even in an extensively developed field with many wells, there are large intervals between the wells, of the order of 1 km. It is impossible from monitoring the well-flow to be sure how much of the hydrocarbon is being extracted from any particular part of the reservoir. Often oil reservoirs are cut by numerous faults and some of them may isolate a volume of the reservoir so that the hydrocarbons cannot flow to the nearby wells. If the location of such isolated 'pools' can be found, additional wells can be drilled to extract these pools and hence increase the overall hydrocarbon recovery from the reservoir.

It is apparent that if the location of such features as the oil–water contact and gas accumulations can be mapped with a seismic survey, then repeated surveys at time intervals during the production of the field offer the prospect of monitoring the extraction of hydrocarbons, and contribute to the management of the production phase of the field operation. This is the rationale for 4D seismic surveys, which essentially consist of the repeated shooting of 3D (and often 3C) surveys over a producing field at regular intervals. The fourth dimension is, of course, time.

The practical implementation of 4D surveying is far from simple (Jack 1997). The essential measurements made by a seismic survey are the values of amplitudes of seismic waves at specific locations and times after a seismic source has been fired. Any factor which affects the location, amplitude or timing of seismic waves must be allowed for when comparing two sets of data recorded in different surveys. Obvious effects would be different geophones in different locations, for each survey. Other effects are much more subtle. The seasonal change in level of the water table may be enough to affect the travel time of seismic waves in the near-surface such that all deep reflections will be systematically mistimed between two surveys in different seasons. As an oilfield develops, the increased plant (pumps, drill-rigs, vehicles) changes (and increases) the background seismic noise with time. In the processing of the raw data to make the final seismic sections for comparison many different mathematical operations change the amplitudes of the data. Each of these must be rigorously checked and identical processing must be carried out for each separate dataset.

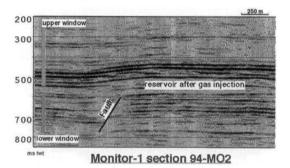

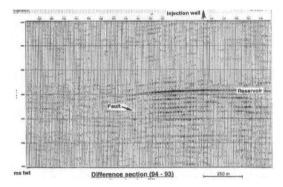

Fig. 4.41 Repeat surveys showing the effect of gas being pumped into a formation for storage. (a) Before gas injection; (b) after gas injection; (c) difference section composed by subtracting (b) from (a). (From Jack 1997.)

The primary properties of the reservoir which change with time as hydrocarbon extraction proceeds are the pore fluid pressure, the nature of the pore fluids, and the temperature. Each of these may have an effect on the seismic response. Changes in fluid pressure will affect the state of stress in the rock matrix combined with temperature, will directly affect factors such as the exsolution of gas from hydrocarbon fluids. That these features can be observed in seismic data has been tested directly by large-scale experiments in producing fields. In some cases gas has been directly pumped into permeable formations to displace pore water, and repeat seismic surveys conducted to monitor the effect (Fig. 4.41). There are also now well-documented case studies of clear location of the more complex effect of steam-flood of reservoirs (Fig. 4.42). In this case pumping of steam into a reservoir has a complex effect of liberating gas dissolved in oil, condensing to form water, and also replacing the oil and gas by uncondensed steam. The figure shows the observable seismic effects of this action over 31 months. The data can be modelled to show that the seismic monitoring is allowing real-time study of the fluid flow in the reservoir (Jack 1997). This ability to monitor producing reservoirs has major importance in allowing sophisticated control of reservoir engineering and production operations.

The economic importance of 4D seismic surveying to the oil industry is apparent. Increasing the oil recovery from a producing oilfield increases the financial return on the huge investment needed to establish a new field and its infrastructure. Relative to this, a 4D seismic survey at perhaps $30 m represents only a marginal cost. Plans are actively being developed to install permanent seismic recording instrumentation over oilfields to facilitate repeat surveys. If all the recording equipment is permanently installed, although this is a large initial expense, much of the difficulty in recording later directly comparable datasets is removed, and only a seismic source is needed. The future prospect is of hydro-

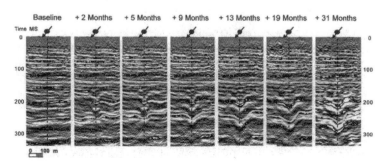

Fig. 4.42 Seismic sections from repeated 3D seismic surveys across an oilfield with steam injection at the well marked on the profile. With increasing duration of steam injection the seismic velocity of the reservoir formation changes progressively, shown by the pull-up, then push-down in the reflection from the base of the reservoir. These changes are due to the changing pore fluid with time. Duri oilfield, Indonesia. (From Jack 1997.)

carbon fields where 4D seismic surveys are routinely used for the management of the production from the field. It can be argued that there is more hydrocarbon resource to be recovered by careful monitoring of known fields, than by exploration for new fields.

4.13 Vertical seismic profiling

Vertical seismic profiling (VSP) is a form of seismic reflection surveying that utilizes boreholes. Shots are normally fired at surface, at the wellhead or offset laterally from it, and recorded at different depths within the borehole using special detectors clamped to the borehole wall. Alternatively, small shots may be fired at different depths within the borehole and recorded at surface using conventional geophones, but in the following account the former configuration is assumed throughout. Typically, for a borehole 1 km or more deep, seismic data are recorded at more than 100 different levels down the borehole. If the surface shot location lies at the wellhead vertically above the borehole detector locations, so that the recorded rays have travelled along vertical ray paths, the method is known as *zero-offset VSP*. If the surface shot locations are offset laterally, so that the recorded rays have travelled along inclined ray paths, the method is known as *offset VSP* (Fig. 4.43).

VSP has several major applications in seismic exploration (Cassell 1984). Perhaps most importantly, reflection events recorded on seismic sections obtained at surface from conventional reflection surveys can be traced by VSP to their point of origin in the subsurface, thus calibrating the seismic sections geologically. Ambiguity as to whether particular events observed on conventional seismic sections represent primary or multiple reflections can be removed by direct comparison of the sections with VSP data. The reflection properties of particular horizons identified in the borehole section can be investigated directly using VSP and it can therefore be determined, for example, whether or not an horizon returns a detectable reflection to the surface.

Uncertainty in interpreting subsurface geology using conventional seismic data is in part due to the surface location of shot points and detectors. VSP recording in a borehole enables the detector to be located in the immediate vicinity of the target zone, thus shortening the overall path length of reflected rays, reducing the effects of attenuation, and reducing the dimensions of the Fresnel zone (Section 4.4.1). By these various means, the overall accuracy of a seismic interpretation may be

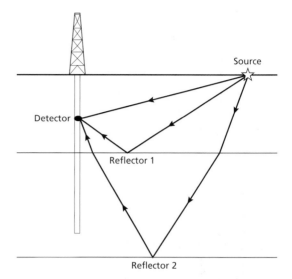

Fig. 4.43 An offset VSP survey configuration.

markedly increased. A particular uncertainty in conventional seismics is the nature of the downgoing pulse that is reflected back to surface from layer boundaries. This uncertainty often reduces the effectiveness of deconvolution of conventional seismic data. By contrast, an intrinsic feature of VSP surveys is that both downgoing and upgoing rays are recorded, and the waveform of the downgoing pulse may be used to optimize the design of a deconvolution operator for inverse filtering of VSP data to enhance resolution. Direct comparison with such VSP data leads to much improved reliability in the geological interpretation of seismic sections recorded at the surface in the vicinity of the borehole.

The nature of VSP data may be considered by reference to Fig. 4.44, which illustrates a synthetic zero-offset VSP dataset for the velocity–depth model shown, each trace being recorded at a different depth. Two sets of events are recorded which have opposite directions of dip in the VSP section. Events whose travel time increases as a function of detector depth represent downgoing rays; the weaker events, whose travel time reduces as a function of detector depth, represent upgoing, reflected rays. Note that the direct downgoing pulse (the first arrival, D0) is followed by other events (DS1, DS2, DS3) with the same dip, representing downgoing near-surface and peg-leg multiples. Each reflected event (U1, U2, U3) terminates at the relevant reflector depth, where it intersects the direct downgoing event.

For most purposes, it is desirable to separate downgo-

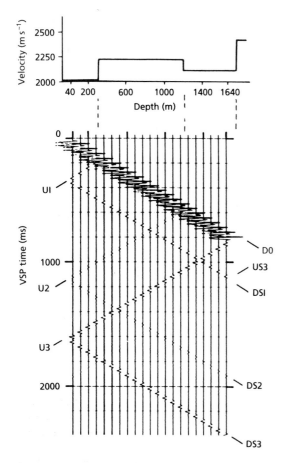

Fig. 4.44 A synthetic zero-offset VSP record section for the velocity–depth model shown. The individual traces are recorded at the different depths shown. D0 is the direct downgoing wave; DS1, DS2 and DS3 are downgoing waves with multiple reflections between the surface and interfaces 1, 2 and 3 respectively. U1, U2 and U3 are primary reflections from the three interfaces; US3 is a reflection from the third interface with multiple reflection in the top layer. (From Cassell 1984.)

ing and upgoing events to produce a VSP section retaining only upgoing, reflected arrivals. The opposite dip of the two types of event in the original VSP section enables this separation to be carried out by *f–k* filtering (see Section 4.8.3). Figure 4.45(a) illustrates a synthetic VSP section after removal of downgoing events. The removal of the stronger downgoing events has enabled representation of the upgoing events at enhanced amplitude, and weak multiple reflection events are now revealed. Note that these terminate at the same depth as the relevant primary event, and therefore do not extend to the point of intersection with the direct downgoing event. It is

now possible to apply a time correction to each trace in the VSP section, based on the travel time of the downgoing direct event, in order to predict the form of seismic trace that would be obtained at surface (Fig. 4.45(b)). By stacking these traces within a time corridor that avoids the multiple events, it is possible to produce a stacked trace containing only primary reflection events. Comparison of this stacked trace with a conventional seismic section from the vicinity of the borehole (Fig. 4.46) enables the geological content of the latter to be identified reliably.

4.14 Interpretation of seismic reflection data

Differing procedures are adopted for the interpretation of two- and three-dimensional seismic data. The results of two-dimensional surveys are presented to the seismic interpreter as non-migrated and migrated seismic sections, from which the geological information is extracted by suitable analysis of the pattern of reflection events. Interpretations are correlated from line to line, and the reflection times of picked events are compared directly at profile intersections. There are two main approaches to the interpretation of seismic sections: *structural analysis*, which is the study of reflector geometry on the basis of reflection times, and *stratigraphical analysis* (or *seismic stratigraphy*), which is the analysis of reflection sequences as the seismic expression of lithologically-distinct depositional sequences. Both structural and stratigraphical analyses are greatly assisted by *seismic modelling*, in which theoretical (synthetic) seismograms are constructed for layered models in order to derive insight into the physical significance of reflection events contained in seismic sections.

In the interpretation of three-dimensional survey data, the interpreter has direct access at a computer work station to all the reflection data contained within the seismic data volume (see Section 4.10), and is able to select various types of data for colour display, for example vertical sections or horizontal sections (time slices) through the data volume. The two most important shortcomings of two-dimensional interpretation are the problem of correlation between adjacent profile lines and the inaccuracy of reflector positioning due to the limitations of two-dimensional migration. The improved coverage and resolution of three-dimensional data often lead to substantial improvements in interpretation as compared with pre-existing two-dimensional interpretation. As with two-dimensional interpretation,

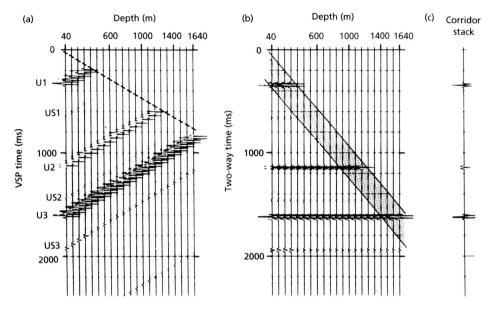

Fig. 4.45 (a) Synthetic VSP section of Fig. 4.44 with downgoing waves removed by filtering. (b) Each trace has been time shifted by the relevant uphole time to simulate a surface recording. (c) Stacked seismogram produced by stacking in the shaded corridor zone of part (b) to avoid multiple events. (From Cassell 1984.)

both structural and stratigraphic analysis may be carried out, and in the following sections examples are taken from both two- and three-dimensional survey applications.

4.14.1 Structural analysis

The main application of structural analysis of seismic sections is in the search for structural traps containing hydrocarbons. Interpretation usually takes place against a background of continuing exploration activity and an associated increase in the amount of information related to the subsurface geology. Reflection events of interest are usually colour-coded initially and labelled as, for example 'red reflector', 'blue reflector', until their geological significance is established. Whereas an initial interpretation of reflections displayed on seismic sections may lack geological control, at some point the geological nature of the reflectors is likely to become established by tracing reflection events back either to outcrop or to an existing borehole for stratigraphic control. Subsurface reflectors may then be referred to by an appropriate stratigraphical indicator such as 'base Tertiary', 'top Lias'.

Most structural interpretation is carried out in units of two-way reflection time rather than depth, and *time-structure* maps are constructed to display the geometry of selected reflection events by means of contours of equal reflection time (Fig. 4.47). *Structural contour maps* can be produced from time-structure maps by conversion of reflection times into depths using appropriate velocity information (e.g. local stacking velocities derived from the reflection survey or sonic log data from boreholes). Time-structure maps obviously bear a close similarity to structural contour maps but are subject to distortion associated with lateral or vertical changes of velocity in the subsurface interval overlying the reflector. Other aspects of structure may be revealed by contouring variations in the reflection time interval between two reflectors, sometimes referred to as *isochron maps*, and these can be converted into *isopach maps* by the conversion of reflection time intervals into thicknesses using the appropriate interval velocity.

Problems often occur in the production of time-structure or isochron maps. The difficulty of correlating reflection events across areas of poor signal-to-noise ratio, structural complexity or rapid stratigraphic transition often leaves the disposition of a reflector poorly resolved. Intersecting survey lines facilitate the checking of an interpretation by comparison of reflection times at intersection points. Mapping reflection times around a

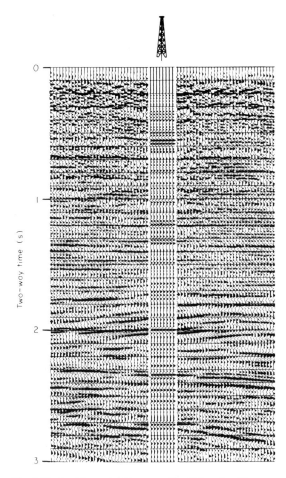

Fig. 4.46 Corridor stack of the zero-offset VSP section (Fig. 4.45(c)) reproduced eight times and spliced into a conventional seismic section based on surface profiling data from the vicinity of the borehole site. Comparison of the VSP stack with the surface recorded data enables the primary events in the seismic section to be reliably distinguished from multiple events. (From Cassell 1984.)

closed loop of survey lines reveals any errors in the identification or correlation of a reflection event across the area of a seismic survey.

Reprocessing of data, or migration, may be employed to help resolve uncertainties of interpretation, but additional seismic lines are often needed to resolve problems associated with an initial phase of interpretation. It is common for several rounds of seismic exploration to be necessary before a prospective structure is sufficiently well defined to locate the optimal position of an exploration borehole.

Structural interpretation of three-dimensional data is able to take advantage of the areal coverage of reflection points, the improved resolution associated with three-dimensional migration and the improved methods of data access, analysis and display provided by dedicated seismic work stations. Examples of the display of geological structures using three-dimensional data volumes are illustrated in Plates 4.1 and 4.2. Interpretation of three-dimensional data is often crucial to the successful development of oilfields with a complex geological structure. An example is the North Cormorant oilfield in the UK Sector of the North Sea, where three-dimensional seismics enabled the mapping of far more fault structures than had been possible using pre-existing two-dimensional data, and revealed a set of NW–SE trending faults that had previously been unsuspected.

4.14.2 Stratigraphical analysis (seismic stratigraphy)

Seismic stratigraphy involves the subdivision of seismic sections into sequences of reflections that are interpreted as the seismic expression of genetically related sedimentary sequences. The principles behind this *seismic sequence analysis* are two-fold. Firstly, reflections are taken to define chronostratigraphical units, since the types of rock interface that produce reflections are stratal surfaces and unconformities; by contrast, the boundaries of diachronous lithological units tend to be transitional and not to produce reflections. Secondly, genetically related sedimentary sequences normally comprise a set of concordant strata that exhibit discordance with underlying and overlying sequences; that is, they are typically bounded by angular unconformities variously representing onlap, downlap, toplap or erosion (Fig. 4.48). A seismic sequence is the representation on a seismic section of a depositional sequence; as such, it is a group of concordant or near-concordant reflection events that terminate against the discordant reflections of adjacent seismic sequences. An example of a seismic sequence identified on a seismic section is illustrated in Plate 4.3.

Having subdivided a seismic section into its constituent sequences, each sequence may be analysed in terms of the internal disposition of reflection events and their character, to obtain insight into the depositional environments responsible for the sequence and into the range of lithofacies that may be represented within it. This use of reflection geometry and character to interpret sedimentary facies is known as *seismic facies analysis*. Individual seismic facies are identified within the seismic sequence illustrated in Plate 4.3. Different types of reflection configuration (Fig. 4.49) are diagnostic of different sedimentary environments. On a regional scale,

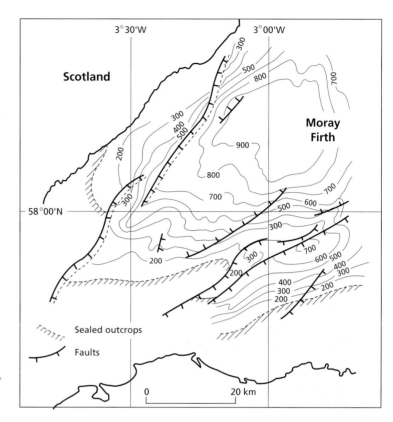

Fig. 4.47 Time-structure map of reflector at the base of the Lower Cretaceous in the Moray Firth off northeast Scotland, UK. Contour values represent two-way travel times of reflection event in milliseconds. (Courtesy British Geological Survey, Edinburgh, UK.)

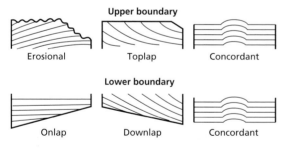

Fig. 4.48 Different types of geological boundary defining seismic sequences. (After Sheriff 1980.)

for example, parallel reflections characterize some shallow-water shelf environments whilst the deeper-water shelf edge and slope environments are often marked by the development of major sigmoidal or oblique cross-bedded units. The ability to identify particular sedimentary environments and predict lithofacies from analysis of seismic sections can be of great value to exploration programmes, providing a pointer to the location of potential source, reservoir and/or seal rocks. Thus, organic-rich basinal muds represent potential source rocks; discrete sand bodies developed in shelf environments represent potential reservoir rocks; and coastal mud and evaporite sequences represent potential seals (Fig. 4.50); the identification of these components in seismic sequences can thus help to focus an exploration programme by identifying areas of high potential.

An example of seismic stratigraphy based on three-dimensional data is illustrated in Plate 4.4. The seiscrop of Plate 4.4(a) shows a meandering stream channel preserved in a Neogene sedimentary sequence in the Gulf of Thailand. The channel geometry and the distinctive lithofacies of the channel fill lead to its clear identification as a distinctive seismic facies. Use of such seiscrops over a wider area enables the regional mapping of a Neogene deltaic environment (Plate 4.4(b)).

Major seismic sequences can often be correlated across broad regions of continental margins and clearly give evidence of being associated with major sea-level changes. The application of seismic stratigraphy in areas of good chronostratigraphical control has led to the development of a model of global cycles of major sea-level change and associated transgressive and regressive depositional sequences throughout the Mesozoic and Cenozoic (Payton 1977). Application of the methods of seismic stratigraphy in offshore sedimentary basins with little or no geological control often enables correlation of locally recognized depositional sequences with the worldwide pattern of sea-level changes (Payton 1977). It also facilitates identification of the major progradational sedimentary sequences which offer the main potential for hydrocarbon generation and accumulation. Stratigraphic analysis therefore greatly enhances the chances of successfully locating hydrocarbon traps in sedimentary basin environments.

Hydrocarbon accumulations are sometimes revealed directly on true-amplitude seismic sections (see below) by localized zones of anomalously strong reflections known as *bright spots*. These high-amplitude reflection events (Fig. 4.51) are attributable to the large reflection coefficients at the top and bottom of gas zones (typically, gas-filled sands) within a hydrocarbon reservoir. In the absence of bright spots, fluid interfaces may nevertheless be directly recognizable by *flat spots* which are horizontal or near-horizontal reflection events discordant to the local geological dip (see also Sections 4.10 and 4.11).

4.14.3 Seismic modelling

Reflection amplitudes may be normalized prior to their presentation on seismic sections so that original distinctions between weak and strong reflections are suppressed. This practice tends to increase the continuity of reflection events across a section and therefore aids their identification and structural mapping. However, much valuable geological information is contained in the true amplitude of a reflection event, which can be recovered from suitably calibrated field recordings. Any lateral variation of reflection amplitude is due to lateral change in the lithology of a rock layer or in its pore fluid content. Thus, whilst the production of normalized-amplitude sections may assist structural mapping of reflectors, it suppresses information that is vital to a full stratigraphic interpretation of the data. With increasing interest centring on stratigraphic interpretation, true-amplitude seismic sections are becoming increasingly important.

In addition to amplitude, the shape and polarity of a reflection event also contain important geological information (Meckel & Nath 1977). Analysis of the significance of lateral changes of shape, polarity and amplitude

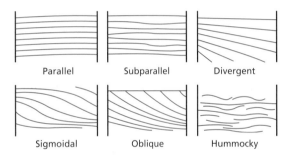

Fig. 4.49 Various internal bedforms that give rise to different seismic facies within sedimentary sequences identified on seismic sections. (After Sheriff 1980.)

Dark grey marls, black organic-rich mudstones	Siltstones, shales, reef limestones	Calcareous sandstones, oolites, bioclastic limestones	Sandstones, mudstones, dolomitic mudstones, evaporites	**Typical lithologies**
Very thin and continuous units	Thin to intermediate tabular bodies with lensoid reef limestones	Intermediate continuous to lensoid bodies	Irregular to discontinuous units	**Bed geometry**
25–50	150–450	100–50	50–25	**Thickness (m)**
Basinal	Outer shelf	Inner shelf	Coastal	**Environment**

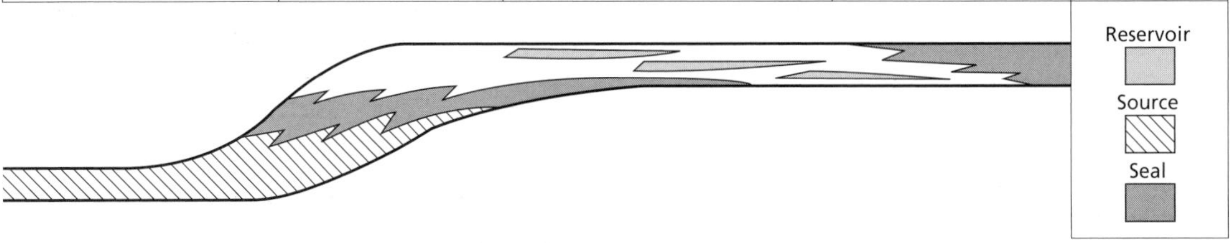

Reservoir

Source

Seal

Fig. 4.50 The overall geometry of a typical depositional sequence and its contained sedimentary facies.

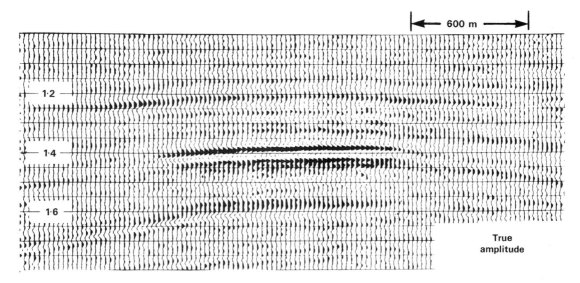

Fig. 4.51 Part of a true-amplitude seismic section containing a seismic bright spot associated with a local hydrocarbon accumulation. (From Sheriff 1980, after Schramm *et al.* 1977.)

observed in true-amplitude seismic sections is carried out by *seismic modelling*, often referred to in this context as *stratigraphic modelling*. Seismic modelling involves the production of synthetic seismograms for layered sequences to investigate the effects of varying the model parameters on the form of the resulting seismograms. Synthetic seismograms and synthetic seismic sections can be compared with observed data, and models can be manipulated in order to simulate the observed data. By this means, valuable insights can be obtained into the subsurface geology responsible for a particular seismic section. The standard type of synthetic seismogram represents the seismic response to vertical propagation of an assumed source wavelet through a model of the subsurface composed of a series of horizontal layers of differing acoustic impedance. Each layer boundary reflects some energy back to the surface, the amplitude and polarity of the reflection being determined by the acoustic impedance contrast. The synthetic seismogram comprises the sum of the individual reflections in their correct travel-time relationships (Fig. 4.52).

In its simplest form, a synthetic seismogram $x(t)$ may be considered as the convolution of the assumed source function $s(t)$ with a reflectivity function $r(t)$ representing the acoustic impedance contrasts in the layered model:

$$x(t) = s(t) \star r(t)$$

However, filtering effects along the downgoing and upgoing ray paths and the overall response of the recording system need to be taken into account. Multiples may or may not be incorporated into the synthetic seismogram.

The acoustic impedance values necessary to compute the reflectivity function may be derived directly from sonic log data (as described in Section 11.8). This is normally achieved assuming density to be constant throughout the model, but it may be important to derive estimates of layer densities in order to compute more accurate impedance values.

Synthetic seismograms can be derived for more complex models using ray-tracing techniques.

Particular stratigraphic features that have been investigated by seismic modelling, to determine the nature of their representation on seismic sections, include thin layers, discontinuous layers, wedge-shaped layers, transitional layer boundaries, variable porosity and type of pore fluid. Figure 4.53 illustrates synthetic seismograms computed across a section of stratigraphic change. These show how the varying pattern of interference between reflection events expresses itself in lateral changes of pulse shape and peak amplitude.

4.14.4 Seismic attribute analysis

Conventional seismic reflection sections are displayed in variable-area format where positive half-cycles of the

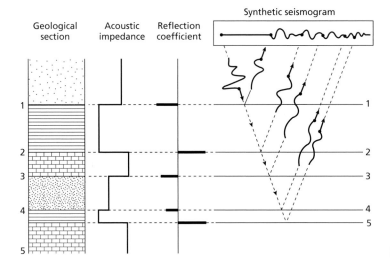

Fig. 4.52 The synthetic seismogram.

waveforms of seismic traces are filled in black. This has the desirable effect of merging the shaded areas from trace to trace to form continuous black lines across the section. These black lines guide the eye of the interpreter to correlate features across the section, and hence make a structural interpretation. The undesirable effect of this display is that the precise amplitude and shape of the waveform, which has been the subject of so much effort during data acquisition and processing, is lost. The amplitude of a normally reflected wave is directly related to the reflection coefficient at the interface, and hence the physical properties (density and velocity) of the formations. Thus, variations in amplitude along a reflector should indicate changes in the properties of the formations.

These properties can be viewed by presenting an image of the seismic section where the amplitude of the seismic wave is displayed as a colour scale. Changes of amplitude along a continuous reflector will then be emphasized by the colour change, rather than hidden in a broad black line. Such amplitude changes may be related to changes in the pore fluid in the rocks, and in favourable circumstances can be *direct hydrocarbon indicators* (DHIs). Amplitude is merely the simplest example of a property (*attribute*) of the seismic wave which can be examined for its geological significance. Others include the seismic wave phase and the frequency content. From the waveform amplitudes the acoustic impedance of each formation can be estimated, and if S-wave data are available Poisson's ratio can be found. On a yet more detailed level, the amplitude variation of reflected wavelets with source–receiver offset (AVO) within each CMP gather can be analysed. This *AVO effect* can be particularly diagnostic in distinguishing between amplitude effects due to rock matrix variation and those due to pore fluids. An excellent review of this complex subject is given in Castagna and Bachus (1993).

4.15 Single-channel marine reflection profiling

Single-channel reflection profiling is a simple but highly effective method of seismic surveying at sea that finds wide use in a variety of offshore applications. It represents reflection surveying reduced to its bare essentials: a marine seismic/acoustic source is towed behind a survey vessel and triggered at a fixed firing rate, and signals reflected from the sea bed and from sub-bottom reflectors are detected by a hydrophone streamer towed in the vicinity of the source (Fig. 4.54). The outputs of the individual hydrophone elements are summed and fed to a single-channel amplifier/processor unit and thence to a chart recorder. This survey procedure is not possible on land because only at sea can the source and detectors be moved forward continuously, and a sufficiently high

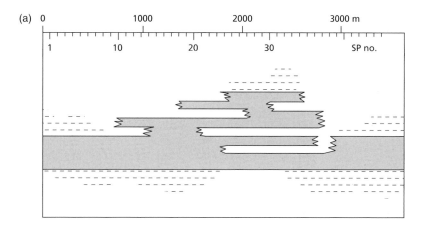

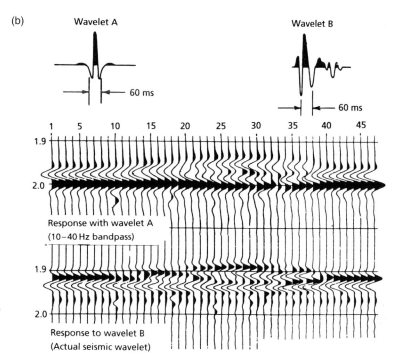

Fig. 4.53 A set of synthetic seismograms simulating a seismic section across a zone of irregular sandstone geometry. (From Neidell & Poggiagliolmi 1977.)

firing rate achieved, to enable surveys to be carried out continuously from a moving vehicle.

The source and hydrophone array are normally towed at shallow depth but some deep-water applications utilize deep-tow systems in which the source and receiver are towed close to the sea bed. Deep-tow systems overcome the transmission losses associated with a long water path, thus giving improved penetration of seismic/ acoustic energy into the sea bed. Moreover, in areas of

rugged bathymetry they produce records that are much simpler to interpret; there is commonly a multiplicity of reflection paths from a rugged sea bed to a surface source–detector location, so that records obtained in deep water using shallow-tow systems commonly exhibit hyperbolic diffraction patterns, bow-tie effects and other undesirable features of non-migrated seismic sections.

In place of the digital recorder used in multichannel

seismic surveying, single-channel profiling typically utilizes an *oceanographic recorder* in which a stylus repeatedly sweeps across the surface of an electrically-conducting recording paper that is continuously moving forward at a slow speed past a strip electrode in contact with the paper. A mark is burnt into the paper whenever an electrical signal is fed to the stylus and passes through the paper to the strip electrode. The seismic/acoustic source is triggered at the commencement of a stylus sweep and all seismic pulses returned during the sweep interval are recorded as a series of dark bands on the

recording paper (Fig. 4.55). The triggering rate and sweep speed are variable over a wide range. For a shallow penetration survey the source may be triggered every 500 ms and the recording interval may be 0–250 ms, whereas for a deep penetration survey in deep water the source may be triggered every 8 s and the recording interval may be 2–6 s.

The analogue recording systems used in single-channel profiling are relatively cheap to operate. There are no processing costs and seismic records are produced in real time by the continuous chart recording of band-pass filtered and amplified signals, sometimes with time variable gain (TVG). When careful consideration is given to source and hydrophone array design and deployment, good basic reflection records may be obtained from a single-channel system, but they cannot compare in quality with the type of seismic record produced by computer processing of multichannel data. Moreover, single-channel recordings cannot provide velocity information so that the conversion of reflection times into reflector depths has to utilize independent estimates of seismic velocity. Nonetheless, single-channel profiling often provides good imaging of subsurface geology and permits estimates of reflector depth and geometry that are sufficiently accurate for many purposes.

The record sections suffer from the presence of multiple reflections, especially multiples of the sea bed reflection, which may obliterate primary reflection events in the later parts of the records. Multiples are a particular problem when surveying in very shallow

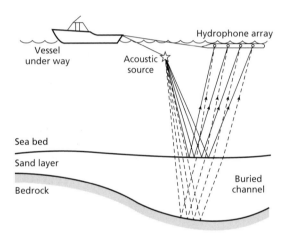

Fig. 4.54 The survey set-up for single-channel seismic reflection profiling.

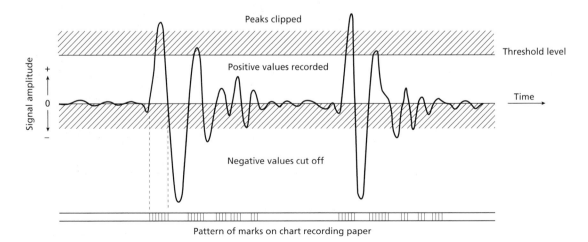

Fig. 4.55 Seismic signals and their representation on the chart recording paper of an oceanographic recorder. (From Le Tirant 1979.)

water, since they then occur at a short time interval after the primary events (Fig. 4.56). Record sections are often difficult to interpret in areas of complex reflector geometry because of the presence of bow-tie effects, diffraction events and other features of non-migrated seismic sections.

4.15.1 Shallow marine seismic sources

As discussed in Chapter 3 there are a variety of marine seismic/acoustic sources, operating at differing energy levels and characterized by different dominant frequencies. Consequently, by selection of a suitable source, single-channel profiling can be applied to a wide range of offshore investigations from high-resolution surveys of near-surface sedimentary layers to surveys of deep geological structure. In general, there is a trade-off between depth of penetration and degree of vertical resolution, since the higher energy sources required to transmit signals to greater depths are characterized by lower dominant frequencies and longer pulse lengths that adversely affect the resolution of the resultant seismic records.

Pingers are low-energy (typically about 5 J), tunable sources that can be operated within the frequency range

from 3 to 12 kHz. The piezoelectric transducers used to generate the pinger signal also serve as receivers for reflected acoustic energy and, hence, a separate hydrophone streamer is not required in pinger surveying. Vertical resolution can be as good as 10–20 cm but depth penetration is limited to a few tens of metres in muddy sediments or several metres in coarse sediments, with virtually no penetration into solid rock. Pinger surveys are commonly used in offshore engineering site investigation and are of particular value in submarine pipeline route surveys. Repeated pinger surveying along a pipeline route enables monitoring of local sediment movement and facilitates location of the pipeline where it has become buried under recent sediments. A typical pinger record is shown in Fig. 4.57.

Boomer sources provide a higher energy output (typically 300–500 J) and operate at lower dominant frequencies (1–5 kHz) than pingers. They therefore provide greater penetration (up to 100 m in bedrock) with good resolution (0.5–1 m). Boomer surveys are useful for mapping thick sedimentary sequences, in connection with channel dredging or sand and gravel extraction, or for high-resolution surveys of shallow geological structures. A boomer record section is illustrated in Fig. 4.58.

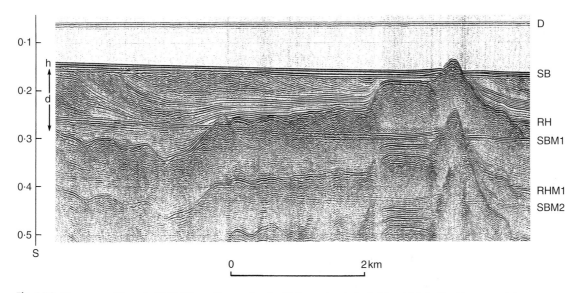

Fig. 4.56 Air gun record from the Gulf of Patras, Greece, showing Holocene hemipelagic (h) and deltaic (d) sediments overlying an irregular erosion surface (rockhead, RH) cut into tectonized Mesozoic and Tertiary rocks of the Hellenide (Alpine) orogenic belt. SB = sea bed reflection; SBM1 and SBM2 = first and second multiples of sea bed reflection; RHM1 = first multiple of rockhead reflection.

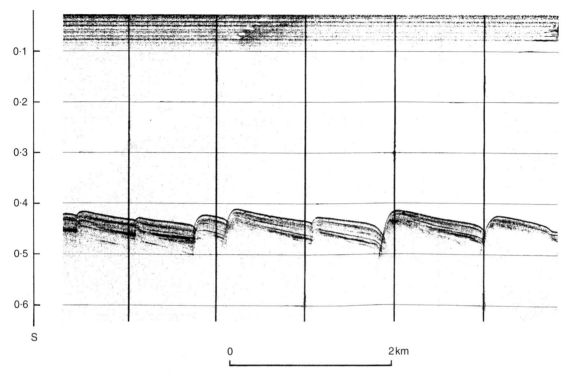

Fig. 4.57 Pinger record from the northern Aegean Sea, Greece, across a zone of active growth faults extending up to the sea bed. The sea floor is underlain by a layered sequence of Holocene muds and silts that can be traced to a depth of about 50 m. Note the diffraction patterns associated with the edges of the individual fault blocks.

Sparker sources can be operated over a wide range of energy levels (300–30 000 J), though the production of spark discharges of several thousand joules every few seconds requires a large power supply and a large bank of capacitors. Sparker surveying therefore represents a versatile tool for a wide range of applications, from shallow penetration surveys (100 m) with moderate resolution (2 m) to deep penetration surveys (>1 km) where resolution is not important. However, sparker surveying cannot match the resolution of precision boomer surveying, and sparkers do not offer as good a source signature as air guns for deeper penetration surveys.

By suitable selection of chamber size and rate of release of compressed air, air gun sources can be tailored to high resolution or deep penetration profiling applications and therefore represent the most versatile source for single-channel profiling. The reflection record shown in Fig. 4.56 was obtained in a shallow water area with a small air gun (40 in^3).

4.15.2 Sidescan sonar systems

Single-channel reflection profiling systems (sometimes referred to as *sub-bottom profiling systems*) are commonly operated in conjunction with a precision echo-sounder, for high-quality bathymetric information, and/or with a sidescan sonar system. *Sidescan sonar* is a sideways-scanning acoustic survey method in which the sea floor to one or both sides of the survey vessel is insonified by beams of high-frequency sound (30–110 kHz) transmitted by hull-mounted or fish-mounted transceiving transducers (Fig. 4.59). Sea bed features facing towards the survey vessel, such as rock outcrops or sedimentary bedforms, reflect acoustic energy back towards the transducers. In the case of features facing away from the vessel, or a featureless sea floor, the acoustic energy is reflected away from the transducers. Signals reflected back to the transducers are fed to the same type of recorder that is used to produce seismic profiling records,

(a)

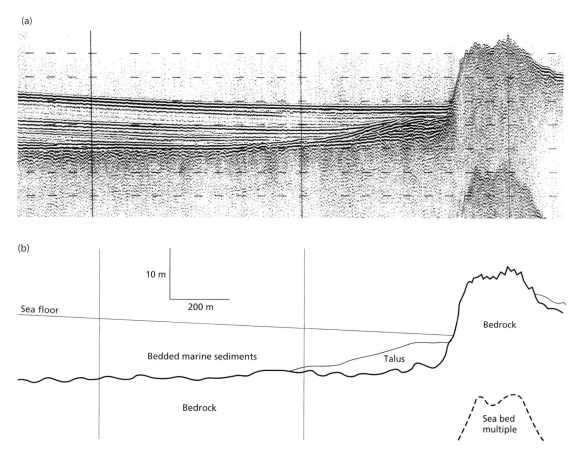

(b)

Fig. 4.58 (a) Precision boomer record from a coastal area of the Irish Sea, UK, and (b) line drawing interpretation showing Holocene sediments up to 10 m thick banked against a reef of Lower Palaeozoic rocks. (Courtesy C.R. Price.)

(a)

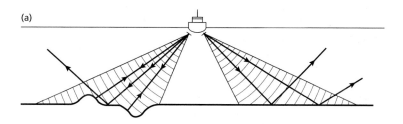

(b)

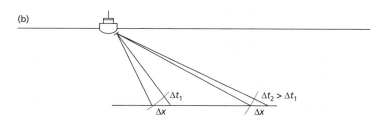

Fig. 4.59 Principles of sidescan sonar. (a) Individual reflected ray paths within the transmitted lobes, showing signal return from topographic features on the sea bed. (b) Scale distortion resulting from oblique incidence: the same widths of sea floor Δx are represented by different time intervals Δt_1 and Δt_2 at the inner and outer edges of the sonograph, respectively.

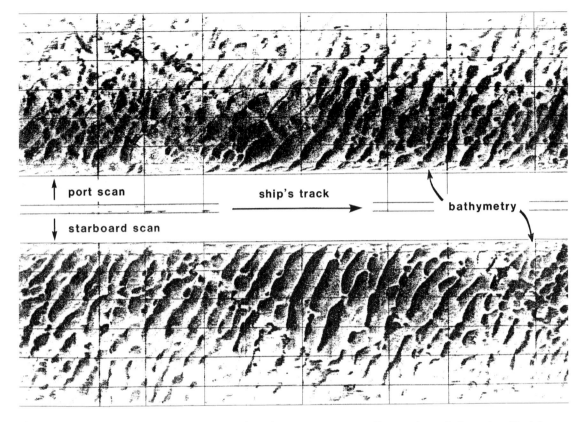

Fig. 4.60 Sonograph obtained from a dual scan survey of a pipeline route across an area of linear sand waves in the southern North Sea. The inner edges of the two swathes define the bathymetry beneath the survey vessel. (Scanning range: 100 m).

and the resulting pattern of returned acoustic energy is known as a *sonograph*. The oblique insonification produces scale distortion resulting from the varying path lengths and angles of incidence of returning rays (Fig. 4.59(b)). This distortion can be automatically corrected prior to display so that the sonograph provides an isometric plan view of sea bed features. A sonograph is shown in Fig. 4.60.

Although not strictly a seismic surveying tool, sidescan sonar provides valuable information on, for example, the configuration and orientation of sedimentary bedforms or on the pattern of rock outcrops. This information is often very useful in complementing the subsurface information derived from shallow seismic reflection surveys. Sidescan sonar is also useful for locating artefacts on the sea floor such as wrecks, cables or pipes. As with sub-bottom profiling systems, results in deep water are much improved by the use of deep-tow systems.

4.16 Applications of seismic reflection surveying

The 1980s and 1990s saw major developments in reflection seismic surveying. Over that period, the general quality of seismic record sections improved markedly due to the move to digital data acquisition systems and the use of increasingly powerful processing techniques. At the same time, the range of applications of the method increased considerably. Previously, reflection surveying was concerned almost exclusively with the search for hydrocarbons and coal, down to depths of a few kilometres. Now, the method is being used increasingly for studies of the entire continental crust and the uppermost mantle to depths of several tens of kilometres. At the other end of the spectrum of target depths the method is increasingly applied for high-resolution onshore mapping of shallow geology to depths of a few tens or hundreds of metres.

The search for hydrocarbons, onshore and offshore, nevertheless remains by far the largest single application of reflection surveying. This reflects the particular strength of the method in producing well-resolved images of sedimentary sequences down to a depth of several kilometres. The method is used at all stages of an exploration programme for hydrocarbons, from the early reconnaissance stage through to the detailed mapping of specific structural targets in preparation for exploration drilling, and on into the field development stage when the overall reservoir geometry requires further detailing.

Because of its relatively high cost, three-dimensional seismic surveying still does not find routine application in hydrocarbon exploration programmes. However, whereas it was originally used only at the field development stage, it now finds widespread application also at the exploration stage in some oilfields. Vertical seismic profiling is another important technique that is being applied increasingly at the stage of oilfield development because of its ability to reveal subsurface detail that is generally unobtainable from surface seismic data alone. In the quest for ever more detailed subsurface information, three component (3C) surveys are becoming more common. The value of repeated surveys during oilfield production is now established and 'time lapse' or 4D surveys are also increasing in usage.

The initial round of seismic exploration for hydrocarbons normally involves speculative surveys along widely-spaced profile lines covering large areas. In this way the major structural or stratigraphic elements of the regional geology are delineated, so enabling the planning of detailed, follow-up reflection surveys in more restricted areas containing the main prospective targets. Where good geological mapping of known sedimentary sequences exists, the need for expenditure on initial speculative seismic surveys is often much reduced and effort can be concentrated from an early stage on the seismic investigation of areas of particular interest.

Detailed reflection surveys involve closely-spaced profile lines and a high density of profile intersection points in order that reflection events can be traced reliably from profile to profile and used to define the prevailing structure. Initial seismic interpretation is likely to in-volve structural mapping, using time-structure and/or isochron maps (Section 4.14.1) in the search for the structural closures that may contain oil or gas. Any closures that are identified may need further delineation by a second round of detailed seismic surveying before the geophysicist is sufficiently confident to select the lo-cation of an exploration borehole from a time-structure map. Three-dimensional seismics may need to be employed when critical structural details are unresolved by interpretation of the two-dimensional survey data.

Exploration boreholes are normally sited on seismic profile lines so that the borehole logs can be correlated directly with the local seismic section. This facilitates precise geological identification of specific seismic reflectors, especially if vertical seismic profiling surveys (Section 4.13) are carried out at the site of the borehole.

Particularly in offshore areas, where the best quality seismic data are generally obtained, the methods of seismic stratigraphy (Section 4.14.2) are increasingly employed on sections displaying seismic sequences to obtain insight into the associated sedimentary lithologies and depositional environments. Such stratigraphic information, derived from seismic facies analysis of the individual sequences, is often of great value to an exploration programme in highlighting the location of potential source rocks (e.g. organic-rich mudstones) and potential reservoir rocks (e.g. a deltaic or reef facies).

The contribution of reflection surveying to the development of hydrocarbon reserves does not end with the discovery of an oil or gas field. Refinement of the seismic interpretation using information from, variously, additional seismic profiles, three-dimensional seismics and vertical seismic profiling data will assist in optimizing the location of production boreholes. In addition, seismic modelling (Section 4.14.3) of amplitude variations and other aspects of reflection character displayed on seismic sections across the producing zone can be used to obtain detailed information on the geometry of the reservoir and on internal lithological variations that may affect the hydrocarbon yield. 4D surveying of producing fields (Section 4.12) has demonstrated that the detection of unexploited areas in a producing field is feasible and adequately repays the cost of the geophysical survey.

Examples of seismic sections from hydrocarbon fields in the North Sea area are shown in Figs 4.61 and 4.62. Figure 4.61 represents a seismic section across the North Viking gas field in the southern North Sea. The gas is trapped in the core of a NW–SE trending anticlinal structure that is extensively faulted at the level of the Lower Permian. A typical combined structural/stratigraphic trap in the northern North Sea is represented by the Brent oilfield structure, and Fig. 4.62 illustrates a seismic section across the field. A tilted fault block containing Upper Palaeozoic, Triassic and Jurassic strata is overlain unconformably by Upper Jurassic, Cretaceous and Tertiary sediments. Two Jurassic sands in the tilted

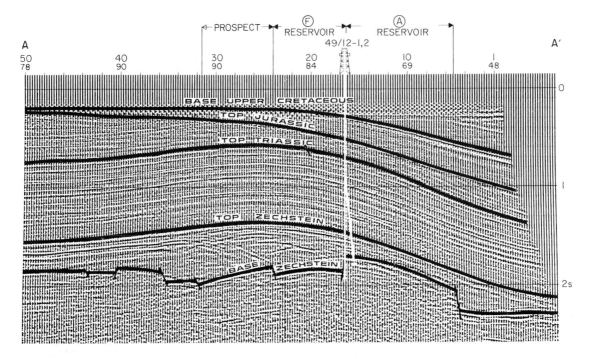

Fig. 4.61 Interpreted seismic section across the North Viking gas field, North Sea. (Courtesy Conoco UK Ltd.)

fault block constitute the main reservoirs, the oil and gas being trapped beneath a capping of unconformably overlying shales of Late Jurassic and Cretaceous age.

Reflection profiling at crustal and lithospheric scale is now being carried out by many developed countries. Following on from the extensive use of multichannel reflection profiling to investigate the crustal structure of oceanic areas, national programmes such as the US COCORP project (Consortium for Continental Reflection Profiling; Brewer & Oliver 1980) and the British BIRPS project (British Institutions Reflection Profiling Syndicate; Brewer 1983) have produced seismic sections through the entire continental crust and the uppermost part of the underlying mantle. These national programmes utilize essentially the same data acquisition systems and processing techniques as the oil industry, whilst increasing the size of source arrays and detector spread lengths; recording times of 15 s are commonly employed, as compared with a standard oil industry recording time of about 4 s. A typical BIRPS section is illustrated in Fig. 4.63.

Crustal reflection profiling results from several different continental areas reveal that the upper part of the continental crust typically has a rather transparent seismic character. Within this, localized bands of dipping reflectors, interpreted as fault zones, pass down into the lower crust (see e.g. Barazangi & Brown (1986) and the special issue of *Tectonophysics* **173** (1990) for a wide range of relevant papers). By contrast, the lower crust is often found to be highly reflective with discontinuous horizontal or gently dipping events giving an overall layered appearance (Fig. 4.63). The origin of this layering is uncertain, but the main possibilities appear to be primary igneous layering, horizontal shear zones and zones of fluid concentration (e.g. Klemperer et al. 1987). All may contribute in some measure to the observed reflectivity. Where refraction and reflection data both exist, the base of the zone of reflectivity is found to coincide with the Mohorovicic discontinuity as defined by refraction interpretation of head wave arrivals from the uppermost mantle (Barton 1986).

The use of reflection seismics for high-resolution studies of shallow geology is a field of growing importance in which developments are linked directly to recent technical advances. Highly portable digital multichannel data acquisition systems, backed up by PC-based processing packages, make it possible to

(a)

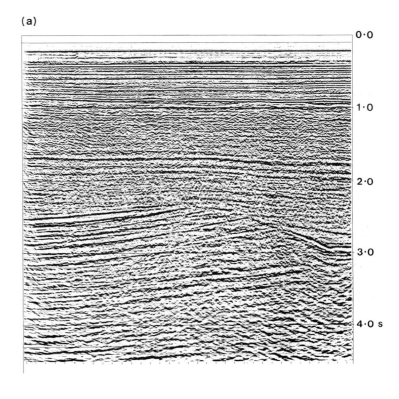

(b)

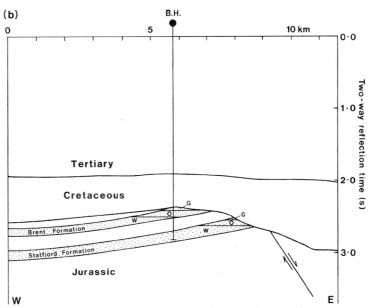

Fig. 4.62 (a) Seismic section (courtesy Shell UK Ltd) and (b) line interpretation across the Brent oilfield, North Sea. G = gas; O = oil; W = water.

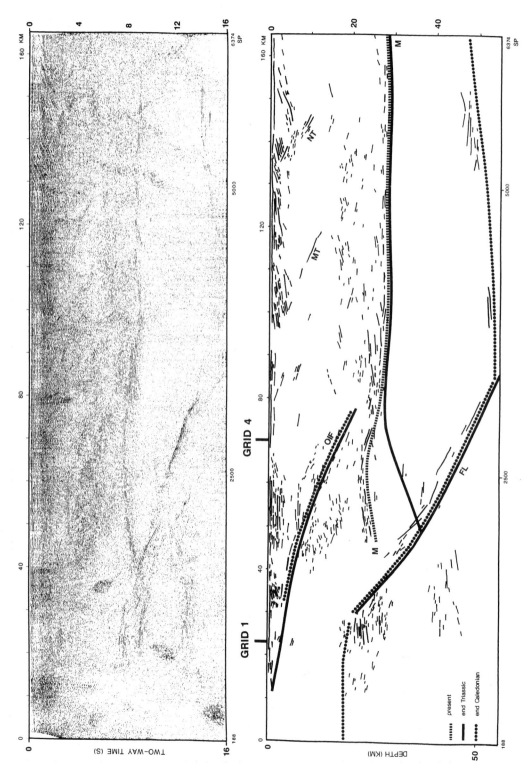

Fig. 4.63 A non-migrated crustal reflection section from the 1986/87 GRID survey of the BIRPS programme, collected along a west–east line about 30 km north of Scotland, UK, and a migrated line drawing of the main reflection events. The main structures are interpreted to be of Caledonian age with later reactivation (FL = Flannan reflection; OIF = Outer Isles fault; MT = Moine thrust; NT = Naver thrust; M = Moho). (From Snyder & Flack 1990.)

produce seismic sections of shallow subsurface geology at reasonable cost. High–resolution reflection seismology is particularly well suited to the investigation of Quaternary sedimentary sequences (Fig. 4.56) and for the detailed mapping of concealed bedrock surfaces of irregular geometry (Fig. 4.64). The contrast between the crustal (Fig. 4.63) and near–surface sections (Fig. 4.64) neatly emphasizes the scalability of the seismic reflection method. In both these applications it is also the geophysical method with the highest resolution, both vertically and horizontally.

Fig. 4.64 A near–surface seismic reflection section showing Mesozoic sediments (reflectors T1–T3 and B) with an angular unconformity (U) against Lower Palaeozoic rocks. (From Ali & Hill 1991.)

Problems

1. A seismic wave is incident normally on a reflector with a reflection coefficient R of 0.01. What proportion of the incident energy is transmitted?

2. What is the root-mean-square velocity in reflection surveying, and how is it related to interval velocity and to stacking velocity?

3. A zero-offset reflection event at 1.000 s has a normal moveout (NMO) of 0.005 s at 200 m offset. What is the stacking velocity?

4. (a) Calculate the approximate dimensions of the Fresnel zone in the following two cases:
(i) Reflection profiling is used to investigate lower crustal structure at a depth of about 30 km. The dominant frequency of the reflected pulse is found to be 10 Hz. Using a typical average crustal velocity of 6.5 km s^{-1}.
(ii) A high-resolution reflection survey is used to map rockhead beneath a Quaternary sediment cover about 100 m thick using a high-frequency source. The dominant frequency of the reflected pulse is found to be 150 Hz. Use a sediment velocity of 2 km s^{-1}.
(b) Discuss the importance of the above Fresnel zone dimensions as indications of the inherent limits on horizontal resolution achievable in different types of reflection survey.
(c) Use the frequency and velocity information to calculate the vertical resolution of the two surveys above and again discuss the general importance of the results obtained to the vertical resolution that is achievable in reflection seismics.

5. In the initial stages of a seismic reflection survey, a noise test indicates a direct wave with a velocity of 3.00 km s^{-1} and a dominant frequency of 100 Hz, and ground roll with a velocity of 1.80 km s^{-1} and a dominant frequency of 30 Hz. What is the optimum spacing of individual geophones in five-element linear arrays in order to suppress these horizontally-travelling phases?

6. In CDP stacking, the method of applying a NMO correction to individual seismic traces creates distortion in seismic pulses recorded at large offset that can degrade the stacking process. Why?

7. Along a two-dimensional marine survey line involving a 48-channel streamer with a hydrophone array interval of 10 m, shots are fired every 40 m.
(a) What is the fold of CMP cover?
(b) If the cover is to be increased to 24-fold, what must the new shot interval be?

8. In single-channel seismic profiling, what is the optimum depth for towing an air gun source with a dominant frequency of 100 Hz such that the reflected ray from the sea surface will interfere constructively with the downgoing primary pulse? (The compressional wave velocity in sea water is 1.505 km s^{-1}.)

9. What is the significance of the curved boundary lines to the typical ground roll sector of the f–k plot illustrated in Fig. 4.25, and how may it be explained?

10. How may three-dimensional seismic survey data be used to study velocity anisotropy?)

Further reading

Ali, J.W. & Hill, I.A. (1991) Reflection seismics for shallow geological applications: a case study from Central England. *J. Geol. Soc. London* **148**, 219–22.

Al-Sadi, H.N. (1980) *Seismic Exploration*. Birhauser Verlag, Basel.

Anstey, N.A. (1982) *Simple Seismics*. IHRDC, Boston.

Bally, A.W. (ed.) (1983) *Seismic Expression of Structural Styles (a picture and work atlas): Vol 1— The layered Earth; Vol 2— Tectonics of extensional provinces; Vol 3— Tectonics of compressional provinces / Strike-slip tectonics*. AAPG Studies in Geology No. 15, American Association of Petroleum Geologists, Tulsa.

Bally, A.W. (ed.) (1987) *Atlas of Seismic Stratigraphy* (3 vols). AAPG Studies in Geology No. 27, American Association of Petroleum Geologists, Tulsa.

Barazangi, M. & Brown, L. (eds) (1986) *Reflection Seismology: The Continental Crust*. AGU Geodynamics Series, No. 14. American Geophysical Union, Washington.

Berg, O.R. & Woolverton, D.G. (eds) (1985) *Seismic Stratigraphy II: An Integrated Approach to Hydrocarbon Exploration*. AAPG Memoir 39, American Association of Petroleum Geologists, Tulsa.

Brown, A.R. (1986) *Interpretation of Three-dimensional Seismic Data*. AAPG Memoir 42, American Association of Petroleum Geologists, Tulsa.

Camina, A.R. & Janacek, G.J. (1984) *Mathematics for Seismic Data Processing and Interpretation*. Graham & Trotman, London.

Cassell, B. (1984) Vertical seismic profiles— an introduction. *First Break*, **2**(11), 9–19.

Castagna, J.P. & Bachus, M.M. (1993) *Offset-Dependent Reflectivity: Theory and Practice of AVO Analysis. Investigations in Geophysics*. Society of Exploration Geophysicists, Tulsa.

Claerbout, J.F. (1985) *Fundamentals of Geophysical Data Processing*. McGraw Hill, New York.

Dobrin, M.B. & Savit, C.H. (1988) *Introduction to Geophysical Prospecting* (4th edn). McGraw Hill, New York.

Hatton, L., Worthington, M.H. & Makin, J. (1986) *Seismic Data Processing*. Blackwell Scientific Publications, Oxford.

Hubral, P. & Krey, T. (1980) *Interval Velocities from Seismic Reflection Time Measurements*. Society of Exploration Geophysicists, Tulsa.

Jack, I. (1997) *Time Lapse Seismic in Reservoir Management*. Society of Exploration Geophysicists, Short Course Notes, Society of Exploration Geophysicists, Tulsa.

Kleyn, A.H. (1983) *Seismic Reflection Interpretation*. Applied Science Publishers, London.

Lavergne, M. (1989) *Seismic Methods*. Editions Technip, Paris.

McQuillin, R., Bacon, M. & Barclay, W. (1979) *An Introduction to Seismic Interpretation*. Graham & Trotman, London.

Payton, C.E. (ed.) (1977) *Seismic Stratigraphy—Application to Hydrocarbon Exploration*. AAPG Memoir 26, American Association of Petroleum Geologists, Tulsa.

Robinson, E.A. (1983) *Migration of Geophysical Data*. IHRDC, Boston.

Robinson, E.A. (1983) *Seismic Velocity Analysis and the Convolutional Model*. IHRDC, Boston.

Robinson, E.S. & Çoruh, C. (1988) *Basic Exploration Geophysics*. Wiley, New York.

Sengbush, R.L. (1983) *Seismic Exploration Methods*. IHRDC, Boston.

Sheriff, R.E. (1980) *Seismic Stratigraphy*. IHRDC, Boston.

Sheriff, R.E. (1982) *Structural Interpretation of Seismic Data*. AAPG Continuing Education Course Note Series No. 23.

Sheriff, R.E. & Geldart, L.P. (1983) *Exploration Seismology, Vol. 2: Data-Processing and Interpretation*. Cambridge University Press, Cambridge.

Waters, K.H. (1978) *Reflection Seismology—A Tool For Energy Resource Exploration*. Wiley, New York.

Ziolkowski, A. (1983) *Deconvolution*. IHRDC, Boston.

5 Seismic refraction surveying

5.1 Introduction

The seismic refraction surveying method uses seismic energy that returns to the surface after travelling through the ground along refracted ray paths. As briefly discussed in Chapter 3, the first arrival of seismic energy at a detector offset from a seismic source always represents either a direct ray or a refracted ray. This fact allows simple refraction surveys to be performed in which attention is concentrated solely on the first arrival (or *onset*) of seismic energy, and time–distance plots of these first arrivals are interpreted to derive information on the depth to refracting interfaces. As is seen later in the chapter, this simple approach does not always yield a full or accurate picture of the subsurface. In such circumstances more complex interpretations may be applied. The method is normally used to locate refracting interfaces (refractors) separating layers of different seismic velocity, but the method is also applicable in cases where velocity varies smoothly as a function of depth or laterally.

Refraction seismograms may also contain reflection events as subsequent arrivals, though generally no special attempt is made to enhance reflected arrivals in refraction surveys. Nevertheless, the relatively high reflection coefficients associated with rays incident on an interface at angles near to the critical angle often lead to strong *wide-angle reflections* which are quite commonly detected at the greater recording ranges that characterize large-scale refraction surveys. These wide-angle reflections often provide valuable additional information on subsurface structure such as, for example, indicating the presence of a low-velocity layer which would not be revealed by refracted arrivals alone.

The vast majority of refraction surveying is carried out along profile lines which are arranged to be sufficiently long to ensure that refracted arrivals from target layers are recorded as first arrivals for at least half the length of the line. Refraction profiles typically need to be between five and ten times as long as the required depth of investigation. A consequence of this requirement is that large seismic sources are needed for the detection of deep refractors in order that sufficient energy is transmitted over the long range necessary for the recording of deep refracted phases as first arrivals. The profile length required in any particular survey depends upon the distribution of velocities with depth at that location. The requirement in refraction surveying for an increase in profile length with increase in the depth of investigation contrasts with the situation in conventional reflection surveying, where near-normal incidence reflections from deep interfaces are recorded at small offset distances.

Refraction seismology is applied to a very wide range of scientific and technical problems, from engineering site investigation surveys to large-scale experiments designed to study the structure of the entire crust or lithosphere. Refraction measurements can provide valuable velocity information for use in the interpretation of reflection surveys, and refracted arrivals recorded during land reflection surveys are used to map the weathered layer, as discussed in Chapter 4. This wide variety of applications leads to an equally wide variety of field survey methods and associated interpretation techniques.

In many geological situations, subsurface refractors may approximate planar surfaces over the linear extent of a refraction line. In such cases the observed travel-time plots are commonly assumed to be derived from a set of planar layers and are analysed to determine depths to, and dips of, individual planar refractors. The geometry of refracted ray paths through planar layer models of the subsurface is considered first, after which consideration is given to methods of dealing with refraction at irregular (non-planar) interfaces.

5.2 Geometry of refracted ray paths: planar interfaces

The general assumptions relating to the ray path

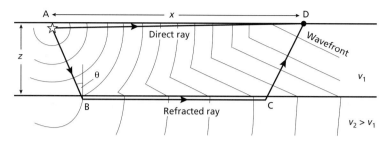

Fig. 5.1 Successive positions of the expanding wavefronts for direct and refracted waves through a two-layer model. Only the wavefront of the first arrival phase is shown. Individual ray paths from source A to detector D are drawn as solid lines.

geometries considered below are that the subsurface is composed of a series of layers, separated by planar and possibly dipping interfaces. Also, within each layer seismic velocities are constant, and the velocities increase with layer depth. Finally, the ray paths are restricted to a vertical plane containing the profile line (i.e. there is no component of cross-dip).

5.2.1 Two-layer case with horizontal interface

Figure 5.1 illustrates progressive positions of the wavefront from a seismic source at A associated with energy travelling directly through an upper layer and energy critically refracted in a lower layer. Direct and refracted ray paths to a detector at D, a distance x from the source, are also shown. The layer velocities are v_1 and v_2 ($>v_1$) and the refracting interface is at a depth z.

The direct ray travels horizontally through the top of the upper layer from A to D at velocity v_1. The refracted ray travels down to the interface and back up to the surface at velocity v_1 along slant paths AB and CD that are inclined at the critical angle θ, and travels along the interface between B and C at the higher velocity v_2. The total travel time along the refracted ray path ABCD is

$$t = t_{AB} + t_{BC} + t_{CD}$$
$$= \frac{z}{v_1 \cos\theta} + \frac{(x - 2z \tan\theta)}{v_2} + \frac{z}{v_1 \cos\theta}$$

Noting that $\sin\theta = v_1/v_2$ (Snell's Law) and $\cos\theta = (1 - v_1^2/v_2^2)^{1/2}$, the travel-time equation may be expressed in a number of different forms, a useful general form being

$$t = \frac{x}{v_2} + \frac{2z \cos\theta}{v_1} \tag{5.1}$$

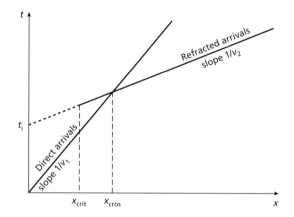

Fig. 5.2 Travel-time curves for the direct wave and the head wave from a single horizontal refractor.

Alternatively

$$t = \frac{x}{v_2} + \frac{2z(v_2^2 - v_1^2)^{1/2}}{v_1 v_2} \tag{5.2}$$

or

$$t = \frac{x}{v_2} + t_i \tag{5.3}$$

where, plotting t against x (Fig. 5.2), t_i is the intercept on the time axis of a travel-time plot or *time–distance plot* having a gradient of $1/v_2$. The *intercept time* t_i, is given by

$$t_i = \frac{2z(v_2^2 - v_1^2)^{1/2}}{v_1 v_2} \tag{from 5.2}$$

Solving for refractor depth

$$z = \frac{t_i v_1 v_2}{2(v_2^2 - v_1^2)^{1/2}}$$

A useful way to consider the equations (5.1) to (5.3) is to note that the total travel time is the time that would have been taken to travel the total range x at the refractor velocity v_2 (that is x/v_2), plus an additional time to allow for the time it takes the wave to travel down to the refractor from the source, and back up to the receiver. The concept of regarding the observed time as a refractor travel-time plus *delay times* at the source and receiver is explored later.

Values of the best-fitting plane layered model parameters, v_1, v_2 and z, can be determined by analysis of the travel-time curves of direct and refracted arrivals:
• v_1 and v_2 can be derived from the reciprocal of the gradient of the relevant travel-time segment, see Fig. 5.2
• the refractor depth, z, can be determined from the intercept time t_i.

At the crossover distance x_{cros} the travel times of direct and refracted rays are equal

$$\frac{x_{cros}}{v_1} = \frac{x_{cros}}{v_2} + \frac{2z(v_2^2 - v_1^2)^{1/2}}{v_1 v_2}$$

Thus, solving for x_{cros}

$$x_{cros} = 2z\left[\frac{v_2 + v_1}{v_2 - v_1}\right]^{1/2} \tag{5.4}$$

From this equation it may be seen that the crossover distance is always greater than twice the depth to the refractor. Also the crossover distance equation (5.4) provides an alternative method of calculating z.

5.2.2 Three-layer case with horizontal interface

The geometry of the ray path in the case of critical refraction at the second interface is shown in Fig. 5.3. The seismic velocities of the three layers are v_1, v_2 ($>v_1$) and v_3 ($>v_2$). The angle of incidence of the ray on the upper interface is θ_{13} and on the lower interface is θ_{23} (critical angle). The thicknesses of layers 1 and 2 are z_1 and z_2 respectively.

By analogy with equation (5.1) for the two-layer case, the travel time along the refracted ray path ABCDEF to

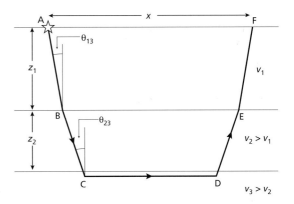

Fig. 5.3 Ray path for a wave refracted through the bottom layer of a three-layer model.

an offset distance x, involving critical refraction at the second interface, can be written in the form

$$t = \frac{x}{v_3} + \frac{2z_1 \cos\theta_{13}}{v_1} + \frac{2z_2 \cos\theta_{23}}{v_2} \tag{5.5}$$

where

$$\theta_{13} = \sin^{-1}(v_1/v_3); \quad \theta_{23} = \sin^{-1}(v_2/v_3)$$

and the notation subscripts for the angles relate directly to the velocities of the layers through which the ray travels at that angle (θ_{13} is the angle of the ray in layer 1 which is critically refracted in layer 3).

Equation (5.5) can also be written

$$t = \frac{x}{v_3} + t_1 + t_2 \tag{5.6}$$

where t_1 and t_2 are the times taken by the ray to travel through layers 1 and 2 respectively (see Fig. 5.4).

The interpretation of travel-time curves for a three-layer case starts with the initial interpretation of the top two layers. Having used the travel-time curve for rays critically refracted at the upper interface to derive z_1 and v_2, the travel-time curve for rays critically refracted at the second interface can be used to derive z_2 and v_3 using equations (5.5) and (5.6) or equations derived from them.

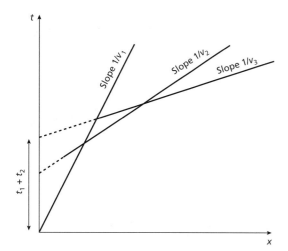

Fig. 5.4 Travel-time curves for the direct wave and the head waves from two horizontal refractors.

5.2.3 Multilayer case with horizontal interfaces

In general the travel time t_n of a ray critically refracted along the top surface of the nth layer is given by

$$t_n = \frac{x}{v_n} + \sum_{i=1}^{n-1} \frac{2z_i \cos \theta_{in}}{v_i} \qquad (5.7)$$

where

$$\theta_{in} = \sin^{-1}(v_i/v_n) .$$

Equation (5.7) can be used progressively to compute layer thicknesses in a sequence of horizontal strata represented by travel-time curves of refracted arrivals. In practice as the number of layers increases it becomes more difficult to identify each of the individual straight-line segments of the travel-time plot. Additionally, with increasing numbers of layers, there is less likelihood that each layer will be bounded by strictly planar horizontal interfaces, and a more complex model may be necessary. It would be unusual to make an interpretation using this method for more than four layers.

5.2.4 Dipping-layer case with planar interfaces

In the case of a dipping refractor (Fig. 5.5(a)) the value of dip enters the travel-time equations as an additional unknown. The reciprocal of the gradient of the travel-time curve no longer represents the refractor velocity but

a quantity known as the *apparent velocity* which is higher than the refractor velocity when recording along a profile line in the updip direction from the shot point and lower when recording downdip.

The conventional method of dealing with the possible presence of refractor dip is to *reverse* the refraction experiment by firing at each end of the profile line and recording seismic arrivals along the line from both shots. In the presence of a component of refractor dip along the profile direction, the *forward* and *reverse* travel time plots for refracted rays will differ in their gradients and intercept times, as shown in Fig. 5.5(b).

The general form of the equation for the travel-time t_n of a ray critically refracted in the nth dipping refractor (Fig. 5.6; Johnson 1976) is given by

$$t_n = \frac{x \sin \beta_1}{v_1} + \sum_{i=1}^{n-1} \frac{h_i(\cos \alpha_i + \cos \beta_i)}{v_i} \qquad (5.8)$$

where h_i is the vertical thickness of the ith layer beneath the shot, v_i is the velocity of the ray in the ith layer, α_i is the angle with respect to the vertical made by the downgoing ray in the ith layer, β_i is the angle with respect to vertical made by the upgoing ray in the ith layer, and x is the offset distance between source and detector.

Equation (5.8) is comparable with equation (5.7), the only differences being the replacement of θ by angles α and β that include a dip term. In the case of shooting downdip, for example (see Fig. 5.6), $\alpha_i = \theta_{in} - \gamma_i$ and $\beta_i = \theta_{in} + \gamma_i$, where γ_i is the dip of the ith layer and $\theta_{in} = \sin^{-1}(v_1/v_n)$ as before. Note that h is the vertical thickness rather than the perpendicular or true thickness of a layer (z).

As an example of the use of equation (5.8) in interpreting travel-time curves, consider the two-layer case illustrated in Fig. 5.5.

Shooting downdip, along the forward profile

$$t_2 = \frac{x \sin \beta_1}{v_1} + \frac{h_1(\cos \alpha + \cos \beta)}{v_1}$$

$$= \frac{x \sin(\theta_{12} + \gamma_1)}{v_1} + \frac{h_1 \cos(\theta_{12} - \gamma_1)}{v_1}$$

$$\quad + \frac{h_1 \cos(\theta_{12} + \gamma_1)}{v_1}$$

$$= \frac{x \sin(\theta_{12} + \gamma_1)}{v_1} + \frac{2h_1 \cos \theta_{12} \cos \gamma_1}{v_1}$$

$$= \frac{x \sin(\theta_{12} + \gamma_1)}{v_1} + \frac{2z \cos \theta_{12}}{v_1} \qquad (5.9)$$

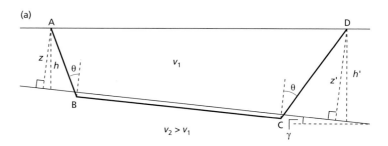

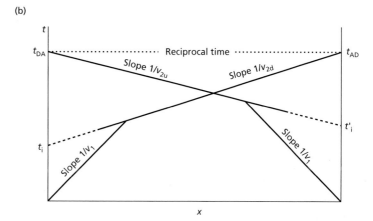

Fig. 5.5 (a) Ray-path geometry and (b) travel-time curves for head wave arrivals from a dipping refractor in the forward and reverse directions along a refraction profile line.

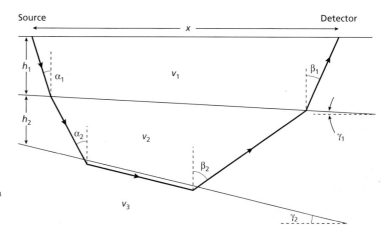

Fig. 5.6 Geometry of the refracted ray path through a multilayer, dipping model. (After Johnson 1976.)

where z is the perpendicular distance to the interface beneath the shot, and $\theta_{12} = \sin^{-1}(v_1/v_2)$.

Equation (5.9) defines a linear plot with a gradient of $\sin(\theta_{12} + \gamma_1)/v_1$ and an intercept time of $2z\cos\theta_{12}/v_1$.

Shooting updip, along the reverse profile

$$t_2' = \frac{x\sin(\theta_{12} - \gamma_1)}{v_1} + \frac{2z'\cos\theta_{12}}{v_1} \qquad (5.10)$$

where z' is the perpendicular distance to the interface beneath the second shot.

The gradients of the travel-time curves of refracted arrivals along the forward and reverse profile lines yield the downdip and updip apparent velocities v_{2d} and v_{2u} respectively (Fig. 5.5(b)). From the forward direction

$$1/v_{2d} = \sin(\theta_{12} + \gamma_1)/v_1 \qquad (5.11)$$

and from the reverse direction

$$1/v_{2u} = \sin(\theta_{12} - \gamma_1)/v_1 \qquad (5.12)$$

Hence

$$\theta_{12} + \gamma_1 = \sin^{-1}(v_1/v_{2d})$$
$$\theta_{12} - \gamma_1 = \sin^{-1}(v_1/v_{2u})$$

Solving for θ and γ yields

$$\theta_{12} = \frac{1}{2}[\sin^{-1}(v_1/v_{2d}) + \sin^{-1}(v_1/v_{2u})]$$

$$\gamma_1 = \frac{1}{2}[\sin^{-1}(v_1/v_{2d}) - \sin^{-1}(v_1/v_{2u})]$$

Knowing v_1, from the gradient of the direct ray travel-time curve, and θ_{12}, the true refractor velocity may be derived using Snell's Law

$$v_2 = v_1/\sin\theta_{12}$$

The perpendicular distances z and z' to the interface under the two ends of the profile are obtained from the intercept times t_i and t_i' of the travel-time curves obtained in the forward and reverse directions

$$t_i = 2z\cos\theta_{12}/v_1$$

$$\therefore \quad z = v_1 t_i/2\cos\theta_{12}$$

and similarly

$$z' = v_1 t_i'/2\cos\theta_{12}$$

By using the computed refractor dip γ_1, the respective perpendicular depths z and z' can be converted into vertical depths h and h' using

$$h = z/\cos\gamma_1$$

and

$$h' = z'/\cos\gamma_1$$

Note that the travel time of a seismic phase from one end of a refraction profile line to the other (i.e. from shot point to shot point) should be the same whether measured in the forward or the reverse direction. Referring to Fig. 5.5(b), this means that t_{AD} should equal t_{DA}. Establishing that there is satisfactory agreement between

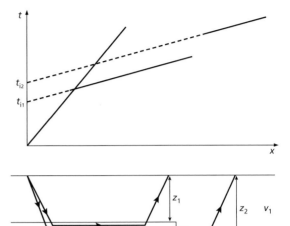

Fig. 5.7   Offset segments of the travel-time curve for refracted arrivals from opposite sides of a fault.

these *reciprocal times* (or *end-to-end* times) is a useful means of checking that travel-time curves have been drawn correctly through a set of refracted ray arrival times derived from a reversed profile.

5.2.5 Faulted planar interfaces

The effect of a fault displacing a planar refractor is to offset the segments of the travel-time plot on opposite sides of the fault (see Fig. 5.7). There are thus two intercept times t_{i1} and t_{i2}, one associated with each of the travel-time curve segments, and the difference between these intercept times ΔT is a measure of the throw of the fault. For example, in the case of the faulted horizontal refractor shown in Fig. 5.7 the throw of the fault Δz is given by

$$\Delta T \approx \frac{\Delta z\cos\theta}{v_1}$$

$$\Delta z \approx \frac{\Delta T v_1}{\cos\theta} = \frac{\Delta T v_1 v_2}{(v_2^2 - v_1^2)^{1/2}}$$

Note that there is some approximation in this formulation, since the ray travelling to the downthrown side of the fault is not the critically refracted ray at A and involves diffraction at the base B of the fault step. However, the error will be negligible where the fault throw is small compared with the refractor depth.

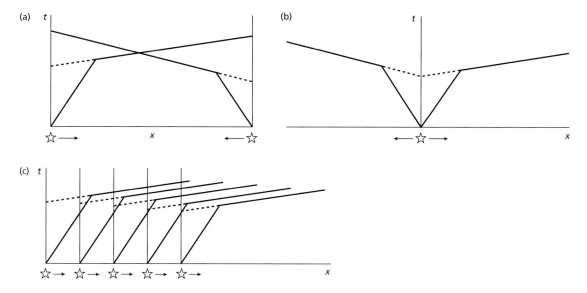

Fig. 5.8 Various types of profile geometry used in refraction surveying. (a) Conventional reversed profile with end shots. (b) Split-profile with central shot. (c) Single-ended profile with repeated shots.

5.3 Profile geometries for studying planar layer problems

The conventional field geometry for a refraction profile involves shooting at each end of the profile line and recording seismic arrivals along the line from both shots. As will be seen with reference to Fig. 5.5(a), only the central portion of the refractor (from B to C) is sampled by refracted rays detected along the line length. Interpreted depths to the refractor under the endpoints of a profile line, using equations given above, are thus not directly measured but are inferred on the basis of the refractor geometry over the shorter length of refractor sampled (BC). Where continuous cover of refractor geometry is required along a series of reversed profiles, individual profile lines should be arranged to overlap in order that all parts of the refractor are directly sampled by critically refracted rays.

In addition to the conventional reversed profile, illustrated schematically in Fig. 5.8(a), other methods of deriving full planar layer interpretations in the presence of dip include the *split-profile* method (Johnson 1976) and the *single-ended profile* method (Cunningham 1974). The split-profile method (Fig. 5.8(b)) involves recording outwards in both directions from a central shot point. Although the interpretation method differs in detail from

that for a conventional reversed profile, it is based on the same general travel-time equation (5.8).

The single-ended profile method (Fig. 5.8(c)) was developed to derive interpretations of low-velocity surface layers represented by refracted arrivals in single-ended reflection spread data, for use in the calculation of static corrections. A simplified treatment is given below.

To obtain a value of refractor dip, estimates of apparent velocity are required in both the forward and reverse directions. The repeated forward shooting of the single-ended profile method enables an apparent velocity in the forward direction to be computed from the gradient of the travel-time curves. For the method of computing the apparent velocity in the reverse direction, consider two refracted ray paths from surface sources S_1 and S_2 to surface detectors D_1 and D_2, respectively (Fig. 5.9). The offset distance is x in both cases, the separation Δx of S_1 and S_2 being the same as that of D_1 and D_2.

Since D_1 is on the downdip side of S_1, the travel time of a refracted ray from S_1 to D_1 is given by equation (5.9), and omitting subscripts to θ and γ in this two-layer case,

$$t_1 = \frac{x\sin(\theta + \gamma)}{v_1} + \frac{2z_1\cos\theta}{v_1} \tag{5.13}$$

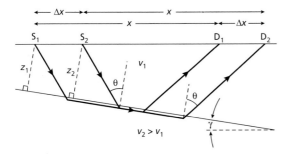

Fig. 5.9 Refraction interpretation using the single-ended profiling method. (After Cunningham 1974.)

and from S_2 to D_2 the travel time is given by

$$t_2 = \frac{x \sin(\theta + \gamma)}{v_1} + \frac{2z_2 \cos \theta}{v_1} \tag{5.14}$$

where z_1 and z_2 are the perpendicular depths to the refractor under shot points S_1 and S_2, respectively. Now,

$$z_2 - z_1 = \Delta x \sin \gamma$$
$$\therefore z_2 = z_1 + \Delta x \sin \gamma \tag{5.15}$$

Substituting equation (5.15) in (5.14) and then subtracting equation (5.13) from (5.14) yields

$$t_2 - t_1 = \Delta t = \frac{\Delta x}{v_1}(2 \sin \gamma \cos \theta)$$
$$= \frac{\Delta x \sin(\theta + \gamma)}{v_1} - \frac{\Delta x \sin(\theta - \gamma)}{v_1}$$

Substituting equations (5.11) and (5.12) in the above equation and rearranging terms

$$\frac{\Delta t}{\Delta x} = \frac{1}{v_{2d}} - \frac{1}{v_{2u}}$$

where v_{2u} and v_{2d} are the updip and downdip apparent velocities, respectively. In the case considered v_{2d} is derived from the single-ended travel-time curves, hence v_{2u} can be calculated from the difference in travel time of refracted rays from adjacent shots recorded at the same offset distance x. With both apparent velocities calculated, interpretation proceeds by the standard methods for conventional reversed profiles discussed in Section 5.2.4.

5.4 Geometry of refracted ray paths: irregular (non-planar) interfaces

The assumption of planar refracting interfaces would often lead to unacceptable error or imprecision in the interpretation of refraction survey data. For example, a survey may be carried out to study the form of the concealed bedrock surface beneath a valley fill of alluvium or glacial drift. Such a surface is unlikely to be modelled adequately by a planar refractor. In such cases the constraint that refracting interfaces be interpreted as planar must be dropped and different interpretation methods must be employed.

The travel-time plot derived from a survey provides a first test of the prevailing refractor geometry. A layered sequence of planar refractors gives rise to a travel-time plot consisting of a series of straight-line segments, each segment representing a particular refracted phase and characterized by a particular gradient and intercept time. Irregular travel-time plots are an indication of irregular refractors (or of lateral velocity variation within individual layers—a complication not discussed here). Methods of interpreting irregular travel-time plots, to determine the non-planar refractor geometry that gives rise to them, are based on the concept of *delay time*.

5.4.1 Delay time

Consider a horizontal refractor separating upper and lower layers of velocity v_1 and v_2 ($>v_1$), respectively (Fig. 5.1). The travel time of a head wave arriving at an offset distance x is given (see equation (5.3)) by

$$t = \frac{x}{v_2} + t_i$$

The intercept time t_i can be considered as composed of two delay times resulting from the presence of the top layer at each end of the ray path. Referring to Fig. 5.10(a), the *delay time* (or *time term*) δ_t is defined as the time difference between the slant path AB through the top layer and the time that would be required for a ray to travel along BC. The equation above clearly shows that the total travel time can be considered as the time a wave would take to travel the whole distance x at refractor velocity v_2, plus additional time t_i taken for the wave to travel down to the refractor at the shot point, and back up to the receiver. These two extra components of time are the *delay times* at the shot and receiver. Each delay time can be calculated in a similar way, referring to Fig. 5.10,

$$\delta_t = t_{AB} - t_{BC}$$

$$= \frac{AB}{v_1} - \frac{BC}{v_2}$$

$$= \frac{z}{v_1 \cos\theta} - \frac{z}{v_2} \tan\theta$$

$$= \frac{z}{v_1 \cos\theta} - \frac{z \sin\theta}{v_1} \frac{\sin\theta}{\cos\theta}$$

$$= \frac{z(1 - \sin^2\theta)}{v_1 \cos\theta} = \frac{z \cos\theta}{v_1}$$

$$= \frac{z(v_2^2 - v_1^2)^{1/2}}{v_1 v_2} \tag{5.16}$$

Solving equation (5.16) for the depth z to the refractor yields

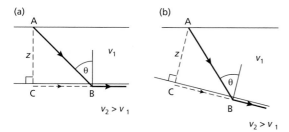

(a)   (b)

$v_2 > v_1$

Fig. 5.10 The concept of delay time.

$$z = \delta_t v_1 / \cos\theta = \delta_t v_1 v_2 / (v_2^2 - v_1^2)^{\frac{1}{2}} \tag{5.17}$$

Thus the delay time can be converted into a refractor depth if v_1 and v_2 are known.

The intercept time t_i in equation (5.3) can be partitioned into two delay times

$$t = x/v_2 + \delta_{ts} + \delta_{td} \tag{5.18}$$

where δ_{ts} and δ_{td} are the delay times at the shot end and detector end of the refracted ray path. Note that in this case of a horizontal refractor,

$$t = \frac{x}{v_2} + \frac{z \cos\theta}{v_1} + \frac{z \cos\theta}{v_1} = \frac{x}{v_2} + \frac{2z \cos\theta}{v_1}$$

This is the same result as derived earlier in equation (5.1), showing that the delay-time concept is implicit even in simple horizontal–lateral interpretation methods.

In the presence of refractor dip the delay time is similarly defined except that the geometry of triangle ABC rotates with the refractor. The delay time is again related to depth by equation (5.17), where z is now the refractor depth at A measured normal to the refractor surface. Using this definition of delay time, the travel time of a ray refracted along a dipping interface (see Fig. 5.11(a)) is given by

$$t = x'/v_2 + \delta_{ts} + \delta_{td} \tag{5.19}$$

where $\delta_{ts} = t_{AB} - t_{BC}$ and $\delta_{td} = t_{DE} - t_{DF}$.

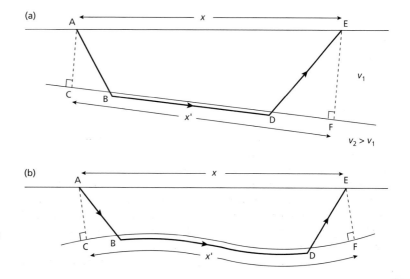

Fig. 5.11 Refracted ray paths associated with (a) a dipping and (b) an irregular refractor.

(a)

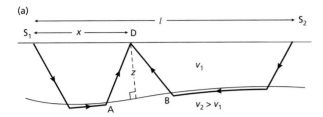

(b)

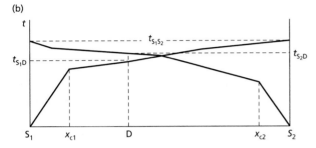

Fig. 5.12 The plus–minus method of refraction interpretation (Hagedoorn 1959). (a) Refracted ray paths from each end of a reversed seismic profile line to an intermediate detector position. (b) Travel-time curves in the forward and reverse directions.

For shallow dips, x' (unknown) is closely similar to the offset distance x (known), in which case equation (5.18) can be used in place of (5.19) and methods applicable to a horizontal refractor employed. This approximation is valid also in the case of an irregular refractor if the relief on the refractor is small in amplitude compared to the average refractor depth (Fig. 5.11(b)).

Delay times cannot be measured directly but occur in pairs in the travel-time equation for a refracted ray from a surface source to a surface detector. The *plus–minus method* of Hagedoorn (1959) provides a means of solving equation (5.18) to derive individual delay time values for the calculation of local depths to an irregular refractor.

5.4.2 The plus–minus interpretation method

Figure 5.12(a) illustrates a two-layer ground model with an irregular refracting interface. Selected ray paths are shown associated with a reversed refraction profile line of length l between end shot points S_1 and S_2. The travel time of a refracted ray travelling from one end of the line to the other is given by

$$t_{S_1 S_2} = l/v_2 + \delta_{t S_1} + \delta_{t S_2} \qquad (5.20)$$

where $\delta_{t S_1}$ and $\delta_{t S_2}$ are the delay times at the shot points. Note that $t_{S_1 S_2}$ is the reciprocal time for this reversed profile (see Fig. 5.12(b)). For rays travelling to an intermediate detector position D from each end of the line, the travel times are, for the forward ray, from shot point S_1

$$t_{S_1 D} = x/v_2 + \delta_{t S_1} + \delta_{t D} \qquad (5.21)$$

for the reverse ray, from shot point S_2

$$t_{S_2 D} = (l - x)/v_2 + \delta_{t S_2} + \delta_{t D} \qquad (5.22)$$

where $\delta_{t D}$ is the delay time at the detector.

v_2 cannot be obtained directly from the irregular travel-time curve of refracted arrivals, but it can be estimated by means of Hagedoorn's *minus* term. This is obtained by taking the difference of equations (5.21) and (5.22)

$$t_{S_1 D} - t_{S_2 D} = 2x/v_2 - l/v_2 + \delta_{t S_1} - \delta_{t S_2}$$
$$= (2x - l)/v_2 + \delta_{t S_1} - \delta_{t S_2}$$

This subtraction eliminates the variable (geophone-station dependent) delay time $\delta_{t D}$ from the above equation. Since the last two terms on the right-hand side of the equation are constant for a particular profile line, plotting the minus term $(t_{S_1 D} - t_{S_2 D})$ against the distance $(2x - l)$ yields a graph of slope $1/v_2$ from which v_2 may be derived. If the assumptions of the plus–minus method are valid, then the minus-time plot will be a straight line. Thus, this plot is a valuable quality control check for the interpretation method. Often it can be difficult to locate the crossover distances in real data, especially if the refracted arrivals line is irregular due to refractor topography. For minus-time points computed from arrival times which are not from the same refractor, the plot will curve away from a central straight section. Also, any lateral change of refractor velocity v_2 along the profile line will show up as a change of gradient in the minus term plot.

For the valid range of detectors determined from the minus-time plot, the delay times can now be calculated. Adding equations (5.21) and (5.22)

$$t_{S_1D} + t_{S_2D} = l/v_2 + \delta_{tS_1} + \delta_{tS_2} + 2\delta_{tD}$$

Substituting equation (5.20) in the above equation yields

$$t_{S_1D} + t_{S_2D} = t_{S_1S_2} + 2\delta_{tD}$$

Hence

$$\delta_{tD} = \frac{1}{2}\left(t_{S_1D} + t_{S_2D} - t_{S_1S_2}\right) \qquad (5.23)$$

This delay time is the *plus* term of the plus–minus method and may be used to compute the perpendicular depth z to the underlying refractor at D using equation (5.17). v_2 is found from the minus-time plot and v_1 is computed from the slope of the direct ray travel-time plot (see Fig. 5.12(b)). Note that the value of all delay times depends on the *reciprocal time*. Errors in this time, which is recorded at maximum range along the profile, and often with the lowest signal-to-noise ratio, introduce a constant error into all delay times. Great care must be taken to check the errors in this value.

A plus term and, hence, a local refractor depth can be computed at all detector positions at which head wave arrivals are recognized from both ends of the profile line. In practice, this normally means the portion of the profile line between the crossover distances; that is, between x_{c1} and x_{c2} in Fig. 5.12(b).

Where a refractor is overlain by more than one layer, equation (5.17) cannot be used directly to derive a refractor depth from a delay time (or plus term). In such a case, either the thickness of each overlying layer is computed separately using refracted arrivals from the shallower interfaces, or an average overburden velocity is used in place of v_1 in equation (5.17) to achieve a depth conversion.

The plus–minus method is only applicable in the case of shallow refractor dips, generally being considered valid for dips of less than 10°. With steeper dips, x' becomes significantly different from the offset distance x. Further, there is an inherent smoothing of the interpreted refractor geometry in the plus–minus method.

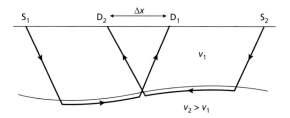

Fig. 5.13 The generalized reciprocal method of refraction interpretation (Palmer 1980).

When computing the plus term for each detector, the refractor is assumed to be planar between the points of emergence from the refractor of the forward and reverse rays, for example between A and B in Fig. 5.12(a) for rays arriving at detector D.

5.4.3 The generalized reciprocal method

This problem of smoothing is solved in the *generalized reciprocal method (GRM)* of refraction interpretation (Palmer 1980) by combining the forward and reverse rays which leave the refractor at approximately the same point and arrive at different detector positions separated by a distance Δx (see Fig. 5.13). The method uses a velocity analysis function t_v given by

$$t_v = \left(t_{S_1D_1} + t_{S_2D_2} - t_{S_1S_2}\right)/2 \qquad (5.24)$$

the values being referred to the mid-point between each pair of detector positions D_1 and D_2. For the case where $D_1 = D_2 = D$ (i.e. $\Delta x = 0$), equation (5.24) reduces to a form similar to Hagedoorn's minus term (see above). The optimal value of Δx for a particular survey is that which produces the closest approach to a linear plot when the velocity analysis function t_v is plotted against distance along the profile line, and is derived by plotting curves for a range of possible Δx values. The overall interpretation method is more complex than the plus–minus method, but can deliver better velocity discrimination, greater lateral resolution and better depth estimates to boundaries. The method also demands denser data coverage than the plus–minus method. The principles of the method, its implementation and example datasets are clearly laid out in Palmer's book (Palmer 1980), but beyond the scope of this one.

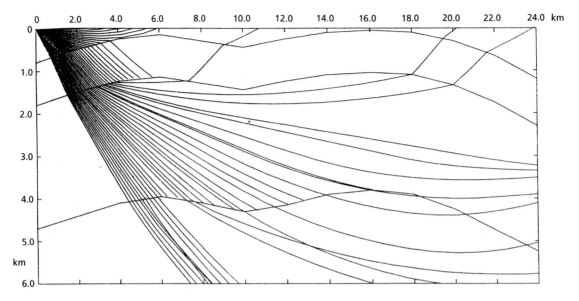

Fig. 5.14 Modelling of complex geology by ray-tracing in the case of a refraction profile between quarries in south Wales, UK. Refracted ray paths from Cornelly Quarry (located in Carboniferous Limestone) are modelled through a layered Palaeozoic sedimentary sequence overlying an irregular Precambrian basement surface at a depth of about 5 km. This model accounts for the measured travel times of refracted arrivals observed along the profile. (From Bayerly & Brooks 1980.)

5.5 Construction of wavefronts and ray-tracing

Given the travel-time plots in the forward and reverse directions along a profile line it is possible to reconstruct the configuration of successive wavefronts in the subsurface and thereby derive, graphically, the form of refracting interfaces. This *wavefront method* (Thornburgh 1930) represents one of the earliest refraction interpretation methods but is no longer widely used.

With the massive expansion in the speed and power of digital computers, and their wide availability, an increasingly important method of refraction interpretation is a modelling technique known as *ray-tracing* (Cerveny *et al.* 1974). In this method structural models are postulated and the travel-times of refracted (and reflected) rays through these models are calculated by computer for comparison with observed travel-times. The model is then adjusted iteratively until the calculated and observed travel-times are in acceptable agreement. This method is especially useful in the case of complex subsurface structures that are difficult to treat analytically. An example of a ray-tracing interpretation is illustrated in Fig. 5.14. The ray-tracing method is particularly valuable in coping with such complexities as horizontal or vertical velocity gradients within layers, highly irregular or steeply dipping refractor interfaces and discontinuous layers.

5.6 The hidden and blind layer problems

It is possible for layers to exist in the Earth, yet not produce any refracted first-arrival waves. In this case the layers will be undetectable in a simple first arrival refraction survey. The observed data could be interpreted using the methods discussed above and yield a self-consistent, but erroneous, solution. For this reason, the possibility of undetected layers should always be considered. In practice, there are two different types of problem. In order to be detected in a first arrival refraction survey, a layer must (a) be underlain by a layer of higher velocity so that head waves are produced, and (b) have a thickness and velocity such that the head waves become first arrivals at some range.

A *hidden layer* is a layer which, whilst producing head waves, does not give rise to first arrivals. Rays travelling to deeper levels arrive before those critically refracted at the top of the layer in question (Fig. 5.15(a)). This may

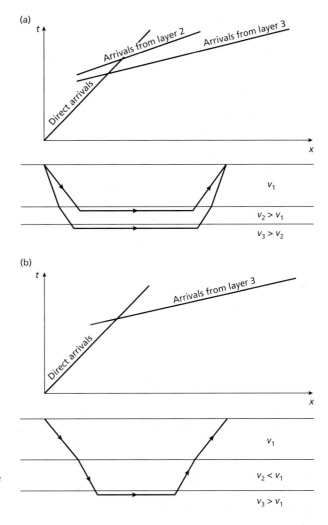

Fig. 5.15 The undetected layer problem in refraction seismology. (a) A hidden layer: a thin layer that does not give rise to first arrivals. (b) A blind layer: a layer of low velocity that does not generate head waves.

result from the thinness of the layer, or from the closeness of its velocity to that of the overlying layer. In such a case, a method of survey involving recognition of only first arrivals will fail to detect the layer. It is good practice to examine the seismic traces for possible arrivals occurring behind the first arrivals. These should then be examined to ensure they are compatible with the structural model derived from the first arrivals.

A *blind layer* presents a more insidious problem, resulting from a low-velocity layer, as illustrated in Fig. 5.15(b). Rays cannot be critically refracted at the top of such a layer and the layer will therefore not give rise to head waves. Hence, a low-velocity layer cannot be detected by refraction surveying, although the top

of the low-velocity layer gives rise to wide-angle reflections that may be detected as later arrivals during a refraction survey.

In the presence of a low-velocity layer, the interpretation of travel-time curves leads to an overestimation of the depth to underlying interfaces. Low-velocity layers are a hazard in all types of refraction seismology. On a small scale, a peat layer in muds and sands above bedrock may escape detection, leading to a false estimation of foundation conditions and rockhead depths beneath a construction site; on a much larger scale, low-velocity zones of regional extent are known to exist within the continental crust and may escape detection in crustal seismic experiments.

5.7 Refraction in layers of continuous velocity change

In some geological situations, velocity varies gradually as a function of depth rather than discontinuously at discrete interfaces of lithological change. In thick clastic sequences, for example, especially clay sequences, velocity increases downwards due to the progressive compaction effects associated with increasing depth of burial. A seismic ray propagating through a layer of gradual velocity change is continuously refracted to follow a curved ray path. For example, in the special case where velocity increases linearly with depth, the seismic ray paths describe arcs of circles. The deepest point reached by a ray travelling on a curved path is known as its *turning point*.

In such cases of continuous velocity change with depth, the travel-time plot for refracted rays that return to the surface along curved ray paths is itself curved, and the geometrical form of the curve may be analysed to derive information on the distribution of velocity as a function of depth (see e.g. Dobrin & Savit 1988).

Velocity increase with depth may be significant in thick surface layers of clay due to progressive compaction and dewatering, but may also be significant in deeply buried layers. Refracted arrivals from such buried layers are not true head waves since the associated rays do not travel along the top surface of the layer but along a curved path in the layer with a turning point at some depth below the interface. Such refracted waves are referred to as *diving waves* (Cerveny & Ravindra 1971). Methods of interpreting refraction data in terms of diving waves are generally complex, but include ray-tracing techniques. Indeed, some ray-tracing programmes require velocity gradients to be introduced into all layers of an interpretation model in order to generate diving waves rather than true head waves.

5.8 Methodology of refraction profiling

Many of the basic principles of refraction surveying have been covered in the preceding sections but in this section several aspects of the design of refraction profile lines are brought together in relation to the particular objectives of a refraction survey.

5.8.1 Field survey arrangements

Although the same principles apply to all scales of refraction profiling, the logistic problems of implementing a profile line increase as the required line length increases. Further, the problems of surveying on land are quite different from those encountered at sea. A consequence of these logistic differences is a very wide variety of survey arrangements for the implementation of refraction profile lines and these differences are illustrated by three examples.

For a small-scale refraction survey of a construction site to locate the water table or rockhead (both of which surfaces are generally good refractors), recordings out to an offset distance of about 100 m normally suffice. Geophones are connected via a multicore cable to a portable 24- or 48-channel seismic recorder. A simple weight-dropping device (even a sledge hammer impacted on to a steel base plate) provides sufficient energy to traverse the short recording range. The dominant frequency of such a source exceeds 100 Hz and the required accuracy of seismic travel times is about 0.5 ms. Such a survey can be easily accomplished by two operators.

The logistic difficulties associated with the cable connection between a detector spread and a recording unit normally limit conventional refraction surveys to maximum shot–detector offsets of about 1 km and, hence, to depths of investigation of a few hundred metres. For larger scale refraction surveys it is necessary to dispense with a cable connection. At sea, such surveys can be carried out by a single vessel in conjunction with free-floating radio-transmitting sonobuoys (Fig. 5.16). Having deployed the sonobuoys, the vessel proceeds along the profile line repeatedly firing explosive charges or an air-gun array. Seismic signals travelling back to the surface through the water layer are detected by a hydrophone suspended beneath each sonobuoy, amplified and transmitted back to the survey vessel where they are recorded along with the shot instant. By this means, refraction lines up to a few tens of kilometres may be implemented.

For large-scale marine surveys, ocean bottom seismographs (OBSs) are deployed on the sea bed. These contain a digital recorder together with a high-precision clock unit to provide an accurate time base for the seismic recordings. Such instruments may be deployed for periods of up to a few days at a time. For the purposes of recovery, the OBSs are 'popped-up' to the surface by remotely triggering a release mechanism. Seabed recording systems provide a better signal-to-noise ratio than hydrophones suspended in the water column and, in deep water, recording on the sea bed allows much better definition of shallow structures. In this type of survey the dominant frequency is typically in the range

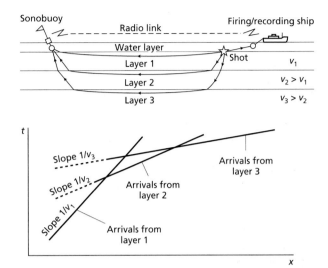

Fig. 5.16 Single-ship seismic refraction profiling.

10–50 Hz and travel times need to be known to about 10 ms.

A large-scale seismic refraction line on land to investigate deep crustal structure is typically 250–300 km long. Seismic events need to be recorded at a series of independently operated recording stations all receiving a standard time signal to provide a common time base for the recordings. Usually this is provided by the signal from the global positioning system (GPS) satellite system. Very large energy sources, such as military depth charges (detonated at sea or in a lake) or large quarry blasts, are required in order that sufficient energy is transmitted over the length of the profile line. The dominant frequency of such sources is less than 10 Hz and the required accuracy of seismic travel times is about 50 ms. Such an experiment requires the active involvement of a large and well-coordinated field crew.

Along extended refraction lines, wide-angle reflection events are often detected together with the refracted phases. These provide an additional source of information on subsurface structure. Wide-angle reflection events are sometimes the most obvious arrivals and may represent the primary interest (e.g. Brooks *et al.* 1984). Surveys specifically designed for the joint study of refracted and wide-angle reflection events are often referred to as *wide-angle surveys*.

5.8.2 Recording scheme

For complete mapping of refractors beneath a seismic line it is important to arrange that head wave arrivals

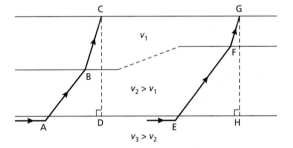

Fig. 5.17 Variation in the travel time of a head wave associated with variation in the thickness of a surface layer.

from all refractors of interest are obtained over the same portion of line. The importance of this can be seen by reference to Fig. 5.17 where it is shown that a change in thickness of a surface low-velocity layer would cause a change in the delay time associated with arrivals from a deeper refractor and may be erroneously interpreted as a change in refractor depth. The actual geometry of the shallow refractor should be mapped by means of shorter reversed profiles along the length of the main profile. These are designed to ensure that head waves from the shallow refractor are recorded at positions where the depth to the basal refractor is required. Knowledge of the disposition of the shallow refractor derived from the shorter profiles would then allow correction of travel times of arrivals from the deeper refractor.

The general design requirement is the formulation of an overall observational scheme as illustrated in

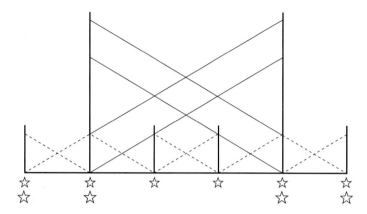

Fig. 5.18 A possible observational scheme to obtain shallow and deeper refraction coverage along a survey line. The inclined lines indicate the range of coverage from the individual shots shown.

Fig. 5.18. Such a scheme might include off-end shots into individual reversed profile lines, since off-end shots extend the length of refractor traversed by recorded head waves and provide insight into the structural causes of any observed complexities in the travel-time curves. Selection of detector spacing along the individual profile lines is determined by the required detail of the refractor geometry, the sampling interval of interpretation points on the refractor being approximately equal to the detector spacing. Thus, the horizontal resolution of the method is equivalent to the detector spacing.

It is often the case that there are insufficient detectors available to cover the full length of the profile with the desired detector spacing. In this case the procedure is to deploy the detectors to cover one segment of the line at the required spacing, then to fire shots at all shot points. The detectors are then moved to another segment of the line and all shot points fired again. The process can be repeated until full data are compiled for the complete profile. At the price of repeating the shots, a profile can thus be recorded of any length with a limited supply of equipment. The same principle is equally applicable to shallow penetration, to detailed refraction surveys for engineering, to environmental and hydrological applications, and to crustal studies.

5.8.3 Weathering and elevation corrections

The type of observational scheme illustrated in Fig. 5.18 is often implemented for the specific purpose of mapping the surface zone of weathering and associated low velocity across the length of a longer profile designed to investigate deeper structure. The velocity and thickness of the weathered layer are highly variable laterally and travel times of rays from underlying refractors need to be

corrected for the variable delay introduced by the layer. This weathering correction is directly analogous to that applied in reflection seismology (see Section 4.6). The weathering correction is particularly important in shallow refraction surveying where the size of the correction is often a substantial percentage of the overall travel time of a refracted ray. In such cases, failure to apply an accurate weathering correction can lead to major error in interpreted depths to shallow refractors.

A weathering correction is applied by effectively replacing the weathered layer of velocity v_w with material of velocity v_1 equal to the velocity of the underlying layer. For a ray critically refracted along the top of the layer immediately underlying the weathered layer, the weathering correction is simply the sum of the delay times at the shot and detector ends of the ray path. Application of this correction replaces the refracted ray path by a direct path from shot to detector in a layer of velocity v_1. For rays from a deeper refractor a different correction is required. Referring to Fig. 5.19, this correction effectively replaces ray path ABCD by ray path AD. For a ray critically refracted in the nth layer the weathering correction t_w is given by

$$t_w = -(z_s + z_d)$$
$$\times \left\{ (v_n^2 - v_1^2)^{1/2} \big/ v_1 v_n - (v_n^2 - v_w^2)^{1/2} \big/ v_w v_n \right\}$$

where z_s and z_d are the thicknesses of the weathered layer beneath the shot and detector respectively, and v_n is the velocity in the nth layer.

In addition to the weathering correction, a correction is also needed to remove the effect of differences in elevation of individual shots and detectors, and an elevation correction is therefore applied to reduce travel times to a

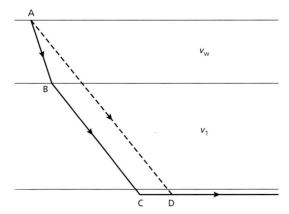

Fig. 5.19 The principle of the weathering correction in refraction seismology.

common datum plane. The elevation correction t_e for rays critically refracted in the nth layer is given by

$$t_e = -\left(h_s + h_d\right)\left\{\left(v_n^2 - v_1^2\right)^{1/2} \big/ v_1 v_n\right\}$$

where h_s and h_d are the heights above datum of the shot point and detector location respectively. It is worth noting that these corrections are more complex than those used for seismic reflection surveys. The difference arises since the assumption of vertical ray paths through the weathered layer used in the reflection case cannot be maintained.

In shallow water marine refraction surveying the water layer is conventionally treated as a weathered layer and a correction applied to replace the water layer by material of velocity equal to the velocity of the sea bed.

5.8.4 Display of refraction seismograms

In small-scale refraction surveys the individual seismograms are conventionally plotted out in their true time relationships in a format similar to that employed to display seismic traces from land reflection spreads (see Fig. 4.8). From such displays, arrival times of refracted waves may be picked and, after suitable correction, used to make the time–distance plots that form the basis of refraction interpretation.

Interpretation of large-scale refraction surveys is often as much concerned with later arriving phases, such as wide-angle reflections or S-wave arrivals, as with first arrivals. To aid recognition of weak coherent phases, the individual seismograms are compiled into an overall record section on which the various seismic phases can be correlated from seismogram to seismogram. The optimal type of display is achieved using a *reduced time* scale in which any event at time t and offset distance x is plotted at the reduced time T where

$$T = t - x\big/v_R$$

and v_R is a scaling factor known as the *reduction velocity*. Thus, for example, a seismic arrival from deep in the Earth's crust with an overall travel time of 30 s to an offset distance of 150 km would, with a reduction velocity of 6 km s^{-1}, have a reduced time of 5 s.

Plotting in reduced time has the effect of progressively reducing travel-time as a function of offset and, therefore, rotating the associated time–distance curves towards the horizontal. For example, a time–distance curve with a reciprocal slope of 6 km s^{-1} on a t–x graph would plot as a horizontal line on a T–x graph using a reduction velocity of 6 km s^{-1}. By appropriate choice of reduction velocity, seismic arrivals from a particular refractor of interest can be arranged to plot about a horizontal datum, so that relief on the refractor will show up directly as departures of the arrivals from a horizontal line. The use of reduced time also enables the display of complete seismograms with an expanded time scale appropriate for the analysis of later arriving phases. An example of a record section from a crustal seismic experiment, plotted in reduced time, is illustrated in Fig. 5.20.

5.9 Other methods of refraction surveying

Although the vast bulk of refraction surveying is carried out along profile lines, other spatial arrangements of shots and detectors may be used for particular purposes. Such arrangements include *fan-shooting* and irregularly distributed shots and recorders as used in the *time term* method.

Fan-shooting (Fig. 5.21) is a convenient method of accurately delineating a subsurface zone of anomalous velocity whose approximate position and size are already known. Detectors are distributed around a segment of arc approximately centred on one or more shot points, and travel-times of refracted rays are measured to each detector. Through a homogeneous medium the travel-times to detectors would be linearly related to range, but

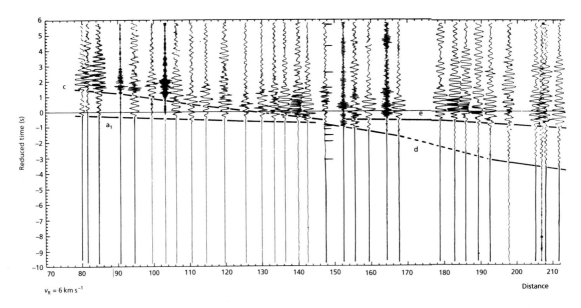

Fig. 5.20 Part of a time section from a large-scale refraction profile, plotted in reduced time using a reduction velocity of 6 km s^{-1}. The section was derived from the LISPB lithospheric seismic profile across Britain established in 1974. Phase a$_1$: head wave arrivals from a shallow crustal refractor with a velocity of about 6.3 km s^{-1}; phases c and e: wide-angle reflections from lower crustal interfaces: phase d: head wave arrivals from the uppermost mantle (the P_n phase of earthquake seismology). (From Bamford *et al.* 1978.)

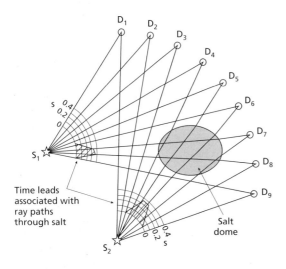

Fig. 5.21 Fan-shooting for the detection of localized zones of anomalous velocity.

any ray path which encounters an anomalous velocity zone will be subject to a time lead or time lag depending upon the velocity of the zone relative to the velocity of the surrounding medium. Localized anomalous zones

capable of detection and delineation by fan-shooting include salt domes, buried valleys and backfilled mine shafts.

An irregular, areal distribution of shots and detectors (Fig. 5.22(a)) represents a completely generalized approach to refraction surveying and facilitates mapping of the three-dimensional geometry of a subsurface refractor using the *time term method* of interpretation (Willmore & Bancroft 1960, Berry & West 1966). Rather than being an intrinsic aspect of the survey design, however, an areal distribution of shot points and recording sites may result simply from an opportunistic approach to refraction surveying in which freely available sources of seismic energy such as quarry blasts are used to derive subsurface information from seismic recordings.

The *time term method* uses the form of the travel–time equation containing delay times (equation (5.18)) and is subject to the same underlying assumptions as other interpretation methods using delay times. However, in the time term method a statistical approach is adopted to deal with a redundancy of data inherent in the method and to derive the best estimate of the interpretation parameters. Introducing an error term into the travel–time equation

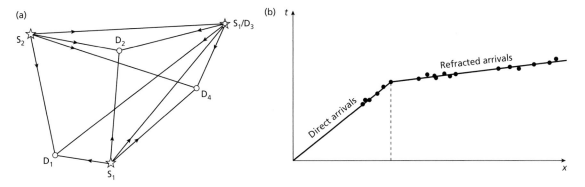

Fig. 5.22 (a) An example of the type of network of shots and detectors from which the travel times of refracted arrivals can be used in a time term analysis of the underlying refractor geometry. (b) The plot of travel time as a function of distance identifies the set of refracted arrivals that may be used in the analysis.

$$t_{ij} = x_{ij}/v + \delta_{ti} + \delta_{tj} + \varepsilon_{ij}$$

where t_{ij} is the travel time of head waves from the ith site to the jth site, x_{ij} is the offset distance between site i and site j, δ_{ti} and δ_{tj} are the delay times (time terms), v is the refractor velocity (assumed constant), and ε_{ij} is an error term associated with the measurement of t_{ij}.

If there are n sites there can be up to $n(n-1)$ observational linear equations of the above type, representing the situation of a shot and detector at each site and all sites sufficiently far apart for the observation of head waves from the underlying refractor. In practice there will be fewer observational equations than this because, normally, only a few of the sites are shot points and head wave arrivals are not recognized along every shot–detector path (Fig. 5.22(b)). There are $(n+1)$ unknowns, namely the individual delay times at the n sites and the refractor velocity v.

If the number m of observational equations equals the number of unknowns, the equations can be solved to derive the unknown quantities, although it is necessary either that at least one shot and detector position should coincide or that the delay time should be known at one site. In fact, with the time term approach to refraction surveying it is normally arranged for m to well exceed $(n+1)$, and for several shot and detector positions to be interchanged. The resulting overdetermined set of equations is solved by deriving values for the individual delay times and refractor velocity that minimize the sum of squares of the errors ε_{ij}. Delay times can then be converted into local refractor depths using the same procedure as in the plus–minus method described earlier.

5.10 Seismic tomography

Although fan-shooting involves surface shots and recorders, the method may be regarded as the historical precursor of an important group of modern exploration methods using shots and detectors located in boreholes. In these methods, known as *seismic tomography*, subsurface zones are systematically investigated by transmitting very large numbers of seismic rays through them. An example is cross-hole seismics (see e.g. Wong *et al.* 1987), in which shots generated at several depths down a borehole are recorded by detector arrays in an adjacent borehole to study variations in the seismic wave transmission through the intervening section of ground. A simple example is shown in Fig. 5.23, where only a limited subset of ray paths are shown.

The volume of ground under investigation is modelled as divided into cubic elements. The seismic sources and receivers are arranged so that multiple seismic rays pass through each element of that volume. If the geological unit under investigation is a near-horizontal bed, then the sources, receivers and volume elements lie in a single horizontal plane and the geometry is directly comparable to the cross-borehole situation. An example of this geometry is the investigation of coal seams prior to long-wall mining techniques. Here the sources and receivers are arranged in the tunnels driven to give access to the seam.

It is theoretically straightforward to develop the method to investigate 3D velocity structures. This is done for medical imaging such as CAT scanning, where X-rays are directed though the investigated volume by moving the source and receiver freely around the

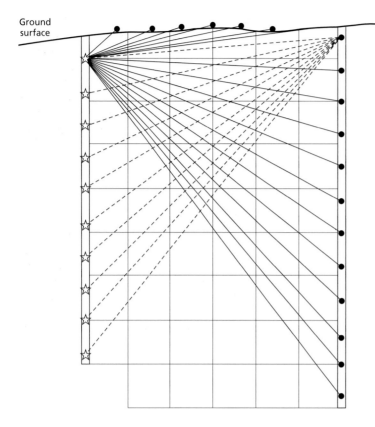

Ground
surface

Fig. 5.23 Idealized observation scheme for a simple cross-hole seismic transmission tomography survey. Dots mark receivers, stars mark sources. For clarity, only the ray paths from one source to all receivers (solid lines), and all sources to one receiver (dashed lines) are shown. Also shown is the regular grid of elements for which velocity values are derived.

perimeter of the volume. In the geological case the difficulty lies in getting access to place sources and receivers at locations distributed uniformly around the volume under investigation. Multiple vertical boreholes merely allow the collection of a number of vertical 2D sections as shown in Fig. 5.23.

The total travel-times for each ray are the basic data used for interpretation. Each cubic element is assigned an initial seismic velocity. Assuming a linear ray path from source and receiver, the time spent by each ray in each element can be calculated. The velocity assigned to each individual element can then be adjusted so that the errors between the observed travel-times and the calculated ones are minimized. A more sophisticated approach is to include in the solution the effect of refraction of the seismic wave as it passes between volume elements of different velocity. Such a solution has more variable parameters and requires a dense pattern of intersecting ray paths within the irradiated section. Note that the calculation of the true ray path is very difficult. It cannot be found by applying Snell's Law at the element boundaries, since these boundaries have no physical

reality. Common methods of solution of the resulting equations are the *algebraic reconstruction technique* (ART) and the *simultaneous reconstruction technique* (SIRT). The details of these techniques are beyond the scope of this book, but are well described by Ivansson (1986).

Use of high-frequency sources permits accurate travel-time determination and consequent high-resolution imaging of the velocity structure. This is necessary since a change in velocity in any one element only has a very small effect on the total travel-time for the ray path. Less commonly, parameters other than the P-wave travel-times can be analysed. Particular examples would be the S-wave travel-times, and the attenuation of the seismic wave. The above discussion has only considered transmission tomography, where the ray path is the simple minimum travel-time path from source to receiver. With additional complications, the same basic approach can be used with more complex ray paths. Reflection tomography involves the application of tomographic principles to reflected seismic waves. While it is considerably more complex than conventional seismic reflection processing, in areas of complex structure, particularly with large velocity variations, it can produce greatly improved seismic images.

The information derived from seismic tomography may be used to predict spatial variations in, for example, lithology, pore fluids, or rock fracturing, and the method is therefore of potential value in a wide range of exploration and engineering applications. As with many geophysical methods, it can also be applied on a variety of spatial scales, from ranges of hundreds of metres, down to engineering or archaeological investigations of single columns in ancient buildings (Cardarelli & de Nardis 2001).

5.11 Applications of seismic refraction surveying

Exploration using refraction methods covers a very wide range of applications. Refraction surveys can provide estimates of the elastic constants of local rock types, which have important engineering applications: use of special sources and geophones allows the separate recording of

shear wave arrivals, and the combination of P- and S-wave velocity information enables calculation of Poisson's ratio (Section 3.3.1). If an estimate of density is available, the bulk modulus and shear modulus can also be calculated from P- and S-wave velocities. Such estimates of the elastic constants, based on the propagation of seismic waves, are referred to as dynamic, in contrast to the static estimates derived from load-testing of rock samples in the laboratory. Dynamic estimates tend to yield slightly higher values than loading tests.

5.11.1 Engineering and environmental surveys

On the local scale, refraction surveys are widely used in foundation studies on construction sites to derive estimates of depth to rockhead beneath a cover of superficial material. Use of the plus–minus method or the generalized reciprocal method (Section 5.4) allows irregular rockhead geometries to be mapped in detail and thus reduces the need for test drilling with its associated high costs. Figure 5.24 shows a typical profile across fluvial

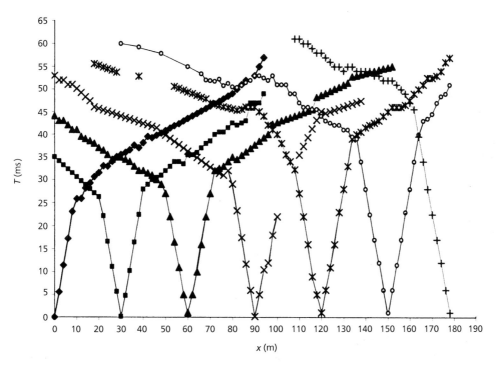

Fig. 5.24 *T–x* graph of a seismic refraction profile recorded over Holocene fluvial sediments overlying Palaeozoic rocks. The geophone separation was 2 m and the shot point separation 30 m. The multiple, overlapping, reversed data allow a continuous plus–minus interpretation of the rockhead interface.

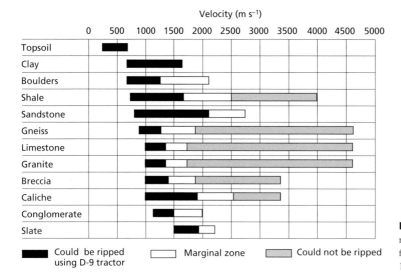

Fig. 5.25 Table showing the variation of rippability with seismic P-wave velocity for a range of lithologies. (After Bell 1993.)

sediments. Here the observation scheme specified a 2 m geophone spacing, and a 30 m shot spacing. The data were recorded with a 48-channel seismograph, with shot points re-fired as the 48 geophones were moved down the profile. The source was a sledgehammer.

The P-wave seismic velocity is related to the elastic constants and the density of the material. It is possible to derive an empirical relationship between the seismic velocity and the 'hardness' of the rock. In engineering usage, an important parameter of rock lithology is its resistance to excavation. If the rock can be removed by mechanical excavation it is termed 'rippable', rather than requiring fracturing by explosives. Empirical tables have been derived relating the 'rippability' of rock units by particular earthmoving equipment to the P-wave seismic velocity. Figure 5.25 shows a typical example of such a table. The range of velocities considered as rippable varies for different lithologies based on empirical averages of such relevant factors as their typical degree of cementation and frequency of jointing. Simple reversed P-wave refraction surveys are sufficient to provide critical information to construction and quarrying operations.

For surveys of near-surface geology, the data collection and interpretation must be efficient and rapid, to make the survey cost-effective against the alternative of direct excavation. The interpretation of seismic refraction profile data is most conveniently carried out using commercial software packages on personal computers. A wide range of good software is available for the plotting,

automatic event picking and interpretation of such data. In some situations the option of excavation instead of geophysical survey is very undesirable. Seismic surveys may be used to define the extent and depth of unrecorded landfill sites, or structures on 'brown-field' redevelopments. Commonly seismic and resistivity surveys may be used together to attempt to 'characterize' the nature of the landfill materials. There is an increasing demand for this sort of investigation in many parts of the world.

5.11.2 Hydrological surveys

The large difference in velocity between dry and wet sediments renders the water table a very effective refractor. Hence, refraction surveys find wide application in exploration programmes for underground water supplies in sedimentary sequences, often employed in conjunction with electrical resistivity methods (see Chapter 8). There can, however, be an ambiguity in interpretation of P-wave refraction data since a layer at depth with a velocity in excess of $1500 \, \text{m} \, \text{s}^{-1}$ could be either the water table, or a layer of more consolidated rock. Recording both P- and S-wave data overcomes this problem, since the water table will affect the P-wave velocity, but not that of the S-waves (Fig. 5.26).

5.11.3 Crustal seismology

The refraction method produces generalized models of subsurface structure with good velocity information,

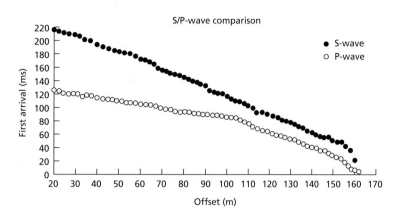

Fig. 5.26 *T–x* graph of a seismic refraction profile with a water-table refractor. The rock unit is the Sherwood Sandstone, having P-wave velocities of 800 and 2000 m s^{-1} for unsaturated and saturated rock respectively (lower line). The equivalent S-wave plot (upper line) shows no effect at the water-table interface.

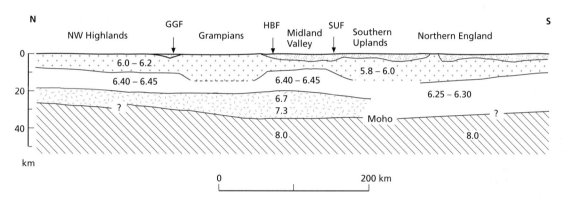

Fig. 5.27 Crustal cross-section across northern Britain based on interpretation of a large-scale seismic refraction experiment. Numbers refer to velocities in km s^{-1}. (After Bamford *et al.* 1978.) Contrast the distance scale with Figs 5.24 and 5.26.

but it is unable to provide the amount of structural detail or the direct imaging of specific structures that are the hallmark of reflection seismology. The occasional need for better velocity information than can be derived from velocity analysis of reflection data alone (see Chapter 4), together with the relative ease of refraction surveying offshore, gives the refraction method an important subsidiary role to reflection surveying in the exploration for hydrocarbons in some offshore areas.

Refraction and wide-angle surveys have been used extensively for regional investigation of the internal constitution and thickness of the Earth's crust. The information derived from such studies is complementary to the direct seismic imaging of crustal structure derived from large-scale reflection surveys of the type discussed in Section 4.16. Interpretation of large-scale refraction and wide-angle surveys is normally carried out by forward modelling of the travel times and amplitudes of recorded

refracted and/or reflected phases using ray-tracing techniques.

Large-scale surveys, using explosives as seismic sources, have been carried out to study crustal structure in most continental areas. An example is the LISPB experiment which was carried out in Britain in 1974 and produced the crustal section for northern Britain reproduced in Fig. 5.27.

Such experiments show that the continental crust is typically 30–40 km thick and that it is often internally layered. It is characterized by major regional variations in thickness and constitution which are often directly related to changes of surface geology. Thus, different orogenic provinces are often characterized by quite different crustal sections. Upper crustal velocities are usually in the range 5.8–6.3 km s^{-1} which, by analogy with velocity measurements of rock samples in the laboratory (see Section 3.4), may be interpreted as representing mainly

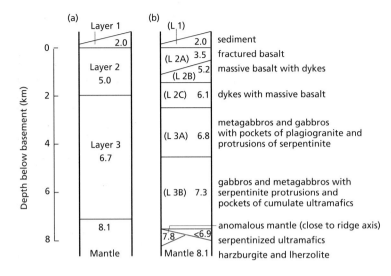

Fig. 5.28 Velocity (km s^{-1}) structure of typical oceanic lithosphere in terms of layered structures proposed in 1965 (a) and 1978 (b), and its geological interpretation. (From Kearey & Vine 1990.)

granitic or granodioritic material. Lower crustal velocities are normally in the range 6.5–7.0 km s^{-1} and may represent any of a variety of igneous and metamorphic rock types, including gabbro, gabbroic anorthosite and basic granulite. The latter rock type is regarded as the most probable major constituent of the lower crust on the basis of experimental studies of seismic velocities (Christensen & Fountain 1975).

5.11.4 Two-ship seismic surveying: combined refraction and reflection surveying

Marine surveys, usually single-ship experiments, have shown the ocean basins to have a crust only 6–8 km thick, composed of three main layers with differing seismic velocities. This thickness and layering is maintained over vast areas beneath all the major oceans. The results of deep-sea drilling, together with the recognition of ophiolite complexes exposed on land as analogues of oceanic lithosphere, have enabled the nature of the individual seismic layers to be identified (Fig. 5.28).

Specialized methods of marine surveying involving the use of two survey vessels and multichannel recording include *expanding spread profiles* and *constant offset profiles* (Stoffa & Buhl 1979). These methods have been developed for the detailed study of the deep structure of the crust and upper mantle under continental margins and oceanic areas.

Expanding spread profiling (ESP) is designed to obtain detailed information relating to a localized region of the crust. The shot-firing vessel and recording vessel travel outwards at the same speed from a central position, obtaining reflected and refracted arrivals from subsurface interfaces out to large offsets. Thus, in addition to near-normal incidence reflections such as would be recorded in a conventional common mid-point (CMP) reflection survey, wide-angle reflections and refracted arrivals are also recorded from the same section of crust. The combined reflection/refraction data allow derivation of a highly-detailed velocity–depth structure for the localized region.

Expanding spread profiles have also been carried out on land to investigate the crustal structure of continental areas (see e.g. Wright *et al.* 1990).

In constant offset profiling (COP), the shot-firing and recording vessels travel along a profile line at a fixed, wide separation. Thus, wide-angle reflections and refractions are continuously recorded along the line. This survey technique facilitates the mapping of lateral changes in crustal structure over wide areas and allows continuous mapping of the types of refracting interface that do not give rise to good near-normal incidence reflections and which therefore cannot be mapped adequately using conventional reflection profiling. Such interfaces include zones of steep velocity gradient, in contrast to the first-order velocity discontinuities that constitute the best reflectors.

Problems

1. A single-ended refraction profile designed to determine the depth to an underlying horizontal refractor reveals a top layer velocity of $3.0\,km\,s^{-1}$ and a refractor velocity of $5.0\,km\,s^{-1}$. The crossover distance is found to be 500 m. What is the refractor depth?

2. What is the delay time for head wave arrivals from layer 3 in the following case?

Layer	Depth (m)	Vel. (km s⁻¹)
1	100	1.5
2	50	2.5
3	–	4.0

3. In order that both the horizontal-layer models given below should produce the same time–distance curves for head wave arrivals, what must be the thickness of the middle layer in Model 2?

	Vel. (km s⁻¹)	Depth (km)
Model 1		
Layer 1	3.0	1.0
Layer 2	5.0	–
Model 2		
Layer 1	3.0	0.5
Layer 2	1.5	?
Layer 3	5.0	–

4. A single-ended refraction survey (Section 5.3) established to locate an underlying planar dipping refractor yields a top layer velocity of $2.2\,km\,s^{-1}$ and a downdip apparent refractor velocity of $4.0\,km\,s^{-1}$. When the shot point and geophones are moved forward by 150 m, in the direction of refractor dip, head wave arrival times to any offset distance are increased by 5 ms. Calculate the dip and true velocity of the refractor. If the intercept time of the refracted ray travel-time curve at the original shot point is 20 ms, what is the vertical depth to the refractor at that location?

5. A split-spread refraction profile (Section 5.3) with a central shot point is established to locate an underlying planar dipping refractor. The resultant time–distance curves yield a top layer veloc-

ity of $2.0\,km\,s^{-1}$ and updip and downdip apparent velocities of $4.5\,km\,s^{-1}$ and $3.5\,km\,s^{-1}$, respectively. The common intercept time is 85 ms. Calculate the true velocity and dip of the refractor and its vertical depth beneath the shot point.

6. The following dataset was obtained from a reversed seismic refraction line 275 m long. The survey was carried out in a level area of alluvial cover to determine depths to the underlying bedrock surface.

Offset (m)	Travel time (ms)
Forward direction:	
12.5	6.0
25	12.5
37.5	19.0
50	25.0
75	37.0
100	42.5
125	48.5
150	53.0
175	57.0
200	61.5
225	66.0
250	71.0
275	76.5
Reverse direction:	
12.5	6.0
25	12.5
37.5	17.0
50	19.5
75	25.0
100	30.5
125	37.5
150	45.5
175	52.0
200	59.0
225	65.5
250	71.0
275	76.5

Carry out a plus–minus interpretation of the data and comment briefly on the resultant bedrock profile.

7. What subsurface structure is responsible for the travel-time curves shown in Fig. 5.29?

Continued

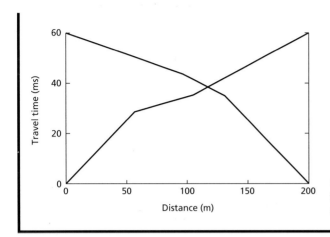

Fig. 5.29 Time–distance curves obtained in the forward and reverse directions along a refraction profile across an unknown subsurface structure.

Further reading

Cardarelli, E. & de Nardis, R. (2001) Seismic refraction, isotropic and anisotropic seismic tomography on an ancient monument (Antonino and Faustina temple AD141). *Geophysical Prospecting*, **49**, 228–41.

Cerveny, V. & Ravindra, R. (1971) *Theory of Seismic Head Waves*. University of Toronto Press.

Dobrin, M.B. & Savit, C.H. (1988) *Introduction to Geophysical Prospecting* (4th edn). McGraw-Hill, New York.

Giese, P., Prodehl, C. & Stein, A. (eds) (1976) *Explosion Seismology in Central Europe*. Springer-verlag, Berlin.

Ivansson, S. (1986) Seismic borehole tomography—theory and computational methods: *Proc. IEEE*, **74**, 328–38.

Palmer, D. (1980) *The Generalised Reciprocal Method of Seismic Refraction Interpretation*. Society of Exploration Geophysicists, Tulsa.

Palmer, D. (1986) *Handbook of Geophysical Exploration: Section 1, Seismic Exploration. Vol. 13: Refraction Seismics*. Enpro Science Publications, Amsterdam.

Sjagren, B. (1984) *Shallow Refraction Seismics*. Chapman & Hall, London.

Stoffa, P.L. & Buhl, P. (1979) Two-ship multichannel seismic experiments for deep crustal studies: expanded spread and constant offset profiles. *J. Geophys. Res.*, **84**, 7645–60.

Willmore, P.L. & Bancroft, A.M. (1960) The time-term approach to refraction seismology. *Geophys. J. R. Astr. Soc.*, **3**, 419–32.

6 Gravity surveying

– Salt domes = negative anomalies (small)
– granite or sediment basins = large scale anomalies
– used to determine shape of bodies

6.1 Introduction

In gravity surveying, subsurface geology is investigated on the basis of variations in the Earth's gravitational field arising from differences of density between subsurface rocks. An underlying concept is the idea of a *causative body*, which is a rock unit of different density from its surroundings. A causative body represents a subsurface zone of anomalous mass and causes a localized perturbation in the gravitational field known as a gravity anomaly. A very wide range of geological situations give rise to zones of anomalous mass that produce significant gravity anomalies. On a small scale, buried relief on a bedrock surface, such as a buried valley, can give rise to measurable anomalies. On a larger scale, small negative anomalies are associated with salt domes, as discussed in Chapter 1. On a larger scale still, major gravity anomalies are generated by granite plutons or sedimentary basins. Interpretation of gravity anomalies allows an assessment to be made of the probable depth and shape of the causative body.

The ability to carry out gravity surveys in marine areas or, to a lesser extent, from the air extends the scope of the method so that the technique may be employed in most areas of the world.

6.2 Basic theory

The basis of the gravity survey method is Newton's Law of Gravitation, which states that the force of attraction F between two masses m_1 and m_2, whose dimensions are small with respect to the distance r between them, is given by

$$F = \frac{Gm_1m_2}{r^2} \qquad (6.1)$$

where G is the Gravitational Constant (6.67×10^{-11} m^3 kg^{-1} s^{-2}).

Consider the gravitational attraction of a spherical, non-rotating, homogeneous Earth of mass M and radius R on a small mass m on its surface. It is relatively simple to show that the mass of a sphere acts as though it were concentrated at the centre of the sphere and by substitution in equation (6.1)

$$F = \frac{GM}{R^2} m = mg \qquad (6.2)$$

Force is related to mass by an acceleration and the term $g = GM/R^2$ is known as the gravitational acceleration or, simply, *gravity*. The weight of the mass is given by mg.

On such an Earth, gravity would be constant. However, the Earth's ellipsoidal shape, rotation, irregular surface relief and internal mass distribution cause gravity to vary over its surface.

The gravitational field is most usefully defined in terms of the *gravitational potential* U:

$$U = \frac{GM}{r} \qquad (6.3)$$

Whereas the gravitational acceleration g is a vector quantity, having both magnitude and direction (vertically downwards), the gravitational potential U is a scalar, having magnitude only. The first derivative of U in any direction gives the component of gravity in that direction. Consequently, a potential field approach provides computational flexibility. Equipotential surfaces can be defined on which U is constant. The sea-level surface, or *geoid*, is the most easily recognized equipotential surface, which is everywhere horizontal, that is, at right angles to the direction of gravity.

geoid is ⊥ to the direction of gravity everywhere

$1 \text{mgal} = 10^{-3} \text{cm/}_{s^2} = 10 \text{gu}$

6.3 Units of gravity

The mean value of gravity at the Earth's surface is about $9.8\,\text{m}\,\text{s}^{-2}$. Variations in gravity caused by density variations in the subsurface are of the order of $100\,\mu\text{m}\,\text{s}^{-2}$. This unit of the micrometre per second per second is referred to as the *gravity unit* (gu). In gravity surveys on land an accuracy of ±0.1 gu is readily attainable, corresponding to about one hundred millionth of the normal gravitational field. At sea the accuracy obtainable is considerably less, about ±10 gu. The c.g.s. unit of gravity is the *milligal* (1 mgal $= 10^{-3}$ gal $= 10^{-3}\,\text{cm}\,\text{s}^{-2}$), equivalent to 10 gu.

6.4 Measurement of gravity

Since gravity is an acceleration, its measurement should simply involve determinations of length and time. However, such apparently simple measurements are not easily achievable at the precision and accuracy required in gravity surveying.

The measurement of an absolute value of gravity is difficult and requires complex apparatus and a lengthy period of observation. Such measurement is classically made using large pendulums or falling body techniques (see e.g. Nettleton 1976, Whitcomb 1987), which can be made with a precision of 0.01 gu. Instruments for measuring absolute gravity in the field were originally bulky, expensive and slow to read (see e.g. Sakuma 1986). A new generation of absolute reading instruments (Brown *et al.* 1999) is now under development which does not suffer from these drawbacks and may well be in more general use in years to come.

The measurement of relative values of gravity, that is, the differences of gravity between locations, is simpler and is the standard procedure in gravity surveying. Absolute gravity values at survey stations may be obtained by reference to the International Gravity Standardization Network (IGSN) of 1971 (Morelli *et al.* 1971), a network of stations at which the absolute values of gravity have been determined by reference to sites of absolute gravity measurements (see Section 6.7). By using a relative reading instrument to determine the difference in gravity between an IGSN station and a field location the absolute value of gravity at that location can be determined.

Previous generations of relative reading instruments were based on small pendulums or the oscillation of torsion fibres and, although portable, took considerable time to read. Modern instruments capable of rapid

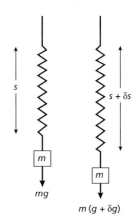

Fig. 6.1 Principle of stable gravimeter operation.

gravity measurements are known as *gravity meters* or *gravimeters*.

Gravimeters are basically spring balances carrying a constant mass. Variations in the weight of the mass caused by variations in gravity cause the length of the spring to vary and give a measure of the change in gravity. In Fig. 6.1 a spring of initial length s has been stretched by an amount δs as a result of an increase in gravity δg increasing the weight of the suspended mass m. The extension of the spring is proportional to the extending force (Hooke's Law), thus

$$m\delta g = k\delta s$$

and

$$\delta s = \frac{m}{k}\delta g \qquad (6.4)$$

where k is the elastic spring constant.

δs must be measured to a precision of $1:10^8$ in instruments suitable for gravity surveying on land. Although a large mass and a weak spring would increase the ratio m/k and, hence, the sensitivity of the instrument, in practice this would make the system liable to collapse. Consequently, some form of optical, mechanical or electronic amplification of the extension is required in practice.

The necessity for the spring to serve a dual function, namely to support the mass and to act as the measuring device, severely restricted the sensitivity of early gravimeters, known as stable or static gravimeters. This

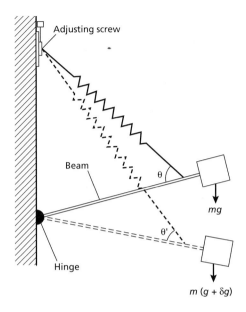

Fig. 6.2 Principle of the LaCoste and Romberg gravimeter.

i requires a certain temperature range

problem is overcome in modern meters (unstable or astatic) which employ an additional force that acts in the same sense as the extension (or contraction) of the spring and consequently amplifies the movement directly.

An example of an unstable instrument is the LaCoste and Romberg gravimeter. The meter consists of a hinged beam, carrying a mass, supported by a spring attached immediately above the hinge (Fig. 6.2). The magnitude of the moment exerted by the spring on the beam is dependent upon the extension of the spring and the sine of the angle θ. If gravity increases, the beam is depressed and the spring further extended. Although the restoring force of the spring is increased, the angle θ is decreased to θ'. By suitable design of the spring and beam geometry the magnitude of the increase of restoring moment with increasing gravity can be made as small as desired. With ordinary springs the working range of such an instrument would be very small. However, by making use of a 'zero-length' spring which is pretensioned during manufacture so that the restoring force is proportional to the physical length of the spring rather than its extension, instruments can be fashioned with a very sensitive response over a wide range. The instrument is read by restoring the beam to the horizontal by altering the vertical location of the spring attachment with a micrometer screw. Thermal effects are removed by a

battery-powered thermostatting system. The range of the instrument is 50 000 gu.

The other unstable instrument in common use is the Worden-type gravimeter. The necessary instability is provided by a similar mechanical arrangement, but in this case the beam is supported by two springs. The first of these springs acts as the measuring device, while the second alters the level of the 2000 gu reading range of the instrument. In certain specialized forms of this instrument the second spring is also calibrated, so that the overall reading range is similar to that of the LaCoste and Romberg gravimeter. Thermal effects are normally minimized by the use of quartz components and a bimetallic beam which compensates automatically for temperature changes. Consequently, no thermostatting is required and it is simply necessary to house the instrument in an evacuated flask. The restricted range of normal forms of the instrument, however, makes it unsuitable for intercontinental gravity ties or surveys in areas where gravity variation is extreme.

Gravimeters for general surveying use are capable of registering changes in gravity with an accuracy of 0.1 gu. A new generation of more efficient zero-length springs has been developed. Microprocessor-controlled instruments are now available which are, within limits, self-levelling, and which allow observations to be made rapidly. Also available for more specialized surveys (Section 6.12) are gravimeters capable of detecting gravity changes as small as 1 microgal (10^{-8} m s^{-2}).

A shortcoming of gravimeters is the phenomenon of drift. This refers to a gradual change in reading with time, observable when the instrument is left at a fixed location. Drift results from the imperfect elasticity of the springs, which undergo anelastic creep with time. Drift can also result from temperature variations which, unless counteracted in some way, cause expansion or contraction of the measuring system and thus give rise to variations in measurements that are unrelated to changes in gravity. Drift is monitored by repeated meter readings at a fixed location throughout the day.

Gravity can be measured at discrete locations at sea using a remote-controlled land gravimeter, housed in a waterproof container, which is lowered over the side of the ship and, by remote operation, levelled and read on the sea bed. Measurements of comparable quality to readings on land can be obtained in this way, and the method has been used with success in relatively shallow waters. The disadvantage of the method is that the meter has to be lowered to the sea bed for each reading so that the rate of surveying is very slow. Moreover, in strong tidal currents, the survey ship needs to be anchored to keep it on station while the gravimeter is on the sea bed.

Gravity measurements can be made continuously at sea using a gravimeter modified for use on ships. Such instruments are known as shipborne, or shipboard, meters. The accuracy of measurements with a shipborne meter is considerably reduced compared to measurements on land because of the severe vertical and horizontal accelerations imposed on the shipborne meter by sea waves and the ship's motion. These external accelerations can cause variations in measured gravity of up to 10^6 gu and represent high-amplitude noise from which a signal of much smaller gravity variations must be extracted. The effects of horizontal accelerations produced by waves, yawing of the ship and changes in its speed and heading can be largely eliminated by mounting the meter on a gyrostabilized, horizontal platform, so that the meter only responds to vertical accelerations. Deviations of the platform from the horizontal produce *off-levelling errors* which are normally less than 10 gu. External vertical accelerations resulting from wave motions cannot be distinguished from gravity but their effect can be diminished by heavily damping the suspension system and by averaging the reading over an interval considerably longer than the maximum period of the wave motions (about 8 s). As the ship oscillates vertically above and below the plane of the mean sea surface, the wave accelerations are equally negative and positive and are effectively removed by averaging over a few minutes. The operation is essentially low-pass filtering in which accelerations with periods of less than 1–5 min are rejected.

With shipborne meters employing a beam-supported sensor, such as the LaCoste and Romberg instrument, a further complication arises due to the influence of horizontal accelerations. The beam of the meter oscillates under the influence of the varying vertical accelerations caused by the ship's motions. When the beam is tilted out of the horizontal it will be further displaced by the turning force associated with any horizontal acceleration. For certain phase relationships between the vertical and horizontal components of motion of the ship, the horizontal accelerations may cause beam displacements that do not average out with time. Consider an example where the position of a meter in space describes a circular motion under the influence of sea waves (Fig. 6.3). At time t_1, as shown in Fig. 6.3, the ship is moving down, displacing the beam upwards, and the horizontal component of motion is to the right, inducing an anticlock-

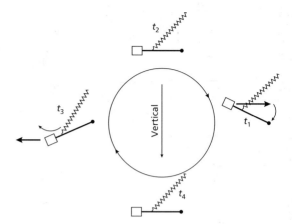

Fig. 6.3 Cross-coupling in a shipborne gravimeter.

wise torque that decreases the upward displacement of the beam. At a slightly later time t_3 the ship is moving up, displacing the beam down, and the horizontal motion is to the left, again inducing an anticlockwise torque which, now, increases the downward displacement of the beam. In such a case, the overall effect of the horizontal accelerations is to produce a systematic error in the beam position. This effect is known as *cross-coupling*, and its magnitude is dependent on the damping characteristics of the meter and the amplitude and phase relationships of the horizontal and vertical motions. It leads to an error known as the *cross-coupling error* in the measured gravity value. In general, the cross-coupling error is small or negligible in good weather conditions but can become very large in high seas. Cross-coupling errors are corrected directly from the outputs of two horizontal accelerometers mounted on the stabilized platform.

The inability to compensate fully for extraneous accelerations reduces the accuracy of these shipborne measurements to 10 gu at best, the actual amount depending on prevailing sea conditions. Instrumental drift monitoring is also less precise as base ties are, of necessity, usually many days apart.

Cross-coupling is one of the major sources of error in measurements of gravity at sea made with instruments utilizing a beam-supported mass, and arises because of the directional nature of the system. No cross-coupling would occur if the sensor were symmetric about a vertical axis, and since the late 1960s new marine meters utilizing this feature have been developed.

The *vibrating string accelerometer* (Bowin *et al.* 1972) is based on the principle that the resonant frequency of a short, vertical string from which a mass is suspended is

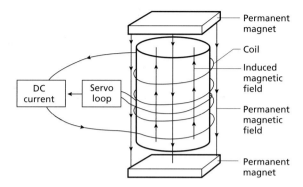

Permanent
magnet

Coil

Induced
magnetic
field

Permanent
magnetic
field

Permanent
magnet

DC
current

Servo
loop

Fig. 6.4 Principle of the accelerometer unit of the Bell marine gravimeter. (After Bell & Watts 1986.)

proportional to the square root of gravity. Changes in this frequency provide a measure of changes in gravity. Gravimeters based on this mechanism have never found much favour because of relatively low reported accuracies and erratic drift.

The most successful axially symmetric instrument to date is the *Bell gravimeter* (Bell & Watts 1986). The sensing element of the meter is the accelerometer shown in Fig. 6.4 which is mounted on a stable platform. The accelerometer, which is about 34 mm high and 23 mm in diameter, consists of a mass, wrapped in a coil, which is constrained to move only vertically between two permanent magnets. A DC current passed through the coil causes the mass to act as a magnet. In the null position, the weight of the mass is balanced by the forces exerted by the permanent magnets. When the mass moves vertically in response to a change in gravity or wave accelerations, the motion is detected by a servo loop which regulates the current in the coil, changing its magnetic moment so that it is driven back to the null position. The varying current is then a measure of changes in the vertical accelerations experienced by the sensor. As with beam-type meters, a weighted average filter is applied to the output in order to separate gravity changes from wave-generated accelerations.

Drift rates of the Bell gravimeter are low and uniform, and it has been demonstrated that the instrument is accurate to just a few gravity units, and is capable of discriminating anomalies with wavelengths of 1–2 km. This accuracy and resolution is considerably greater than that of earlier instruments, and it is anticipated that much smaller gravity anomalies will be detected than was

previously possible. The factor preventing more widespread deployment of the meter is its large cost.

The measurement of gravity from aircraft is complex because of the large possible errors in applying corrections. Eötvös corrections (Section 6.8.5) may be as great as 16 000 gu at a speed of 200 knots, a 1% error in velocity or heading producing maximum errors of 180 gu and 250 gu, respectively. Vertical accelerations associated with the aircraft's motion with periods longer than the instrumental averaging time cannot readily be corrected. In spite of these difficulties, tests undertaken in small aircraft (Halpenny & Darbha 1995) equipped with radar altimeters and GPS navigation have achieved results which differ from those obtained with underwater meters by an average of −2 gu and standard deviation 27 gu. Bell *et al.* (1999) describe a more modern set-up for airborne gravity surveying, which is now in use commercially. A system is also available for use with a helicopter (Seigel & McConnell 1998) in which the gravimeter is lowered to the ground by a cable, levelled and read remotely, so that measurements can be made where landing the aircraft is impossible.

The calibration constants of gravimeters may vary with time and should be checked periodically. The most common procedure is to take readings at two or more locations where absolute or relative values of gravity are known. In calibrating Worden-type meters, these readings would be taken for several settings of the coarse adjusting screw so that the calibration constant is checked over as much of the full range of the instrument as possible. Such a procedure cannot be adopted for the LaCoste and Romberg gravimeter, where each different dial range has its own calibration constant. In this case checking can be accomplished by taking readings at different inclinations of the gravimeter on a tilt table, a task usually entrusted to the instrument's manufacturer.

6.5 Gravity anomalies

Gravimeters effectively respond only to the vertical component of the gravitational attraction of an anomalous mass. Consider the gravitational effect of an anomalous mass δg, with horizontal and vertical components δg_x and δg_z, respectively, on the local gravity field g and its representation on a vector diagram (Fig. 6.5).

Solving the rectangle of forces gives

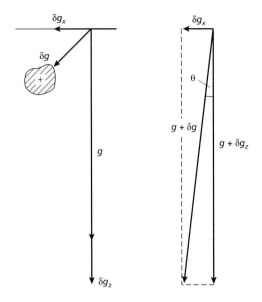

Fig. 6.5 Relationship between the gravitational field and the components of the gravity anomaly of a small mass.

$$g + \delta g = \sqrt{\left((g + \delta g_z)^2 + \delta g_x^{\ 2}\right)}$$
$$= \sqrt{\left(g^2 + 2g\delta g_z + \delta g_z^{\ 2} + \delta g_x^{\ 2}\right)}$$

Terms in δ^2 are insignificantly small and can thus be ignored. Binomial expansion of the equation then gives

$$g + \delta g \approx g + \delta g_z$$

so that

$$\delta g \approx \delta g_z$$

Consequently, measured perturbations in gravity effectively correspond to the vertical component of the attraction of the causative body. The local deflection of the vertical θ is given by

$$\theta = \tan^{-1}\left(\frac{\delta g_x}{g}\right) \tag{6.5}$$

and since $\delta g_z \ll g$, θ is usually insignificant. Very large mass anomalies such as mountain ranges can, however, produce measurable local vertical deflections.

6.6 Gravity anomalies of simple-shaped bodies

Consider the gravitational attraction of a point mass m at a distance r from the mass (Fig. 6.6). The gravitational attraction Δg_r in the direction of the mass is given by

$$\Delta g_r = \frac{Gm}{r^2} \text{ from Newton's Law.}$$

Since only the vertical component of the attraction Δg_z is measured, the gravity anomaly Δg caused by the mass is

$$\Delta g = \frac{Gm}{r^2} \cos \theta$$

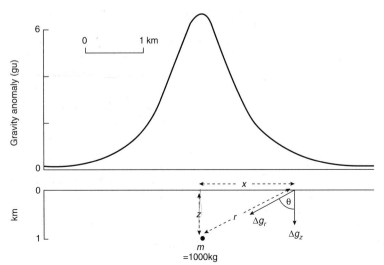

Fig. 6.6 The gravity anomaly of a point mass or sphere.

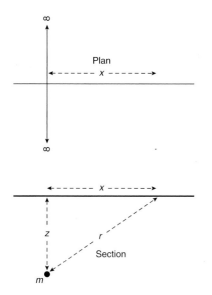

Fig. 6.7 Coordinates describing an infinite horizontal line mass.

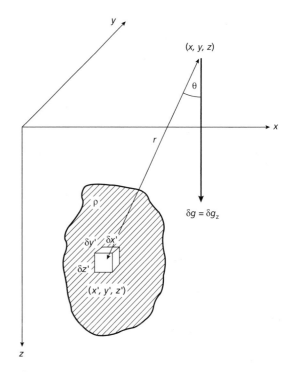

Fig. 6.8 The gravity anomaly of an element of a mass of irregular shape.

or

$$\Delta g = \frac{Gmz}{r^3} \tag{6.6}$$

Since a sphere acts as though its mass were concentrated at its centre, equation (6.6) also corresponds to the gravity anomaly of a sphere whose centre lies at a depth z.

Equation (6.6) can be used to build up the gravity anomaly of many simple geometric shapes by constructing them from a suite of small elements which correspond to point masses, and then summing (integrating) the attractions of these elements to derive the anomaly of the whole body.

Integration of equation (6.6) in a horizontal direction provides the equation for a line mass (Fig. 6.7) extending to infinity in this direction

$$\Delta g = \frac{2Gmz}{r^2} \tag{6.7}$$

Equation (6.7) also represents the anomaly of a horizontal cylinder, whose mass acts as though it is concentrated along its axis.

Integration in the second horizontal direction provides the gravity anomaly of an infinite horizontal sheet, and a further integration in the vertical direction between fixed limits provides the anomaly of an infinite horizontal slab

$$\Delta g = 2\pi G\rho t \tag{6.8}$$

where ρ is the density of the slab and t its thickness. Note that this attraction is independent of both the location of the observation point and the depth of the slab.

A similar series of integrations, this time between fixed limits, can be used to determine the anomaly of a right rectangular prism.

In general, the gravity anomaly of a body of *any* shape can be determined by summing the attractions of all the mass elements which make up the body. Consider a small prismatic element of such a body of density ρ, located at x', y', z', with sides of length $\delta x'$, $\delta y'$, $\delta z'$ (Fig. 6.8). The mass δm of this element is given by

$$\delta m = \rho\, \delta x'\, \delta y'\, \delta z'$$

Consequently, its attraction δg at a point outside the body (x, y, z), a distance r from the element, is derived from equation (6.6):

$$\delta g = G\rho \frac{(z' - z)}{r^3} \delta x' \delta y' \delta z'$$

The anomaly of the whole body Δg is then found by summing all such elements which make up the body

$$\Delta g = \Sigma\Sigma\Sigma G\rho \frac{(z'-z)}{r^3}\delta x'\delta y'\delta z' \tag{6.9}$$

If $\delta x'$, $\delta y'$ and $\delta z'$ are allowed to approach zero, then

$$\Delta g = \int\int\int G\rho \frac{(z'-z)}{r^3}dx'\,dy'\,dz' \tag{6.10}$$

where

$$r = \sqrt{(x'-x)^2 + (y'-y)^2 + (z'-z)^2}$$

As shown before, the attraction of bodies of regular geometry can be determined by integrating equation (6.10) analytically. The anomalies of irregularly shaped bodies are calculated by numerical integration using equations of the form of equation (6.9).

6.7 Gravity surveying

The station spacing used in a gravity survey may vary from a few metres in the case of detailed mineral or geotechnical surveys to several kilometres in regional reconnaissance surveys. The station density should be greatest where the gravity field is changing most rapidly, as accurate measurement of gravity gradients is critical to subsequent interpretation. If absolute gravity values are required in order to interface the results with other gravity surveys, at least one easily accessible base station must be available where the absolute value of gravity is known. If the location of the nearest IGSN station is inconvenient, a gravimeter can be used to establish a local base by measuring the difference in gravity between the IGSN station and the local base. Because of instrumental drift this cannot be accomplished directly and a procedure known as *looping* is adopted. A series of alternate readings at recorded times is made at the two stations and drift curves constructed for each (Fig. 6.9). The differences in ordinate measurements (Δg_{1-4}) for the two stations then may be averaged to give a measure of the drift-corrected gravity difference.

During a gravity survey the gravimeter is read at a base station at a frequency dependent on the drift characteristics of the instrument. At each survey station, location, time, elevation/water depth and gravimeter reading are recorded.

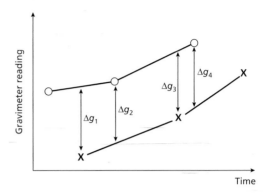

Fig. 6.9 The principle of looping. Crosses and circles represent alternate gravimeter readings taken at two base stations. The vertical separations between the drift curves for the two stations (Δg_{1-4}) provide an estimate of the gravity difference between them.

In order to obtain a reduced gravity value accurate to ± 1 gu, the reduction procedure described in the following section indicates that the gravimeter must be read to a precision of ± 0.1 gu, the latitude of the station must be known to ± 10 m and the elevation of the station must be known to ± 10 mm. The latitude of the station must consequently be determined from maps at a scale of $1:10\,000$ or smaller, or by the use of electronic position-fixing systems. Uncertainties in the elevations of gravity stations probably account for the greatest errors in reduced gravity values on land; at sea, water depths are easily determined with a precision depth recorder to an accuracy consistent with the gravity measurements. In well-surveyed land areas, the density of accurately-determined elevations at bench marks is normally sufficiently high that gravity stations can be sited at bench marks or connected to them by levelling surveys. Reconnaissance gravity surveys of less well-mapped areas require some form of independent elevation determination. Many such areas have been surveyed using aneroid altimeters. The accuracy of heights determined by such instruments is dependent upon the prevailing climatic conditions and is of the order of 1–5 m, leading to a relatively large uncertainty in the elevation corrections applied to the measured gravity values. The optimal equipment at present is the global positioning system (GPS) (Davis *et al.* 1989), whose constellation of 24 satellites is now complete and an unadulterated signal is broadcast. Signals from these can be monitored by a small, inexpensive receiver. Use of differential GPS, that is, the comparison between GPS signals between a base set at a known elevation and a mobile field set, can provide elevations to an accuracy of some 25 mm.

6.8 Gravity reduction

Before the results of a gravity survey can be interpreted it is necessary to correct for all variations in the Earth's gravitational field which do not result from the differences of density in the underlying rocks. This process is known as *gravity reduction* (LaFehr 1991) or *reduction to the geoid*, as sea-level is usually the most convenient datum level.

6.8.1 Drift correction

Correction for instrumental drift is based on repeated readings at a base station at recorded times throughout the day. The meter reading is plotted against time

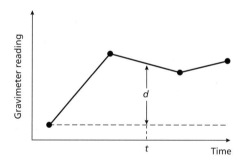

Fig. 6.10 A gravimeter drift curve constructed from repeated readings at a fixed location. The drift correction to be subtracted for a reading taken at time *t* is *d*.

(Fig. 6.10) and drift is assumed to be linear between consecutive base readings. The drift correction at time *t* is *d*, which is subtracted from the observed value.

After drift correction the difference in gravity between an observation point and the base is found by multiplication of the difference in meter reading by the calibration factor of the gravimeter. Knowing this difference in gravity, the absolute gravity at the observation point g_{obs} can be computed from the known value of gravity at the base. Alternatively, readings can be related to an arbitrary datum, but this practice is not desirable as the results from different surveys cannot then be tied together.

6.8.2 Latitude correction

Gravity varies with latitude because of the non-spherical shape of the Earth and because the angular velocity of a point on the Earth's surface decreases from a maximum at the equator to zero at the poles (Fig. 6.11(a)). The centripetal acceleration generated by this rotation has a negative radial component that consequently causes gravity to decrease from pole to equator. The true shape of the Earth is an oblate spheroid or polar flattened ellipsoid (Fig. 6.11(b)) whose difference in equatorial and polar radii is some 21 km. Consequently, points near the equator are farther from the centre of mass of the Earth than those near the poles, causing gravity to increase from the equator to the poles. The amplitude of this effect is reduced by the differing subsurface mass distributions resulting from the equatorial bulge, the mass underlying equatorial regions being greater than that underlying polar regions.

The net effect of these various factors is that gravity at

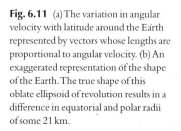

Fig. 6.11 (a) The variation in angular velocity with latitude around the Earth represented by vectors whose lengths are proportional to angular velocity. (b) An exaggerated representation of the shape of the Earth. The true shape of this oblate ellipsoid of revolution results in a difference in equatorial and polar radii of some 21 km.

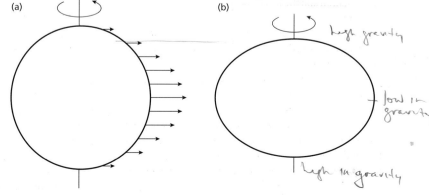

— FAC corrects for ↓ in g w/ height (↑ distance from centre of earth) → + on land

— Bouger approximates rock layer below → — on land

the poles exceeds gravity at the equator by some 51 860 gu, with the north–south gravity gradient at latitude ϕ being $8.12 \sin 2\phi$ gu km^{-1}.

Clairaut's formula relates gravity to latitude on the reference spheroid according to an equation of the form

$$g_\phi = g_0(1 + k_1 \sin^2 \phi - k_2 \sin^2 2\phi) \qquad (6.11)$$

where g_ϕ is the predicted value of gravity at latitude ϕ, g_0 is the value of gravity at the equator and k_1, k_2 are constants dependent on the shape and speed of rotation of the Earth. Equation (6.11) is, in fact, an approximation of an infinite series. The values of g_0, k_1 and k_2 in current use define the International Gravity Formula 1967 ($g_0 = 9\,780\,318$ gu, $k_1 = 0.0053024$, $k_2 = 0.0000059$; IAG 1971). Prior to 1967 less accurate constants were employed in the International Gravity Formula (1930). Results deduced using the earlier formula must be modified before incorporation into survey data reduced using the Gravity Formula 1967 by using the relationship $g_\phi(1967) - g_\phi(1930) = (136 \sin^2 \phi - 172)$ gu.

An alternative, more accurate, representation of the Gravity Formula 1967 (Mittermayer 1969), in which the constants are adjusted so as to minimize errors resulting from the truncation of the series, is

$$g_\phi = 9\,780\,318.5\,(1 + 0.005278895 \sin^2 \phi + 0.000023462 \sin^4 \phi)\text{ gu}$$

This form, however, is less suitable if the survey results are to incorporate pre-1967 data made compatible with the Gravity Formula 1967 using the above relationship.

The value g_ϕ gives the predicted value of gravity at sea-level at any point on the Earth's surface and is subtracted from the observed gravity to correct for latitude variation.

6.8.3 Elevation corrections

Correction for the differing elevations of gravity stations is made in three parts. The free-air correction (FAC) cor-

rects for the decrease in gravity with height in free air resulting from increased distance from the centre of the Earth, according to Newton's Law. To reduce to datum an observation taken at height h (Fig. 6.12(a)),

$$\text{FAC} = 3.086h \text{ gu } (h \text{ in metres})$$

The FAC is positive for an observation point above datum to correct for the decrease in gravity with elevation.

The free-air correction accounts solely for variation in the distance of the observation point from the centre of the Earth; no account is taken of the gravitational effect of the rock present between the observation point and datum. The *Bouguer correction* (BC) removes this effect by approximating the rock layer beneath the observation point to an infinite horizontal slab with a thickness equal to the elevation of the observation above datum (Fig. 6.12(b)). If ρ is the density of the rock, from equation (6.8)

$$\text{BC} = 2\pi G \rho h = 0.4191 \rho h \text{ gu}$$
$$(h \text{ in metres}, \rho \text{ in Mg m}^{-3})$$

On land the Bouguer correction must be subtracted, as the gravitational attraction of the rock between observation point and datum must be removed from the observed gravity value. The Bouguer correction of sea surface observations is positive to account for the lack of rock between surface and sea bed. The correction is equivalent to the replacement of the water layer by material of a specified rock density ρ_r. In this case

$$\text{BC} = 2\pi G(\rho_r - \rho_w)z$$

where z is the water depth and ρ_w the density of water.

The free-air and Bouguer corrections are often applied together as the *combined elevation correction*.

The Bouguer correction makes the assumption that the topography around the gravity station is flat. This is

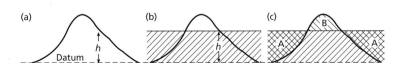

Fig. 6.12 (a) The free-air correction for an observation at a height h above datum. (b) The Bouguer correction. The shaded region corresponds to a slab of rock of thickness h extending to infinity in both horizontal directions. (c) The terrain correction.

[handwritten annotation at top: — positive terrain correction corrects for Bouguer "A" level]

rarely the case and a further correction, the *terrain correction* (TC), must be made to account for topographic relief in the vicinity of the gravity station. This correction is always positive as may be appreciated from consideration of Fig. 6.12(c). The regions designated A form part of the Bouguer correction slab although they do not consist of rock. Consequently, the Bouguer correction has overcorrected for these areas and their effect must be restored by a positive terrain correction. Region B consists of rock material that has been excluded from the Bouguer correction. It exerts an upward attraction at the observation point causing gravity to decrease. Its attraction must thus be corrected by a positive terrain correction.

Classically, terrain corrections are applied using a circular graticule known, after its inventor, as a Hammer chart (Fig. 6.13), divided by radial and concentric lines into a large number of compartments. The outermost zone extends to almost 22 km, beyond which topographic effects are usually negligible. The graticule is laid on a topographic map with its centre on the gravity station and the average topographic elevation of each compartment is determined. The elevation of the gravity station is subtracted from these values, and the gravitational effect of each compartment is determined by reference to tables constructed using the formula for the gravitational effect of a sector of a vertical cylinder at its axis. The terrain correction is then computed by summing the gravitational contribution of all compartments. Table 6.1 shows the method of computation. Such operations are time consuming as the topography of over 130 compartments has to be averaged for each station, but terrain correction is the one operation in gravity reduction that cannot be fully automated. Labour can be reduced by averaging topography within a rectangular grid. Only a single digitization is required as the topographic effects may be calculated at any point within the grid by summing the effects of the right rectangular prisms defined by the grid squares and their elevation difference with the gravity station. This operation can effectively correct for the topography of areas distant

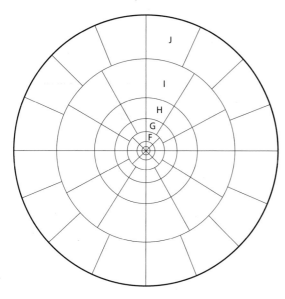

Fig. 6.13 A typical graticule used in the calculation of terrain corrections. A series of such graticules with zones varying in radius from 2 m to 21.9 km is used with topographic maps of varying scale.

Table 6.1 Terrain corrections.

Zone	r_1	r_2	n	Zone	r_1	r_2	n
B	2.0	16.6	4	H	1529.4	2614.4	12
C	16.6	53.3	6	I	2614.4	4468.8	12
D	53.3	170.1	6	J	4468.8	6652.2	16
E	170.1	390.1	8	K	6652.2	9902.5	16
F	390.1	894.8	8	L	9902.5	14740.9	16
G	894.8	1529.4	12	M	14740.9	21943.3	16

$$T = 0.4191\frac{\rho}{n}\left(r_2 - r_1 + \sqrt{r_1^2 + z^2} - \sqrt{r_2^2 + z^2}\right)$$

where T = terrain correction of compartment (gu); ρ = Bouguer correction density (Mg m^{-3}); n = number of compartments in zone; r_1 = inner radius of zone (m); r_2 = outer radius of zone (m); and z = modulus of elevation difference between observation point and mean elevation of compartment (m).

from the gravity station and can be readily computerized. Such an approach is likely to be increasingly adopted as digital elevation models for large regions become available (Cogbill 1990). Correction for inner zones, however, must still be performed manually as any reasonable digitization scheme for a complete survey area and its environs must employ a sampling interval that is too large to provide an accurate representation of the terrain close to the station.

Terrain effects are low in areas of subdued topography, rarely exceeding 10 gu in flat-lying areas. In areas of rugged topography terrain effects are considerably greater, being at a maximum in steep-sided valleys, at the base or top of cliffs and at the summits of mountains.

Where terrain effects are considerably less than the desired accuracy of a survey, the terrain correction may be ignored. Sprenke (1989) provides a means of assessing the distance to which terrain corrections are necessary. However, the usual necessity for this correction accounts for the bulk of time spent on gravity reduction and is thus a major contributor to the cost of a gravity survey.

6.8.4 Tidal correction

Gravity measured at a fixed location varies with time because of periodic variation in the gravitational effects of the Sun and Moon associated with their orbital motions, and correction must be made for this variation in a high-precision survey. In spite of its much smaller mass, the gravitational attraction of the Moon is larger than that of the Sun because of its proximity. Also, these gravitational effects cause the shape of the solid Earth to vary in much the same way that the celestial attractions cause tides in the sea. These *solid Earth tides* are considerably smaller than oceanic tides and lag farther behind the lunar motion. They cause the elevation of an observation point to be altered by a few centimetres and thus vary its distance from the centre of mass of the Earth. The periodic gravity variations caused by the combined effects of Sun and Moon are known as *tidal variations*. They have a maximum amplitude of some 3 gu and a minimum period of about 12 h.

If a gravimeter with a relatively high drift rate is used, base ties are made at an interval much smaller than the minimum Earth tide period and the tidal variations are automatically removed during the drift correction. If a meter with a low drift rate is employed, base ties are normally made only at the start and end of the day so that the tidal variation has undergone a full cycle. In such a case, a separate tidal correction may need to be made. The tidal

effects are predictable and can be computed by a small computer program.

6.8.5 Eötvös correction

The Eötvös correction (EC) is applied to gravity measurements taken on a moving vehicle such as a ship or an aircraft. Depending on the direction of travel, vehicular motion will generate a centripetal acceleration which either reinforces or opposes gravity. The correction required is

$$EC = 75.03V \sin \alpha \cos \phi + 0.04154V^2 \text{ gu}$$

where V is the speed of the vehicle in knots, α the heading and ϕ the latitude of the observation. In mid-latitudes the Eötvös correction is about +75 gu for each knot of E to W motion so that speed and heading must be accurately known.

6.8.6 Free-air and Bouguer anomalies

The *free-air anomaly* (FAA) and *Bouguer anomaly* (BA) may now be defined

$$FAA = g_{obs} - g_\phi + FAC \ (\pm EC) \tag{6.12}$$

$$BA = g_{obs} - g_\phi + FAC \pm BC + TC \ (\pm EC) \tag{6.13}$$

The Bouguer anomaly forms the basis for the interpretation of gravity data on land. In marine surveys Bouguer anomalies are conventionally computed for inshore and shallow water areas as the Bouguer correction removes the local gravitational effects associated with local changes in water depth. Moreover, the computation of the Bouguer anomaly in such areas allows direct comparison of gravity anomalies offshore and onshore and permits the combination of land and marine data into gravity contour maps. These may be used, for example, in tracing geological features across coastlines. The Bouguer anomaly is not appropriate for deeper water surveys, however, as in such areas the application of a Bouguer correction is an artificial device that leads to very large positive Bouguer anomaly values without significantly enhancing local gravity features of geological origin. Consequently, the free-air anomaly is frequently used for interpretation in such areas. Moreover, the FAA provides a broad assessment of the degree of isostatic compensation of an area (e.g. Bott 1982).

Gravity anomalies are conventionally displayed on

profiles or as contoured (isogal) maps. Interpretation of the latter may be facilitated by utilizing digital image processing techniques similar to those used in the display of remotely sensed data. In particular, colour and shaded relief images may reveal structural features that may not be readily discernible on unprocessed maps (Plate 5.1a). This type of processing is equally appropriate to magnetic anomalies (Plate 5.1b; see for example Lee *et al.* 1990).

6.9 Rock densities

Gravity anomalies result from the difference in density, or *density contrast*, between a body of rock and its surroundings. For a body of density ρ_1 embedded in material of density ρ_2, the density contrast $\Delta\rho$ is given by

$$\Delta\rho = \rho_1 - \rho_2$$

The sign of the density contrast determines the sign of the gravity anomaly.

Rock densities are among the least variable of all geophysical parameters. Most common rock types have densities in the range between 1.60 and 3.20 $Mg\,m^{-3}$. The density of a rock is dependent on both its mineral composition and porosity.

Variation in porosity is the main cause of density variation in sedimentary rocks. Thus, in sedimentary rock sequences, density tends to increase with depth, due to compaction, and with age, due to progressive cementation.

Most igneous and metamorphic rocks have negligible porosity, and composition is the main cause of density variation. Density generally increases as acidity decreases; thus there is a progression of density increase from acid through basic to ultrabasic igneous rock types. Density ranges for common rock types and ores are presented in Table 6.2.

A knowledge of rock density is necessary both for application of the Bouguer and terrain corrections and for the interpretation of gravity anomalies.

Density is commonly determined by direct measurements on rock samples. A sample is weighed in air and in water. The difference in weights provides the volume of the sample and so the dry density can be obtained. If the rock is porous the saturated density may be calculated by following the above procedure after saturating the rock with water. The density value employed in interpretation then depends upon the location of the rock above or below the water table.

Table 6.2 Approximate density ranges ($Mg\,m^{-3}$) of some common rock types and ores.

Alluvium (wet)	1.96–2.00
Clay	1.63–2.60
Shale	2.06–2.66
Sandstone	
Cretaceous	2.05–2.35
Triassic	2.25–2.30
Carboniferous	2.35–2.55
Limestone	2.60–2.80
Chalk	1.94–2.23
Dolomite	2.28–2.90
Halite	2.10–2.40
Granite	2.52–2.75
Granodiorite	2.67–2.79
Anorthosite	2.61–2.75
Basalt	2.70–3.20
Gabbro	2.85–3.12
Gneiss	2.61–2.99
Quartzite	2.60–2.70
Amphibolite	2.79–3.14
Chromite	4.30–4.60
Pyrrhotite	4.50–4.80
Magnetite	4.90–5.20
Pyrite	4.90–5.20
Cassiterite	6.80–7.10
Galena	7.40–7.60

NB. The lower end of the density range quoted in many texts is often unreasonably extended by measurements made on samples affected by physical or chemical weathering.

It should be stressed that the density of any particular rock type can be quite variable. Consequently, it is usually necessary to measure several tens of samples of each particular rock type in order to obtain a reliable mean density and variance.

As well as these direct methods of density determination, there are several indirect (or *in situ*) methods. These usually provide a mean density of a particular rock unit which may be internally quite variable. *In situ* methods do, however, yield valuable information where sampling is hampered by lack of exposure or made impossible because the rocks concerned occur only at depth.

The measurement of gravity at different depths beneath the surface using a special borehole gravimeter (see Section 11.11) or, more commonly, a standard gravimeter in a mineshaft, provides a measure of the mean density of the material between the observation levels. In Fig. 6.14 gravity has been measured at the surface and at a point underground at a depth h immediately below. If g_1 and g_2 are the values of gravity obtained at the two levels, then, applying free-air and Bouguer corrections, one obtains

$$g_1 - g_2 = 3.086h - 4\pi G\rho h \qquad (6.14)$$

The Bouguer correction is double that employed on the surface as the slab of rock between the observation levels exerts both a downward attraction at the surface

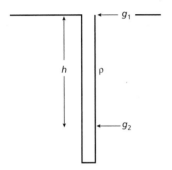

Fig. 6.14 Density determination by subsurface gravity measurements. The measured gravity difference $g_1 - g_2$ over a height difference h can be used to determine the mean density ρ of the rock separating the measurements.

location and an upward attraction at the underground location. The density ρ of the medium separating the two observations can then be found from the difference in gravity. Density may also be measured in boreholes using a density (gamma–gamma) logger as discussed in Section 11.7.2.

Nettleton's method of density determination involves taking gravity observations over a small isolated topographic prominence. Field data are reduced using a series of different densities for the Bouguer and terrain corrections (Fig. 6.15). The density value that yields a Bouguer anomaly with the least correlation (positive or negative) with the topography is taken to represent the density of the prominence. The method is useful in that no borehole or mineshaft is required, and a mean density of the material forming the prominence is provided. A disadvantage of the method is that isolated relief features may be formed of anomalous materials which are not representative of the area in general.

Density information is also provided from the P-wave velocities of rocks obtained in seismic surveys.

Figure 6.16 shows graphs of the logarithm of P-wave velocity against density for various rock types (Gardner *et al.* 1974), and the best-fitting linear relationship. Other

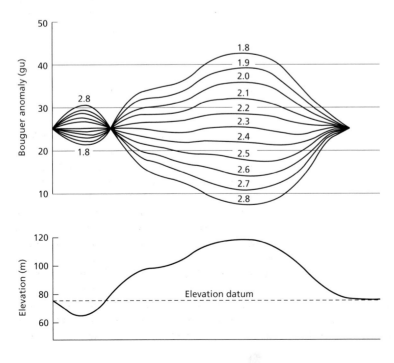

Fig. 6.15 Nettleton's method of density determination over an isolated topographic feature. Gravity reductions have been performed using densities ranging from 1.8 to 2.8 Mg m^{-3} for both Bouguer and terrain corrections. The profile corresponding to a value of 2.3 Mg m^{-3} shows least correlation with topography so this density is taken to represent the density of the feature. (After Dobrin & Savit 1988.)

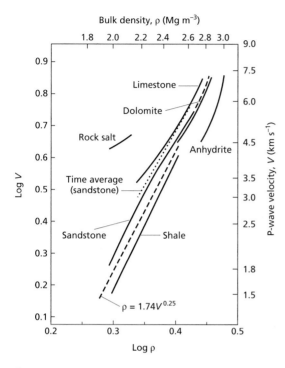

Fig. 6.16 Graphs of the logarithm of P-wave velocity against density for various rock types. Also shown is the best-fitting linear relationship between density and log velocity (after Gardner *et al.* 1974).

6.10 Interpretation of gravity anomalies

6.10.1 The inverse problem

The interpretation of potential field anomalies (gravity, magnetic and electrical) is inherently ambiguous. The ambiguity arises because any given anomaly could be caused by an infinite number of possible sources. For example, concentric spheres of constant mass but differing density and radius would all produce the same anomaly, since their mass acts as though located at the centre of the

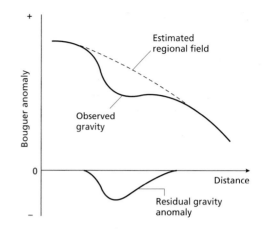

Fig. 6.17 The separation of regional and residual gravity anomalies from the observed Bouguer anomaly.

sphere. This ambiguity represents the *inverse problem* of potential field interpretation, which states that, although the anomaly of a given body may be calculated uniquely, there are an infinite number of bodies that could give rise to any specified anomaly. An important task in interpretation is to decrease this ambiguity by using all available external constraints on the nature and form of the anomalous body. Such constraints include geological information derived from surface outcrops, boreholes and mines, and from other, complementary, geophysical techniques (see e.g. Lines *et al.* 1988).

6.10.2 Regional fields and residual anomalies

Bouguer anomaly fields are often characterized by a broad, gently varying, regional anomaly on which may be superimposed shorter wavelength local anomalies (Fig. 6.17). Usually in gravity surveying it is the local anomalies that are of prime interest and the first step in interpretation is the removal of the *regional field* to isolate the *residual anomalies*. This may be performed graphically by sketching in a linear or curvilinear field by eye. Such a method is biased by the interpreter, but this is not necessarily disadvantageous as geological knowledge can be incorporated into the selection of the regional field. Several analytical methods of regional field analysis are available and include trend surface analysis (fitting a polynomial to the observed data, see Beltrão *et al.* (1991)) and low-pass filtering (Section 6.12). Such procedures must be used critically as fictitious residual anomalies can sometimes arise when the regional field is

workers (e.g. Birch 1960, 1961, Christensen & Fountain 1975) have derived similar relationships. The empirical velocity–density curve of Nafe and Drake (1963) indicates that densities estimated from seismic velocities are probably no more accurate than about $\pm 0.10\,\mathrm{Mg\,m^{-3}}$. This, however, is the only method available for the estimation of densities of deeply buried rock units that cannot be sampled directly.

subtracted from the observed data due to the mathematical procedures employed.

It is necessary before carrying out interpretation to differentiate between two-dimensional and three-dimensional anomalies. Two-dimensional anomalies are elongated in one horizontal direction so that the anomaly length in this direction is at least twice the anomaly width. Such anomalies may be interpreted in terms of structures which theoretically extend to infinity in the elongate direction by using profiles at right angles to the strike. Three-dimensional anomalies may have any shape and are considerably more difficult to interpret quantitatively.

Gravity interpretation proceeds via the methods of direct and indirect interpretation.

6.10.3 Direct interpretation

Direct interpretation provides, directly from the gravity anomalies, information on the anomalous body which is largely independent of the true shape of the body. Various methods are discussed below.

Limiting depth

Limiting depth refers to the maximum depth at which the top of a body could lie and still produce an observed gravity anomaly. Gravity anomalies decay with the inverse square of the distance from their source so that anomalies caused by deep structures are of lower amplitude and greater extent than those caused by shallow sources. This wavenumber–amplitude relationship to depth may be quantified to compute the maximum depth (or limiting depth) at which the top of the anomalous body could be situated.

(a) Half-width method. The half-width of an anomaly $(x_{1/2})$ is the horizontal distance from the anomaly maximum to the point at which the anomaly has reduced to half of its maximum value (Fig. 6.18(a)).

If the anomaly is three-dimensional, the initial assumption is made that it results from a point mass. Manipulation of the point mass formula (equation (6.6)) allows its depth to be determined in terms of the half-width

$$z = \frac{x_{1/2}}{\sqrt{\sqrt[3]{4} - 1}}$$

Here, z represents the actual depth of the point mass or the centre of a sphere with the same mass. It is an overestimate of the depth to the top of the sphere, that is, the limiting depth. Consequently, the limiting depth for any three-dimensional body is given by

$$z < \frac{x_{1/2}}{\sqrt{\sqrt[3]{4} - 1}} \tag{6.15}$$

A similar approach is adopted for a two-dimensional anomaly, with the initial assumption that the anomaly

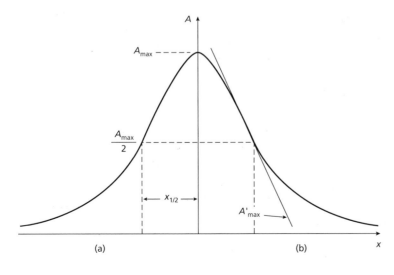

Fig. 6.18 Limiting depth calculations using (a) the half-width method and (b) the gradient–amplitude ratio.

results from a horizontal line mass (equation (6.7)). The depth to a line mass or to the centre of a horizontal cylinder with the same mass distribution is given by

$$z = x_{1/2}$$

For any two-dimensional body, the limiting depth is then given by

$$z < x_{1/2} \tag{6.16}$$

(b) *Gradient–amplitude ratio method.* This method requires the computation of the maximum anomaly amplitude (A_{max}) and the maximum horizontal gravity gradient (A'_{max}) (Fig. 6.18(b)). Again the initial assumption is made that a three-dimensional anomaly is caused by a point mass and a two-dimensional anomaly by a line mass. By differentiation of the relevant formulae, for any three-dimensional body

$$z < 0.86 \left| \frac{A_{max}}{A'_{max}} \right| \tag{6.17}$$

and for any two-dimensional body

$$z < 0.65 \left| \frac{A_{max}}{A'_{max}} \right| \tag{6.18}$$

(c) *Second derivative methods.* There are a number of limiting depth methods based on the computation of the maximum second horizontal derivative, or maximum rate of change of gradient, of a gravity anomaly (Smith 1959). Such methods provide rather more accurate limiting depth estimates than either the half-width or gradient–amplitude ratio methods if the observed anomaly is free from noise.

Excess mass

The excess mass of a body can be uniquely determined from its gravity anomaly without making any assumptions about its shape, depth or density. Excess mass refers to the difference in mass between the body and the mass of country rock that would otherwise fill the space occupied by the body. The basis of this calculation is a formula derived from Gauss' theorem, and it involves a surface integration of the residual anomaly over the area in which it occurs. The survey area is divided into n grid squares of area Δa and the mean residual anomaly Δg found for each square. The excess mass M_e is then given

by

$$M_e = \frac{1}{2\pi G} \sum_{i=1}^{n} \Delta g_i \Delta a_i \tag{6.19}$$

Before using this procedure it is important that the regional field is removed so that the anomaly tails to zero. The method only works well for isolated anomalies whose extremities are well defined. Gravity anomalies decay slowly with distance from source and so these tails can cover a wide area and be important contributors to the summation.

To compute the actual mass M of the body, the densities of both anomalous body (ρ_1) and country rock (ρ_2) must be known:

$$M = \frac{\rho_1 M_e}{(\rho_1 - \rho_2)} \tag{6.20}$$

The method is of use in estimating the tonnage of ore bodies. It has also been used, for example, in the estimation of the mass deficiency associated with the Chicxulub crater, Yucatan (CamposEnriquez *et al.* 1998), whose formation due to meteorite or asteroid impact has been associated with the extinction of the dinosaurs.

Inflection point

The locations of inflection points on gravity profiles, i.e. positions where the horizontal gravity gradient changes most rapidly, can provide useful information on the nature of the edge of an anomalous body. Over structures with outward dipping contacts, such as granite bodies (Fig. 6.19(a)), the inflection points (arrowed) lie near the base of the anomaly. Over structures with inward dipping contacts such as sedimentary basins (Fig. 6.19(b)), the inflection points lie near the uppermost edge of the anomaly.

Approximate thickness

If the density contrast $\Delta\rho$ of an anomalous body is known, its thickness t may be crudely estimated from its maximum gravity anomaly Δg by making use of the Bouguer slab formula (equation (6.8)):

$$t \approx \frac{\Delta g}{2\pi G \Delta \rho} \tag{6.21}$$

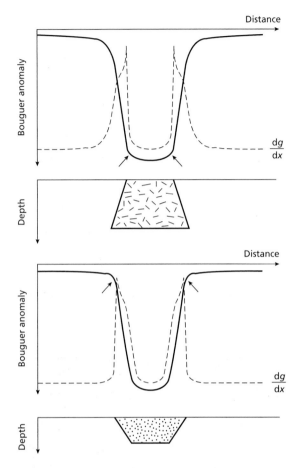

Fig. 6.19 Bouguer anomaly profiles across (a) a granite body, and (b) a sedimentary basin. The inflection points are marked with an arrow. The broken lines represent the horizontal derivative (rate of change of gradient) of the gravity anomaly, which is at a maximum at the inflection points.

This thickness will always be an underestimate for a body of restricted horizontal extent. The method is commonly used in estimating the throw of a fault from the difference in the gravity fields of the upthrown and downthrown sides.

The technique of source depth determination by Euler deconvolution, described in Section 7.10.2, is also applicable to gravity anomalies (Keating 1998).

6.10.4 Indirect interpretation

In indirect interpretation, the causative body of a gravity anomaly is simulated by a model whose theoretical anomaly can be computed, and the shape of the model is altered until the computed anomaly closely matches the observed anomaly. Because of the inverse problem this model will not be a unique interpretation, but ambiguity can be decreased by using other constraints on the nature and form of the anomalous body.

A simple approach to indirect interpretation is the comparison of the observed anomaly with the anomaly computed for certain standard geometrical shapes whose size, position, form and density contrast are altered to improve the fit. Two-dimensional anomalies may be compared with anomalies computed for horizontal cylinders or half-cylinders, and three-dimensional anomalies compared with those of spheres, vertical cylinders or right rectangular prisms. Combinations of such shapes may also be used to simulate an observed anomaly.

Figure 6.20(a) shows a large, circular gravity anomaly situated near Darnley Bay, NWT, Canada. The anomaly is radially symmetrical and a profile across the anomaly (Fig. 6.20(b)) can be simulated by a model constructed from a suite of coaxial cylinders whose diameters decrease with depth so that the anomalous body has the overall form of an inverted cone. This study illustrates well the non-uniqueness of gravity interpretation. The nature of the causative body is unknown and so no information is available on its density. An alternative interpretation, again in the form of an inverted cone, but with an increased density contrast, is presented in Fig. 6.20(b). Both models provide adequate simulations of the observed anomaly, and cannot be distinguished using the information available.

The computation of anomalies over a model of irregular form is accomplished by dividing the model into a series of regularly-shaped compartments and calculating the combined effect of these compartments at each observation point. At one time this operation was performed by the use of graticules, but nowadays the calculations are invariably performed by computers.

A two-dimensional gravity anomaly may be represented by a profile normal to the direction of elongation. This profile can be interpreted in terms of a model which maintains a constant cross-section to infinity in the horizontal directions perpendicular to the profile.

The basic unit for constructing the anomaly of a two-dimensional model is the semi-infinite slab with a sloping edge shown in Fig. 6.21, which extends to infinity into and out of the plane of the figure. The gravity anomaly of this slab Δg is given by

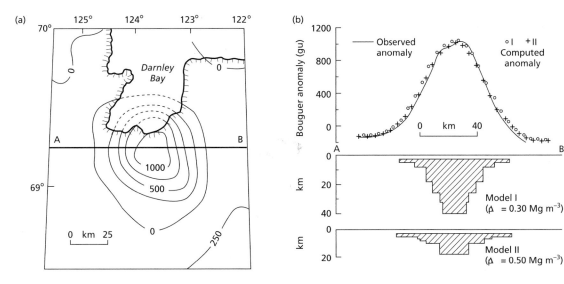

Fig. 6.20 (a) The circular gravity anomaly at Darnley Bay, NWT, Canada. Contour interval 250 gu. (b) Two possible interpretations of the anomaly in terms of a model constructed from a suite of coaxial vertical cylinders. (After Stacey 1971.)

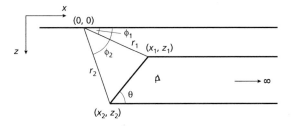

Fig. 6.21 Parameters used in defining the gravity anomaly of a semi-infinite slab with a sloping edge.

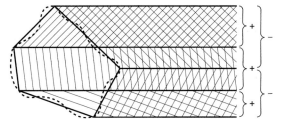

Fig. 6.22 The computation of gravity anomalies of two-dimensional bodies of irregular cross-section. The body (dashed line) is approximated by a polygon and the effects of semi-infinite slabs with sloping edges defined by the sides of the polygon are progressively added and subtracted until the anomaly of the polygon is obtained.

$$\Delta g = 2G\Delta\rho[-\{x_1 \sin\theta + z_1 \cos\theta\}$$
$$\times \{\sin\theta \log_e(r_2/r_1) + \cos\theta(\phi_2 - \phi_1)\}$$
$$+ z_2\phi_2 - z_1\phi_1] \qquad (6.22)$$

where $\Delta\rho$ is the density contrast of the slab, angles are expressed in radians and other parameters are defined as in Fig. 6.21 (Talwani *et al.* 1959). To calculate the anomaly of a two-dimensional body of irregular cross-section, the body is approximated by a polygon as shown in Fig. 6.22. The anomaly of the polygon is then found by proceeding around it summing the anomalies of the slabs bounded by edges where the depth increases and subtracting those where the depth decreases.

Figure 6.23 illustrates a two-dimensional interpretation, in terms of a model of irregular geometry represented by a polygonal outline, of the Bodmin Moor granite of southwest England. The shape of the uppermost part of the model is controlled by the surface outcrop of granite, while the density contrasts employed are based on density measurements on rock samples. The interpretation shows unambiguously that the contacts of the granite slope outwards. Ambiguity is evident, however, in the interpretation of the gravity gradient over the northern flank of the granite. The model presented in Fig. 6.23 interprets the cause of this gradient as a

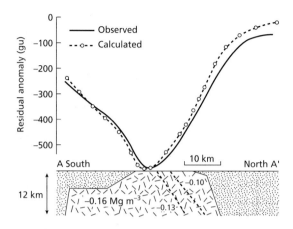

Fig. 6.23 A two-dimensional interpretation of the gravity anomaly of the Bodmin Moor granite, southwest England. See Fig. 6.27 for location. (After Bott & Scott 1964.)

northerly increase in the density of the granite; a possible alternative, however, would be a northerly thinning of a granite body of constant density contrast.

Two-dimensional methods can sometimes be extended to three-dimensional bodies by applying end-correction factors to account for the restricted extent of the causative body in the strike direction (Cady 1980). The end-correction factors are, however, only approximations and full three-dimensional modelling is preferable.

The gravity anomaly of a three-dimensional body may be calculated by dividing the body into a series of horizontal slices and approximating each slice by a polygon (Talwani & Ewing 1960). Alternatively the body may be constructed out of a suite of right rectangular prisms.

However a model calculation is performed, indirect interpretation involves four steps:
1. Construction of a reasonable model.
2. Computation of its gravity anomaly.
3. Comparison of computed with observed anomaly.
4. Alteration of model to improve correspondence of observed and calculated anomalies and return to step 2.

The process is thus iterative and the goodness of fit between observed and calculated anomalies is gradually improved. Step 4 can be performed manually for bodies of relatively simple geometry so that an interpretation is readily accomplished using interactive routines on a personal computer (Götze & Lahmeyer 1988). Bodies of complex geometry in two- or three-dimensions are not so simply dealt with and in such cases it is advantageous to employ techniques which perform the iteration automatically.

The most flexible of such methods is *non-linear optimization* (Al-Chalabi 1972). All variables (body points, density contrasts, regional field) may be allowed to vary within defined limits. The method then attempts to minimize some function F which defines the goodness of fit, for example

$$F = \sum_{i=1}^{n} \left(\Delta g_{obs_i} - \Delta g_{calc_i} \right)^2$$

where Δg_{obs} and Δg_{calc} are series of n observed and calculated values.

The minimization proceeds by altering the values of the variables within their stated limits to produce a successively smaller value for F for each iteration. The technique is elegant and successful but expensive in computer time.

Other such automatic techniques involve the simulation of the observed profile by a thin layer of variable density. This *equivalent layer* is then progressively expanded so that the whole body is of a uniform, specified density contrast. The body then has the form of a series of vertical prisms in either two or three dimensions which extend either above, below or symmetrically around the original equivalent layer. Such methods are less flexible than the non-linear optimization technique in that usually only a single density contrast may be specified and the model produced must either have a specified base or top or be symmetrical about a central horizontal plane.

6.11 Elementary potential theory and potential field manipulation

Gravitational and magnetic fields are both potential fields. In general the potential at any point is defined as the work necessary to move a unit mass or pole from an infinite distance to that point through the ambient field. Potential fields obey Laplace's equation which states that the sum of the rates of change of the field gradient in three orthogonal directions is zero. In a normal Cartesian coordinate system with horizontal axes x, y and a vertical axis z, Laplace's equation is stated

$$\frac{\partial^2 A}{\partial x^2} + \frac{\partial^2 A}{\partial y^2} + \frac{\partial^2 A}{\partial z^2} = 0 \tag{6.23}$$

where A refers to a gravitational or magnetic field and is a function of (x, y, z).

In the case of a two-dimensional field there is no variation along one of the horizontal directions so that A is a function of x and z only and equation (6.23) simplifies to

$$\frac{\partial^2 A}{\partial x^2} + \frac{\partial^2 A}{\partial z^2} = 0 \qquad (6.24)$$

Solution of this partial differential equation is easily performed by separation of variables

$$A_k(x, z) = (a\cos kx + b\sin kx)e^{kz} \qquad (6.25)$$

where a and b are constants, the positive variable k is the spatial frequency or wavenumber, A_k is the potential field amplitude corresponding to that wavenumber and z is the level of observation. Equation (6.25) shows that a potential field can be represented in terms of sine and cosine waves whose amplitude is controlled exponentially by the level of observation.

Consider the simplest possible case where the two-dimensional anomaly measured at the surface $A(x, 0)$ is a sine wave

$$A(x, 0) = A_0 \sin kx \qquad (6.26)$$

where A_0 is a constant and k the wavenumber of the sine wave. Equation (6.25) enables the general form of the equation to be stated for any value of z

$$A(x, z) = (A_0 \sin kx)e^{kz} \qquad (6.27)$$

The field at a height h above the surface can then be determined by substitution in equation (6.27)

$$A(x, -h) = (A_0 \sin kx)e^{-kh} \qquad (6.28)$$

and the field at depth d below the surface

$$A(x, d) = (A_0 \sin kx)e^{kd} \qquad (6.29)$$

The sign of h and d is important as the z-axis is normally defined as positive downwards.

Equation (6.27) is an over-simplification in that a potential field is never a function of a single sine wave. Invariably such a field is composed of a range of wavenumbers. However, the technique is still valid as long as the field can be expressed in terms of all its component wavenumbers, a task easily performed by use of

the Fourier transform (Section 2.3). If, then, instead of the terms $(a\cos kx + b\sin kx)$ in equation (6.25) or $(A_0 \sin kx)$ in equation (6.27), the full Fourier spectrum, derived by Fourier transformation of the field into the wavenumber domain, is substituted, the results of equations (6.28) and (6.29) remain valid.

These latter equations show that the field measured at the surface can be used to predict the field at any level above or below the plane of observation. This is the basis of the upward and downward field continuation methods in which the potential field above or below the original plane of measurement is calculated in order to accentuate the effects of deep or shallow structures respectively.

Upward continuation methods are employed in gravity interpretation to determine the form of regional gravity variation over a survey area, since the regional field is assumed to originate from relatively deep-seated structures. Figure 6.24(a) is a Bouguer anomaly map of the Saguenay area in Quebec, Canada, and Fig. 6.24(b) represents the field continued upward to an elevation of 16 km. Comparison of the two figures clearly illustrates how the high-wavenumber components of the observed field have been effectively removed by the continuation process. The upward continued field must result from relatively deep structures and consequently represents a valid regional field for the area. Upward continuation is also useful in the interpretation of magnetic anomaly fields (see Chapter 7) over areas containing many near-surface magnetic sources such as dykes and other intrusions. Upward continuation attenuates the high-wavenumber anomalies associated with such features and enhances, relatively, the anomalies of the deeper-seated sources.

Downward continuation of potential fields is of more restricted application. The technique may be used in the resolution of the separate anomalies caused by adjacent structures whose effects overlap at the level of observation. On downward continuation, high-wavenumber components are relatively enhanced and the anomalies show extreme fluctuations if the field is continued to a depth greater than that of its causative structure. The level at which these fluctuations commence provides an estimate of the limiting depth of the anomalous body. The effectiveness of this method is diminished if the potential field is contaminated with noise, as the noise is accentuated on downward continuation.

The selective enhancement of the low- or high-wavenumber components of potential fields may be achieved in a different but analogous manner by the

(a)

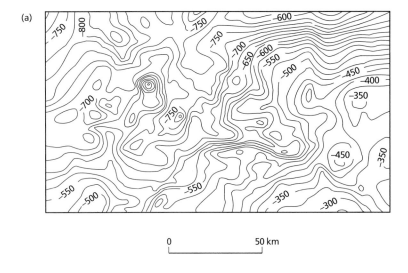

0 ⊢————————————⊣ 50 km

(b)

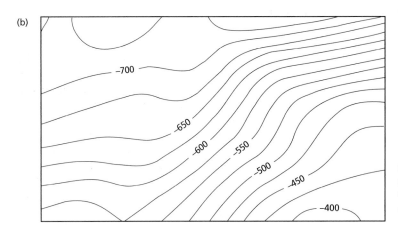

Fig. 6.24 (a) Observed Bouguer anomalies (gu) over the Saguenay area, Quebec, Canada. (b) The gravity field continued upward to an elevation of 16 km. (After Duncan & Garland 1977.)

application of *wavenumber filters*. Gravitational and magnetic fields may be processed and analysed in a similar fashion to seismic data, replacing frequency by wavenumber. Such processing is more complex than the equivalent seismic filtering as potential field data are generally arranged in two horizontal dimensions, that is, contour maps, rather than a single dimension. However, it is possible to devise two-dimensional filters for the selective removal of high- or low-wavenumber components from the observed anomalies. The consequence of the application of such techniques is similar to upward or downward continuation in that shallow structures are mainly responsible for the high-wavenumber compo-

nents of anomalies and deep structures for the low wavenumbers. However, it is not possible fully to isolate local or regional anomalies by wavenumber filtering because the wavenumber spectra of deep and shallow sources overlap.

Other manipulations of potential fields may be accomplished by the use of more complex filter operators (e.g. Gunn 1975, Cooper 1997). Vertical or horizontal derivatives of any order may be computed from the observed field. Such computations are not widely employed, but second horizontal derivative maps are occasionally used for interpretation as they accentuate anomalies associated with shallow bodies.

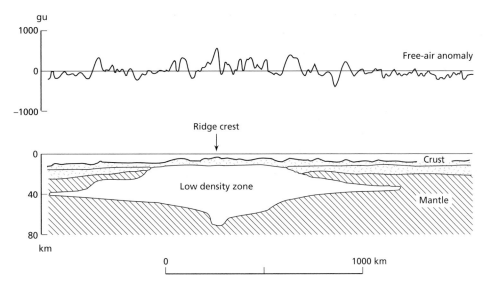

Fig. 6.25 Free-air anomaly profile across the mid-Atlantic ridge. (After Talwani *et al.* 1965.)

6.12 Applications of gravity surveying

Gravity studies are used extensively in the investigation of large- and medium-scale geological structures (Paterson & Reeves 1985). Early marine surveys, performed from submarines, indicated the existence of large positive and negative gravity anomalies associated with island arcs and oceanic trenches, respectively; subsequent shipborne work has demonstrated their lateral continuity and has shown that most of the major features of the Earth's surface can be delineated by gravity surveying. Gravity anomalies have also shown that most of these major relief features are in isostatic equilibrium, suggesting that the lithosphere is not capable of sustaining significant loads and yields isostatically to any change in surface loading. Figure 6.25 shows the near-zero free-air anomalies over an ocean ridge which suggest that it is in isostatic equilibrium. The gravity interpretation, which is constrained by seismic refraction results, indicates that this compensation takes the form of a zone of mass deficiency in the underlying mantle. Its low seismic velocity and the high heat flow at the surface suggest that this is a region of partial melting and, perhaps, hydration. Gravity surveying can also be used in the study of ancient suture zones, which are interpreted as the sites of former plate boundaries within the continental lithosphere. These zones are often characterized by major linear gravity anomalies resulting from the different crustal sections juxtaposed across the sutures (Fig. 6.26).

On the medium scale, gravity anomalies can reveal the subsurface form of igneous intrusions such as granite batholiths and anorthosite massifs. For example, gravity surveys in southwest England (Bott *et al.* 1958) have revealed a belt of large-amplitude, negative Bouguer anomalies overlying a region of outcropping granites (Fig. 6.27). Modelling of the gravity anomalies (Fig. 6.23) has led to the postulation of a continuous batholith some 10–15 km thick underlying southwest England (see e.g. Brooks *et al.* 1983). Studies such as these have provided important constraints on the mechanism of emplacement, composition and origin of igneous bodies. Similarly, gravity surveying has been extensively used in the location of sedimentary basins, and their interpreted structures have provided important information on mechanisms of basin formation.

The gravity method was once extensively used by the petroleum industry for the location of possible hydrocarbon traps, but the subsequent vast improvement in efficiency and technology of seismic surveying has led to the demise of gravity surveying as a primary exploration tool.

In commercial applications, gravity surveying is rarely used in reconnaissance exploration. This is because the method is relatively slow to execute, and therefore expensive, due to the necessity of accurately determined elevations and the length of the reduction procedure. Gravity methods do find application, however, as a follow-up technique used on a target defined by

_– used in O&G mainly as a follow up target &
identification of traps by ∅_

_– used in hydrogeological to det shape/geometry of
aquifer → gravity lows_

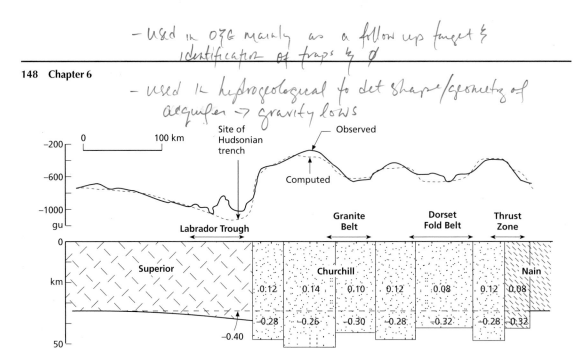

Fig. 6.26 Bouguer anomaly profile across a structural province boundary in the Canadian Shield. Density contrasts in Mg m^{-3}. (After Thomas & Kearey 1980.)

∗ Couldnt it be used in O&G to image oil/water contact

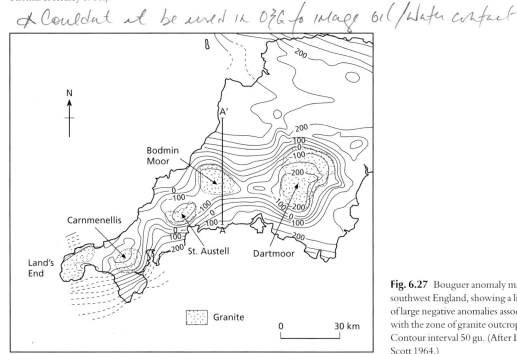

Fig. 6.27 Bouguer anomaly map of southwest England, showing a linear belt of large negative anomalies associated with the zone of granite outcrops. Contour interval 50 gu. (After Bott & Scott 1964.)

another, more cost-effective method. An important application of this type in mineral exploration is the determination of ore tonnage by the excess mass method described in Section 6.10.3.

Gravity surveying may be used in hydrogeological investigations to determine the geometry of potential aquifers. Figure 6.28 shows a Bouguer anomaly map of an area near Taltal, Chile (Van Overmeeren 1975). The region is extremely arid, with groundwater supply and storage controlled by deep geological features. The gravity minima revealed by the contours probably represent two buried valleys in the alluvium overlying the granodioritic bedrock. Figure 6.29 shows an interpretation of a profile over the minima. The bedrock topography was controlled by the results from a seismic refraction line which had been interpreted using the plus–minus

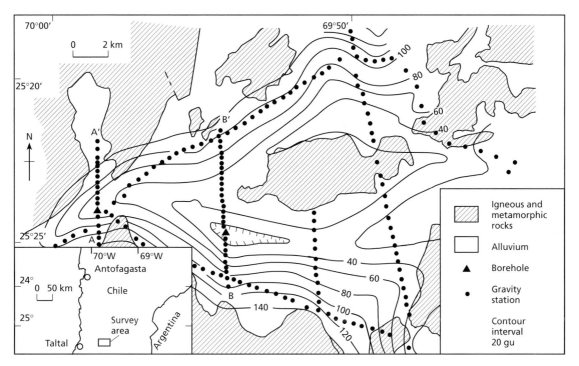

Fig. 6.28 Geological map of an area near Taltal, Chile, showing location of gravity stations and contoured Bouguer anomalies. (After Van Overmeeren 1975.)

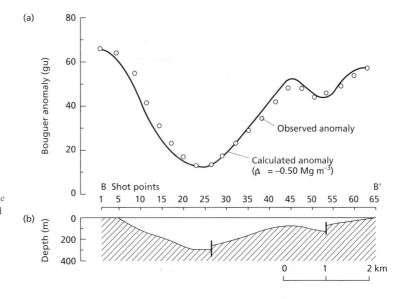

Fig. 6.29 Profile B–B′, Taltal area, Chile (see Fig. 6.28 for location). (a) Observed Bouguer anomaly and calculated anomaly for a model with a density contrast ($\Delta\rho$) of $-0.50\,\mathrm{Mg\,m^{-3}}$. (b) Gravity interpretation. (After Van Overmeeren 1975.)

Micrography uses
↳ Void detection
↳ archeological

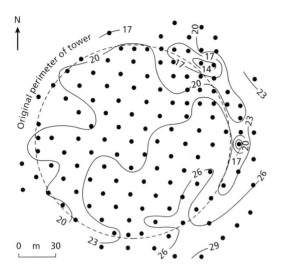

Fig. 6.30 Bouguer anomalies, uncorrected for topographic effects, over the cooling tower area. Contour interval 0.3 gu. (After Arzi 1975.)

method (see Section 5.4). The seismic control allowed a mean density of the highly variable valley-fill deposits to be determined. On the basis of the geophysical results, two boreholes (Fig. 6.28) were sunk in the deepest parts of the valley fill and located groundwater ponded in the bedrock depressions.

In engineering and geotechnical applications, gravity surveying is sometimes used in the location of subsurface voids. Void detection has been made possible by the development of microgravimetric techniques which can detect gravity changes as small as a microgal. Arzi (1975) described a microgravity survey of the proposed site of a cooling tower serving a nuclear power plant, where it was suspected that solution cavities might be present in the dolomitic bedrock. Measurements were made on a

15 m grid at points whose elevations had been determined to ±3 mm, with base readings at 40 min intervals. The soil thickness had been determined so that its effects could be computed and 'stripped' from the observations to remove gravity variations caused by undulating bedrock topography. The resulting Bouguer anomaly map is shown in Fig. 6.30. In the NE part of the site there are two minima near the proposed perimeter of the cooling tower, and subsequent drilling confirmed that they originated from buried cavities. Remedial work entailed the injection of grouting material into the cavities. A check on the effectiveness of the grouting was provided by a repeat gravity survey which, by an excess mass calculation (Section 6.10.3), showed that the change in the gravity field before and after grouting was caused by the replacement of voids by grouting material. Casten and Gram (1989) have described microgravity surveys performed underground to locate cavities which might pose a threat to the safety of mine workings.

Microgravity surveys also find application in archaeological investigations, where they may be used in the detection of buried buildings, tombs and other artefacts. The technique has also been used to study the temporal movement of groundwater through a region.

An important recent development in gravity surveying is the design of portable instruments capable of measuring absolute gravity with high precision. Although the cost of such instruments is high it is possible that they will be used in the future to investigate large-scale mass movements in the Earth's interior and small cyclic gravity variations associated with neotectonic phenomena such as earthquakes and postglacial uplift.

Gravitational studies, both of the type described in this chapter and satellite observations, are important in geodesy, the study of the shape of the Earth. Gravity surveying also has military significance, since the trajectory of a missile is affected by gravity variation along its flight path.

Problems

1. Compare and contrast the LaCoste–Romberg and Worden-type gravimeters. State also the advantages and disadvantages of the two types of instrument.

2. What are the magnitudes of the terrain correction at gravity stations (a) at the top, (b) at the base, and (c) half-way up a vertical cliff 100 m high?

3. The table shows data collected along a north–south gravity profile. Distances are measured from the south end of the profile, whose latitude is 51°12′24″N. The calibration constant of the Worden gravimeter used on the survey is 3.792 gu per dial unit. Before, during and after the survey, readings (marked BS) were taken at a base station where the value of gravity is 9 811 442.2 gu. This was done in order to monitor instrumental drift and to allow the absolute value of gravity to be determined at each observation point.

Station	Time	Dist. (m)	Elev. (m)	Reading
BS	0805			2934.2
1	0835	0	84.26	2946.3
2	0844	20	86.85	2941.0
3	0855	40	89.43	2935.7
4	0903	60	93.08	2930.4
1	0918			2946.5
BS	0940			2934.7
1	1009			2946.3
5	1024	80	100.37	2926.6
6	1033	100	100.91	2927.9
7	1044	120	103.22	2920.0
X	1053	140	107.35	2915.1
1	1111			2946.5
BS	1145			2935.2
1	1214			2946.2
9	1232	160	110.10	2911.5
10	1242	180	114.89	2907.2
11	1300	200	118.96	2904.0
1	1315			2946.3
BS	1350			2935.5

(a) Perform a gravity reduction of the survey data and comment on the accuracy of each step. Use a density of 2.70 Mg m^{-3} for the Bouguer correction.

(b) Draw a series of sections illustrating the variation in topography, observed gravity, free-air anomaly and Bouguer anomaly along the profile. Comment on the sections.

(c) What further information would be required before a full interpretation could be made of the Bouguer anomaly?

4. Two survey vessels with shipborne gravity meters are steaming at 6 knots in opposite directions along an east–west course. If the difference in gravity read by the two meters is 635 gu as the ships pass, what is the latitude?

5. The gravity anomaly Δg of an infinite horizontal slab of thickness t and density contrast $\Delta\rho$ is given by

$$\Delta g = 2\pi G \Delta\rho t$$

where the gravitational constant G is $6.67 \times 10^{-11}\,\text{m}^3\,\text{kg}^{-1}\,\text{s}^{-2}$.

(a) Scale this equation to provide Δg in gu when $\Delta\rho$ is expressed in Mg m^{-3} and t in m.

(b) This equation is used to provide a preliminary estimate of the gravity anomaly of a body of specified thickness. Using this equation, calculate the gravity anomaly of (i) a granite 12 km thick of density 2.67 Mg m^{-3}; and (ii) a sandstone body 4 km thick of density 2.30 Mg m^{-3}, where the density of the surrounding metamorphic rocks is 2.80 Mg m^{-3}. Are the anomalies so calculated liable to be over- or underestimates?

6. Show that the half-width of the gravity anomaly caused by a horizontal cylinder is equal to the depth of the axis of the cylinder.

7. Figure 6.31 is a Bouguer anomaly map, contoured at an interval of 50 gu, of a drift-covered area.

(a) On the map, sketch in what you consider to be the regional field and then remove it from the observed field to isolate residual anomalies,

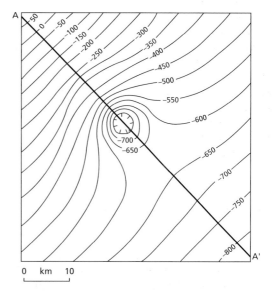

0   km   10

Fig. 6.31 Bouguer anomaly map pertaining to Question 7. Contour interval 50 gu.

Continued

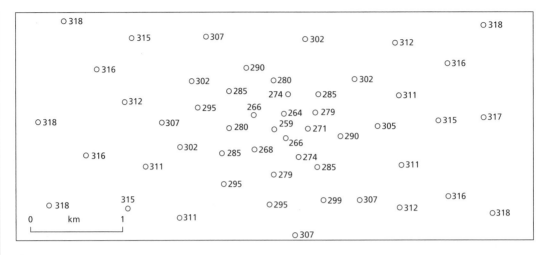

Fig. 6.32 Bouguer anomaly observations pertaining to Question 8. Values in gu.

which can be represented on the map as contours drawn in a different colour.

(b) Construct gravity profiles along line A–A′ illustrating the observed, regional and residual anomalies.

(c) Perform a direct interpretation of the residual anomaly, obtaining as much information as possible on the depth, thickness and shape of the source.

(d) The bedrock constitutes part of a Precambrian shield. Speculate on the nature of the anomalous body, giving reasons for your ideas.

8. Contour the gravity data on the map shown in Fig. 6.32 using an interval of 10 gu. Draw a representative profile.

(a) Use limiting depth calculations based on the half-width and gradient–amplitude methods to determine the depth to the centre of mass of the anomalous body. Comment on any difference between the depth estimates provided by the two methods.

(b) Determine the mass deficiency present using the formula for the gravity anomaly of a point mass. If the anomaly is caused by a salt dome of density $2.22\,Mg\,m^{-3}$ within sediments of density $2.60\,Mg\,m^{-3}$, calculate the volume and mass of salt present and the depth to the top of the salt dome. Compute the actual gravity anomaly of

the salt and comment on any differences with the observed anomaly.

(c) What is the lowest possible density contrast of the anomalous body?

(d) Determine the mass deficiency present using a method based on Gauss' Theorem. Comment on the accuracy of the value obtained and compare it with the answer to (b). Calculate the actual mass present assuming the same densities as in (a).

9. The map in Fig. 6.33 shows Bouguer anomalies over a gabbro intrusion in a schist terrain. In the eastern part of the map, horizontally bedded Mesozoic sediments unconformably overlie the schists. A seismic refraction line has been established over the sediments in the location shown. Time–distance data and typical velocities and densities are given below.

Interpret the geophysical results using the following scheme:

(a) Use the refraction data to determine the thickness and possible nature of the Mesozoic rocks beneath the seismic line.

(b) Use this interpretation to calculate the gravity anomaly of the Mesozoic rocks at this location. Correct the observed gravity anomaly for the effect of the Mesozoic rocks.

(c) Determine the maximum gravity anomaly of

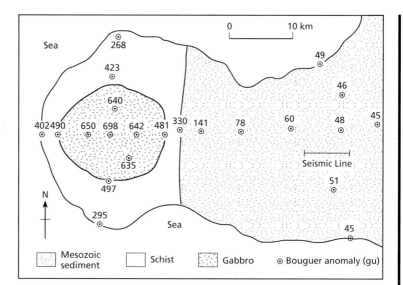

Fig. 6.33 Map of geophysical observations pertaining to Question 9. Bouguer anomaly values in gu.

the gabbro. Assuming the gabbro to have the form of a vertical cylinder, determine the depth to its base.

The gravity anomaly Δg of a vertical cylinder of density contrast $\Delta \rho$, radius r, length L, depth to top z_U and depth to base z_L is given by

$$\Delta g = 2\pi G \Delta \rho \left(L - \sqrt{z_L^2 + r^2} + \sqrt{z_U^2 + r^2} \right)$$

where G is the gravitational constant.

State any assumptions and possible causes of error in your interpretation.

Typical densities and seismic velocities

	ρ (Mg m^{-3})	Veloc. (km s^{-1})
Jur./Cret.	2.15	1.20–1.80
Trias	2.35	2.40–3.00
Schist	2.75	3.60–4.90
Gabbro	2.95	

Jur. = Jurassic; Cret. = Cretaceous.

Seismic data

Dist. (m)	Time (s)
530	0.349
600	0.391
670	0.441
1130	0.739
1200	0.787
1270	0.831
1800	1.160
1870	1.177
1940	1.192
2730	1.377
2800	1.393
2870	1.409
3530	1.563
3600	1.582
3670	1.599

10. Over a typical ocean spreading centre, the free-air gravity anomaly is approximately zero and the Bouguer anomaly is large and negative. Why?

Further reading

Baranov, W. (1975) *Potential Fields and Their Transformations in Applied Geophysics*. Gebrüder Borntraeger, Berlin.

Blakely, R.J. (1995) *Potential Theory in Gravity and Magnetic Applications*. Cambridge University Press, Cambridge.

Bott, M.H.P. (1973) Inverse methods in the interpretation of magnetic and gravity anomalies. In: Alder, B., Fernbach, S. & Bolt, B.A. (eds), *Methods in Computational Physics*, **13**, 133–62.

Dehlinger, P. (1978) *Marine Gravity*. Elsevier, Amsterdam.

Gibson, R.I. & Millegan, P.S. (eds) (1998) *Geologic Applications of*

Gravity and Magnetics: Case Histories. SEG Reference Series 8 & AAPG Studies in Geology 43, Tulsa.

LaCoste, L.J.B. (1967) Measurement of gravity at sea and in the air. *Rev. Geophys.*, **5**, 477–526.

LaCoste, L.J.B., Ford, J., Bowles, R. & Archer, K. (1982) Gravity measurements in an airplane using state-of-the-art navigation and altimetry. *Geophysics*, **47**, 832–7.

Milsom, J. (1989) *Field Geophysics.* Open University Press, Milton Keynes.

Nettleton, L.L. (1971) *Elementary Gravity and Magnetics for Geologists and Seismologists.* Monograph Series No. 1. Society of Exploration Geophysicists, Tulsa.

Nettleton, L.L. (1976) *Gravity and Magnetics in Oil Exploration.* McGraw-Hill, New York.

Ramsey, A.S. (1964) *An Introduction to the Theory of Newtonian Attraction.* Cambridge University Press, Cambridge.

Torge, W. (1989) *Gravimetry.* Walter de Gruyter, Berlin.

Tsuboi, C. (1983) *Gravity.* Allen & Unwin, London.

7 Magnetic surveying

7.1 Introduction

The aim of a magnetic survey is to investigate subsurface geology on the basis of anomalies in the Earth's magnetic field resulting from the magnetic properties of the underlying rocks. Although most rock-forming minerals are effectively non-magnetic, certain rock types contain sufficient magnetic minerals to produce significant magnetic anomalies. Similarly, man-made ferrous objects also generate magnetic anomalies. Magnetic surveying thus has a broad range of applications, from small-scale engineering or archaeological surveys to detect buried metallic objects, to large-scale surveys carried out to investigate regional geological structure.

Magnetic surveys can be performed on land, at sea and in the air. Consequently, the technique is widely employed, and the speed of operation of airborne surveys makes the method very attractive in the search for types of ore deposit that contain magnetic minerals.

7.2 Basic concepts

Within the vicinity of a bar magnet a magnetic flux is developed which flows from one end of the magnet to the other (Fig. 7.1). This flux can be mapped from the directions assumed by a small compass needle suspended within it. The points within the magnet where the flux converges are known as the poles of the magnet. A freely-suspended bar magnet similarly aligns in the flux of the Earth's magnetic field. The pole of the magnet which tends to point in the direction of the Earth's north pole is called the north-seeking or positive pole, and this is balanced by a south-seeking or negative pole of identical strength at the opposite end of the magnet.

The force F between two magnetic poles of strengths m_1 and m_2 separated by a distance r is given by

$$F = \frac{\mu_0 m_1 m_2}{4\pi \mu_R r^2} \qquad (7.1)$$

where μ_0 and μ_R are constants corresponding to the *magnetic permeability of vacuum* and the *relative magnetic permeability* of the medium separating the poles (see later). The force is attractive if the poles are of different sign and repulsive if they are of like sign.

The *magnetic field B* due to a pole of strength m at a distance r from the pole is defined as the force exerted on a unit positive pole at that point

$$B = \frac{\mu_0 m}{4\pi \mu_R r^2} \qquad (7.2)$$

Magnetic fields can be defined in terms of *magnetic potentials* in a similar manner to gravitational fields. For a single pole of strength m, the magnetic potential V at a distance r from the pole is given by

$$V = \frac{\mu_0 m}{4\pi \mu_R r} \qquad (7.3)$$

The magnetic field component in any direction is then given by the partial derivative of the potential in that direction.

In the SI system of units, magnetic parameters are defined in terms of the flow of electrical current (see e.g. Reilly 1972). If a current is passed through a coil consisting of several turns of wire, a *magnetic flux* flows through and around the coil annulus which arises from a *magnetizing force H*. The magnitude of H is proportional to the number of turns in the coil and the strength of the current, and inversely proportional to the length of the wire, so that H is expressed in $A\,m^{-1}$. The density of the magnetic flux, measured over an area perpendicular to the direction of flow, is known as the *magnetic induction* or *magnetic field B* of the coil. B is proportional to H and the constant of proportionality μ is known as the *magnetic permeability*. Lenz's law of induction relates the rate of change of magnetic flux in a circuit to the voltage devel-

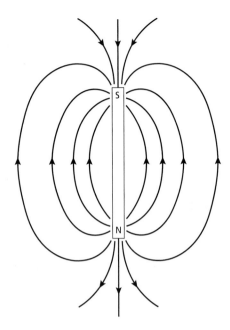

Fig. 7.1 The magnetic flux surrounding a bar magnet.

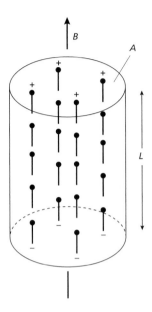

Fig. 7.2 Schematic representation of an element of material in which elementary dipoles align in the direction of an external field B to produce an overall induced magnetization.

oped within it, so that B is expressed in $Vs\,m^{-2}$ (Weber (Wb) m^{-2}). The unit of the $Wb\,m^{-2}$ is designated the *tesla* (T). Permeability is consequently expressed in $Wb\,A^{-1}m^{-1}$ or Henry (H) m^{-1}. The c.g.s. unit of magnetic field strength is the *gauss* (G), numerically equivalent to $10^{-4}\,T$.

The tesla is too large a unit in which to express the small magnetic anomalies caused by rocks, and a subunit, the *nanotesla* (nT), is employed $(1\,nT = 10^{-9}\,T)$. The c.g.s. system employs the numerically equivalent *gamma* (γ), equal to $10^{-5}\,G$.

Common magnets exhibit a pair of poles and are therefore referred to as dipoles. The *magnetic moment M* of a dipole with poles of strength m a distance l apart is given by

$$M = ml \qquad (7.4)$$

The magnetic moment of a current-carrying coil is proportional to the number of turns in the coil, its cross-sectional area and the magnitude of the current, so that magnetic moment is expressed in $A\,m^2$.

When a material is placed in a magnetic field it may acquire a magnetization in the direction of the field which is lost when the material is removed from the field. This phenomenon is referred to as *induced magneti-*

zation or *magnetic polarization*, and results from the alignment of elementary dipoles (see below) within the material in the direction of the field. As a result of this alignment the material has magnetic poles distributed over its surface which correspond to the ends of the dipoles (Fig. 7.2). The intensity of induced magnetization J_i of a material is defined as the dipole moment per unit volume of material:

$$J_i = \frac{M}{LA} \qquad (7.5)$$

where M is the magnetic moment of a sample of length L and cross-sectional area A. J_i is consequently expressed in $A\,m^{-1}$. In the c.g.s. system intensity of magnetization is expressed in $emu\,cm^{-3}$ (emu = electromagnetic unit), where $1\,emu\,cm^{-3} = 1000\,A\,m^{-1}$.

The induced intensity of magnetization is proportional to the strength of the magnetizing force H of the inducing field:

$$J_i = kH \qquad (7.6)$$

where k is the *magnetic susceptibility* of the material. Since J_i and H are both measured in $A\,m^{-1}$, susceptibility is dimensionless in the SI system. In the c.g.s. system susceptibility is similarly dimensionless, but a consequence of rationalizing the SI system is that SI susceptibility values are a factor 4π greater than corresponding c.g.s. values.

In a vacuum the magnetic field strength B and magnetizing force H are related by $B = \mu_0 H$ where μ_0 is the permeability of vacuum $(4\pi \times 10^{-7}\,\mathrm{H\,m^{-1}})$. Air and water have very similar permeabilities to μ_0 and so this relationship can be taken to represent the Earth's magnetic field when it is undisturbed by magnetic materials. When a magnetic material is placed in this field, the resulting magnetization gives rise to an additional magnetic field in the region occupied by the material, whose strength is given by $\mu_0 J_i$. Within the body the total magnetic field, or magnetic induction, B is given by

$$B = \mu_0 H + \mu_0 J_i$$

Substituting equation (7.6)

$$B = \mu_0 H + \mu_0 k H = (1 + k)\mu_0 H = \mu_R \mu_0 H$$

where μ_R is a dimensionless constant known as the *relative magnetic permeability*. The magnetic permeability μ is thus equal to the product of the relative permeability and the permeability of vacuum, and has the same dimensions as μ_0. For air and water μ_R is thus close to unity.

All substances are magnetic at an atomic scale. Each atom acts as a dipole due to both the spin of its electrons and the orbital path of the electrons around the nucleus. Quantum theory allows two electrons to exist in the same state (or electron shell) provided that their spins are in opposite directions. Two such electrons are called paired electrons and their spin magnetic moments cancel. In *diamagnetic* materials all electron shells are full and no unpaired electrons exist. When placed in a magnetic field the orbital paths of the electrons rotate so as to produce a magnetic field in opposition to the applied field. Consequently, the susceptibility of diamagnetic substances is weak and negative. In *paramagnetic* substances the electron shells are incomplete so that a magnetic field results from the spin of their unpaired electrons. When placed in an external magnetic field the dipoles corresponding to the unpaired electron spins rotate to produce a field in the same sense as the applied field so that the susceptibility is positive. This is still, however, a relatively weak effect.

In small grains of certain paramagnetic substances whose atoms contain several unpaired electrons, the dipoles associated with the spins of the unpaired electrons are magnetically coupled between adjacent atoms. Such a grain is then said to constitute a single *magnetic do-*

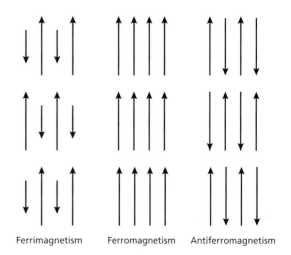

Ferrimagnetism   Ferromagnetism   Antiferromagnetism

Fig. 7.3 Schematic representation of the strength and orientation of elementary dipoles within ferrimagnetic, ferromagnetic and antiferromagnetic domains.

main. Depending on the degree of overlap of the electron orbits, this coupling may be either parallel or antiparallel. In *ferromagnetic* materials the dipoles are parallel (Fig. 7.3), giving rise to a very strong spontaneous magnetization which can exist even in the absence of an external magnetic field, and a very high susceptibility. Ferromagnetic substances include iron, cobalt and nickel, and rarely occur naturally in the Earth's crust. In *antiferromagnetic* materials such as haematite, the dipole coupling is antiparallel with equal numbers of dipoles in each direction. The magnetic fields of the dipoles are self-cancelling so that there is no external magnetic effect. However, defects in the crystal lattice structure of an antiferromagnetic material may give rise to a small net magnetization, called *parasitic antiferromagnetism.* In *ferrimagnetic* materials such as magnetite, the dipole coupling is similarly antiparallel, but the strength of dipoles in each direction are unequal. Consequently ferrimagnetic materials can exhibit a strong spontaneous magnetization and a high susceptibility. Virtually all the minerals responsible for the magnetic properties of common rock types (Section 7.3) fall into this category.

The strength of the magnetization of ferromagnetic and ferrimagnetic substances decreases with temperature and disappears at the *Curie temperature.* Above this temperature interatomic distances are increased to separations which preclude electron coupling, and the material behaves as an ordinary paramagnetic substance.

In larger grains, the total magnetic energy is decreased if the magnetization of each grain subdivides into individual volume elements (magnetic domains) with diameters of the order of a micrometre, within which there is parallel coupling of dipoles. In the absence of any external magnetic field the domains become oriented in such a way as to reduce the magnetic forces between adjacent domains. The boundary between two domains, the *Bloch wall*, is a narrow zone in which the dipoles cant over from one domain direction to the other.

When a multidomain grain is placed in a weak external magnetic field, the Bloch wall unrolls and causes a growth of those domains magnetized in the direction of the field at the expense of domains magnetized in other directions. This induced magnetization is lost when the applied field is removed as the domain walls rotate back to their original configuration. When stronger fields are applied, domain walls unroll irreversibly across small imperfections in the grain so that those domains magnetized in the direction of the field are permanently enlarged. The inherited magnetization remaining after removal of the applied field is known as *remanent*, or *permanent*, *magnetization* J_r. The application of even stronger magnetic fields causes all possible domain wall movements to occur and the material is then said to be magnetically saturated.

Primary remanent magnetization may be acquired either as an igneous rock solidifies and cools through the Curie temperature of its magnetic minerals (thermoremanent magnetization, TRM) or as the magnetic particles of a sediment align within the Earth's field during sedimentation (detrital remanent magnetization, DRM). Secondary remanent magnetizations may be impressed later in the history of a rock as magnetic minerals recrystallize or grow during diagenesis or metamorphism (chemical remanent magnetization, CRM). Remanent magnetization may develop slowly in a rock standing in an ambient magnetic field as the domain magnetizations relax into the direction of the field (viscous remanent magnetization, VRM).

Any rock containing magnetic minerals may possess both induced and remanent magnetizations J_i and J_r. The relative intensities of induced and remanent magnetizations are commonly expressed in terms of the *Königsberger ratio*, $J_r : J_i$. These may be in different directions and may differ significantly in magnitude. The magnetic effects of such a rock arise from the resultant J of the two magnetization vectors (Fig. 7.4). The magnitude of J controls the amplitude of the magnetic anomaly and the orientation of J influences its shape.

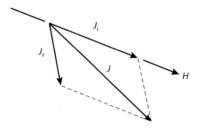

Fig. 7.4 Vector diagram illustrating the relationship between induced (J_i), remanent (J_r) and total (J) magnetization components.

7.3 Rock magnetism

Most common rock-forming minerals exhibit a very low magnetic susceptibility and rocks owe their magnetic character to the generally small proportion of magnetic minerals that they contain. There are only two geochemical groups which provide such minerals. The iron–titanium–oxygen group possesses a solid solution series of magnetic minerals from magnetite (Fe_3O_4) to ulvöspinel (Fe_2TiO_4). The other common iron oxide, haematite (Fe_2O_3), is antiferromagnetic and thus does not give rise to magnetic anomalies (see Section 7.12) unless a parasitic antiferromagnetism is developed. The iron–sulphur group provides the magnetic mineral pyrrhotite (FeS_{1+x}, $0 < x < 0.15$) whose magnetic susceptibility is dependent upon the actual composition.

By far the most common magnetic mineral is magnetite, which has a Curie temperature of 578°C. Although the size, shape and dispersion of the magnetite grains within a rock affect its magnetic character, it is reasonable to classify the magnetic behaviour of rocks according to their overall magnetite content. A histogram illustrating the susceptibilities of common rock types is presented in Fig. 7.5.

Basic igneous rocks are usually highly magnetic due to their relatively high magnetite content. The proportion of magnetite in igneous rocks tends to decrease with increasing acidity so that acid igneous rocks, although variable in their magnetic behaviour, are usually less magnetic than basic rocks. Metamorphic rocks are also variable in their magnetic character. If the partial pressure of oxygen is relatively low, magnetite becomes resorbed and the iron and oxygen are incorporated into other mineral phases as the grade of metamorphism increases. Relatively high oxygen partial pressure can, however, result in the formation of magnetite as an accessory mineral in metamorphic reactions.

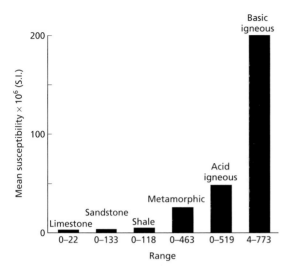

Fig. 7.5 Histogram showing mean values and ranges in susceptibility of common rock types. (After Dobrin & Savit 1988).

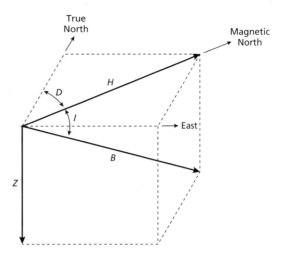

Fig. 7.6 The geomagnetic elements.

In general the magnetite content and, hence, the susceptibility of rocks is extremely variable and there can be considerable overlap between different lithologies. It is not usually possible to identify with certainty the causative lithology of any anomaly from magnetic information alone. However, sedimentary rocks are effectively non-magnetic unless they contain a significant amount of magnetite in the heavy mineral fraction. Where magnetic anomalies are observed over sediment-covered areas the anomalies are generally caused by an underlying igneous or metamorphic basement, or by intrusions into the sediments.

Common causes of magnetic anomalies include dykes, faulted, folded or truncated sills and lava flows, massive basic intrusions, metamorphic basement rocks and magnetite ore bodies. Magnetic anomalies range in amplitude from a few tens of nT over deep metamorphic basement to several hundred nT over basic intrusions and may reach an amplitude of several thousand nT over magnetite ores.

7.4 The geomagnetic field

Magnetic anomalies caused by rocks are localized effects superimposed on the normal magnetic field of the Earth (geomagnetic field). Consequently, knowledge of the behaviour of the geomagnetic field is necessary both in the reduction of magnetic data to a suitable datum and in the interpretation of the resulting anomalies. The geomagnetic field is geometrically more complex than the gravity field of the Earth and exhibits irregular variation in both orientation and magnitude with latitude, longitude and time.

At any point on the Earth's surface a freely suspended magnetic needle will assume a position in space in the direction of the ambient geomagnetic field. This will generally be at an angle to both the vertical and geographic north. In order to describe the magnetic field vector, use is made of descriptors known as the geomagnetic elements (Fig. 7.6). The *total field vector B* has a vertical component Z and a horizontal component H in the direction of magnetic north. The dip of B is the *inclination I* of the field and the horizontal angle between geographic and magnetic north is the *declination D*. B varies in strength from about 25 000 nT in equatorial regions to about 70 000 nT at the poles.

In the northern hemisphere the magnetic field generally dips downward towards the north and becomes vertical at the north magnetic pole (Fig. 7.7). In the southern hemisphere the dip is generally upwards towards the north. The line of zero inclination approximates the geographic equator, and is known as the magnetic equator.

About 90% of the Earth's field can be represented by the field of a theoretical magnetic dipole at the centre of the Earth inclined at about 11.5° to the axis of rotation. The magnetic moment of this fictitious *geocentric dipole* can be calculated from the observed field. If this dipole field is subtracted from the observed magnetic field, the

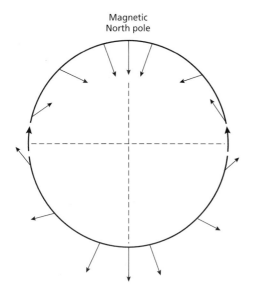

Magnetic
North pole

Fig. 7.7 The variation of the inclination of the total magnetic field with latitude based on a simple dipole approximation of the geomagnetic field. (After Sharma 1976.)

residual field can then be approximated by the effects of a second, smaller, dipole. The process can be continued by fitting dipoles of ever decreasing moment until the observed geomagnetic field is simulated to any required degree of accuracy. The effects of each fictitious dipole contribute to a function known as a harmonic and the technique of successive approximations of the observed field is known as spherical harmonic analysis – the equivalent of Fourier analysis in spherical polar coordinates. The method has been used to compute the formula of the International Geomagnetic Reference Field (IGRF) which defines the theoretical undisturbed magnetic field at any point on the Earth's surface. In magnetic surveying, the IGRF is used to remove from the magnetic data those magnetic variations attributable to this theoretical field. The formula is considerably more complex than the equivalent Gravity Formula used for latitude correction (see Section 6.8.2) as a large number of harmonics is employed (Barraclough & Malin 1971, Peddie 1983).

The geomagnetic field cannot in fact result from permanent magnetism in the Earth's deep interior. The required dipolar magnetic moments are far greater than is considered realistic and the prevailing high temperatures are far in excess of the Curie temperature of any known magnetic material. The cause of the geomagnetic field is attributed to a dynamo action produced by the circula-tion of charged particles in coupled convective cells within the outer, fluid, part of the Earth's core. The exchange of dominance between such cells is believed to produce the periodic changes in polarity of the geomagnetic field revealed by palaeomagnetic studies. The circulation patterns within the core are not fixed and change slowly with time. This is reflected in a slow, progressive, temporal change in all the geomagnetic elements known as *secular variation*. Such variation is predictable and a well-known example is the gradual rotation of the north magnetic pole around the geographic pole.

Magnetic effects of external origin cause the geomagnetic field to vary on a daily basis to produce *diurnal variations*. Under normal conditions (Q or quiet days) the diurnal variation is smooth and regular and has an amplitude of about 20–80 nT, being at a maximum in polar regions. Such variation results from the magnetic field induced by the flow of charged particles within the ionosphere towards the magnetic poles, as both the circulation patterns and diurnal variations vary in sympathy with the tidal effects of the Sun and Moon.

Some days (D or disturbed days) are distinguished by far less regular diurnal variations and involve large, short-term disturbances in the geomagnetic field, with amplitudes of up to 1000 nT, known as *magnetic storms*. Such days are usually associated with intense solar activity and result from the arrival in the ionosphere of charged solar particles. Magnetic surveying should be discontinued during such storms because of the impossibility of correcting the data collected for the rapid and high-amplitude changes in the magnetic field.

7.5 Magnetic anomalies

All magnetic anomalies caused by rocks are superimposed on the geomagnetic field in the same way that gravity anomalies are superimposed on the Earth's gravitational field. The magnetic case is more complex, however, as the geomagnetic field varies not only in amplitude, but also in direction, whereas the gravitational field is everywhere, by definition, vertical.

Describing the normal geomagnetic field by a vector diagram (Fig. 7.8(a)), the geomagnetic elements are related

$$B^2 = H^2 + Z^2 \qquad (7.7)$$

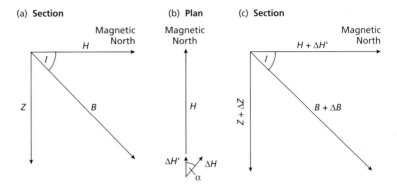

Fig. 7.8 Vector representation of the geomagnetic field with and without a superimposed magnetic anomaly.

A magnetic anomaly is now superimposed on the Earth's field causing a change ΔB in the strength of the total field vector B. Let the anomaly produce a vertical component ΔZ and a horizontal component ΔH at an angle α to H (Fig. 7.8(b)). Only that part of ΔH in the direction of H, namely $\Delta H'$, will contribute to the anomaly

$$\Delta H' = \Delta H \cos \alpha \qquad (7.8)$$

Using a similar vector diagram to include the magnetic anomaly (Fig. 7.8(c))

$$(B + \Delta B)^2 = (H + \Delta H')^2 + (Z + \Delta Z)^2$$

If this equation is expanded, the equality of equation (7.7) substituted and the insignificant terms in Δ^2 ignored, the equation reduces to

$$\Delta B = \Delta Z \frac{Z}{B} + \Delta H' \frac{H}{B}$$

Substituting equation (7.8) and angular descriptions of geomagnetic element ratios gives

$$\Delta B = \Delta Z \sin I + \Delta H \cos I \cos \alpha \qquad (7.9)$$

where I is the inclination of the geomagnetic field.

This approach can be used to calculate the magnetic anomaly caused by a small isolated magnetic pole of strength m, defined as the effect of this pole on a unit positive pole at the observation point. The pole is situated at depth z, a horizontal distance x and radial distance r from the observation point (Fig. 7.9). The force of repulsion ΔB_r on the unit positive pole in the direction r is given by substitution in equation (7.1), with $\mu_R = 1$,

$$\Delta B_r = \frac{Cm}{r^2}$$

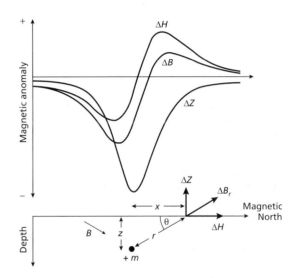

Fig. 7.9 The horizontal (ΔH), vertical (ΔZ) and total field (ΔB) anomalies due to an isolated positive pole.

where $\quad C = \dfrac{\mu_0}{4\pi}$

If it is assumed that the profile lies in the direction of magnetic north so that the horizontal component of the anomaly lies in this direction, the horizontal (ΔH) and vertical (ΔZ) components of this force can be computed by resolving in the relevant directions

$$\Delta H = \frac{Cm}{r^2} \cos \theta = \frac{Cmx}{r^3} \qquad (7.10)$$

$$\Delta Z = \frac{-Cm}{r^2} \sin \theta = \frac{-Cmz}{r^3} \qquad (7.11)$$

The vertical field anomaly is negative as, by convention, the z-axis is positive downwards. Plots of the form of these anomalies are shown in Fig. 7.9. The horizontal field anomaly is a positive/negative couplet and the vertical field anomaly is centred over the pole.

The total field anomaly ΔB is then obtained by substituting the expressions of equations (7.10) and (7.11) in equation (7.9), where $\alpha = 0$. If the profile were not in the direction of magnetic north, the angle α would represent the angle between magnetic north and the profile direction.

7.6 Magnetic surveying instruments

7.6.1 Introduction

Since the early 1900s a variety of surveying instruments have been designed that is capable of measuring the geomagnetic elements Z, H and B. Most modern survey instruments, however, are designed to measure B only. The precision normally required is ± 0.1 nT which is approximately one part in 5×10^6 of the background field, a considerably lower requirement of precision than is necessary for gravity measurements (see Chapter 6).

In early magnetic surveys the geomagnetic elements were measured using *magnetic variometers*. There were several types, including the torsion head magnetometer and the Schmidt vertical balance, but all consisted essentially of bar magnets suspended in the Earth's field. Such devices required accurate levelling and a stable platform for measurement so that readings were time consuming and limited to sites on land.

7.6.2 Fluxgate magnetometer

Since the 1940s, a new generation of instruments has been developed which provides virtually instantaneous readings and requires only coarse orientation so that magnetic measurements can be taken on land, at sea and in the air.

The first such device to be developed was the *fluxgate magnetometer*, which found early application during the second world war in the detection of submarines from the air. The instrument employs two identical ferromagnetic cores of such high permeability that the geomagnetic field can induce a magnetization that is a substantial proportion of their saturation value (see Section 7.2). Identical primary and secondary coils are wound in opposite directions around the cores (Fig. 7.10). An alternating current of 50–1000 Hz is passed through the primary coils (Fig. 7.10(a)), generating an alternating magnetic field. In the absence of any external magnetic

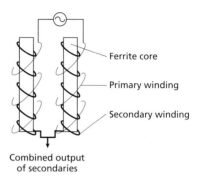

Combined output
of secondaries

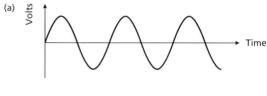

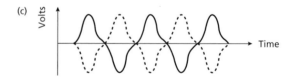

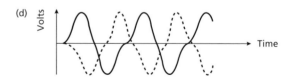

Fig. 7.10 Principle of the fluxgate magnetometer. Solid and broken lines in (b)–(d) refer to the responses of the two cores.

field, the cores are driven to saturation near the peak of each half-cycle of the current (Fig. 7.10(b)). The alternating magnetic field in the cores induces an alternating voltage in the secondary coils which is at a maximum when the field is changing most rapidly (Fig. 7.10(c)). Since the coils are wound in opposite directions, the voltage in the coils is equal and of opposite sign so that their combined output is zero. In the presence of an external magnetic field, such as the Earth's field, which has a component parallel to the axis of the cores, saturation occurs earlier for the core whose primary field is reinforced by the external field and later for the core opposed by the external field. The induced voltages are now out of phase as the cores reach saturation at different times (Fig. 7.10(d)). Consequently, the combined output of the secondary coils is no longer zero but consists of a series of voltage pulses (Fig. 7.10(e)), the magnitude of which can be shown to be proportional to the amplitude of the external field component.

The instrument can be used to measure Z or H by aligning the cores in these directions, but the required accuracy of orientation is some eleven seconds of arc to achieve a reading accuracy of ± 1 nT. Such accuracy is difficult to obtain on the ground and impossible when the instrument is mobile. The total geomagnetic field can, however, be measured to ± 1 nT with far less precise orientation as the field changes much more slowly as a function of orientation about the total field direction. Airborne versions of the instrument employ orienting mechanisms of various types to maintain the axis of the instrument in the direction of the geomagnetic field. This is accomplished by making use of the feedback signal generated by additional sensors whenever the instrument moves out of orientation to drive servomotors which realign the cores into the desired direction.

The fluxgate magnetometer is a continuous reading instrument and is relatively insensitive to magnetic field gradients along the length of the cores. The instrument may be temperature sensitive, requiring correction.

7.6.3 Proton magnetometer

The most commonly used magnetometer for both survey work and observatory monitoring is currently the *nuclear precession* or *proton magnetometer*. The sensing device of the proton magnetometer is a container filled with a liquid rich in hydrogen atoms, such as kerosene or water, surrounded by a coil (Fig. 7.11(a)). The hydrogen nuclei (protons) act as small dipoles and normally align parallel to the ambient geomagnetic field B_e (Fig. 7.11(b)). A current is passed through the coil to generate a magnetic field B_p 50–100 times larger than the geomagnetic field, and in a different direction, causing the protons to realign in this new direction (Fig. 7.11(c)). The current to the coil is then switched off so that the

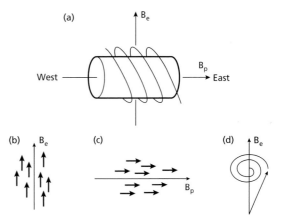

Fig. 7.11 Principle of the proton magnetometer.

polarizing field is rapidly removed. The protons return to their original alignment with B_e by spiralling, or precessing, in phase around this direction (Fig. 7.11(d)) with a period of about 0.5 ms, taking some 1–3 s to achieve their original orientation. The frequency f of this precession is given by

$$f = \frac{\gamma_p B_e}{2\pi}$$

where γ_p is the gyromagnetic ratio of the proton, an accurately known constant. Consequently, measurement of f, about 2 kHz, provides a very accurate measurement of the strength of the total geomagnetic field. f is determined by measurement of the alternating voltage of the same frequency induced to flow in the coil by the precessing protons.

Field instruments provide absolute readings of the total magnetic field accurate to ± 0.1 nT although much greater precision can be attained if necessary. The sensor does not have to be accurately oriented, although it should ideally lie at an appreciable angle to the total field vector. Consequently, readings may be taken by sensors towed behind ships or aircraft without the necessity of orienting mechanisms. Aeromagnetic surveying with proton magnetometers may suffer from the slight disadvantage that readings are not continuous due to the finite cycle period. Small anomalies may be missed since an aircraft travels a significant distance between the discrete measurements, which may be spaced at intervals of a few seconds. This problem has been largely obviated by modern instruments with recycling periods of the order of a second. The proton magnetometer is sensitive to acute magnetic gradients which may cause protons in different parts of the sensor to precess at different rates with a consequent adverse effect on precession signal strength.

Many modern proton magnetometers make use of the *Overhauser Effect*. To the sensor fluid is added a liquid containing some free electrons in 'unpaired' orbits. The protons are then polarized indirectly using radio-frequency energy near 60 MHz. The power consumption of such instruments is only some 25% of classical proton magnetometers, so that the instruments are lighter and more compact. The signal generated by the fluid is about 100 times stronger, so there is much lower noise; gradient tolerance is some three times better; sampling rates are faster.

7.6.4 Optically pumped magnetometer

Optically pumped or *alkali vapour magnetometers* have a significantly higher precision than other types. They comprise a glass cell containing an evaporated alkali metal such as caesium, rubidium or potassium which is energized by light of a particular wavelength. In these alkali atoms there exist valence electrons partitioned into two energy levels 1 and 2. The wavelength of the energizing light is selected to excite electrons from level 2 to the higher level 3, a process termed polarization. Electrons at level 3 are unstable and spontaneously decay back to levels 1 and 2. As this process is repeated, level 1 becomes fully populated at the expense of level 2 becoming underpopulated. This process is known as optical pumping and leads to the stage in which the cell stops absorbing light and turns from opaque to transparent. The energy difference between levels 1 and 2 is proportional to the strength of the ambient magnetic field. Depolarization then takes place by the application of radiofrequency power. The wavelength corresponding to the energy difference between levels 1 and 2 depolarizes the cell and is a measure of the magnetic field strength. A photodetector is used to balance the cell between transparent and opaque states. The depolarization is extremely rapid so that readings are effectively instantaneous. The sensitivity of optically pumped magnetometers can be as high as ± 0.01 nT. This precision is not required for surveys involving total field measurements, where the level of background 'noise' is of the order of 1 nT. The usual application is in the magnetic gradiometers described below, which rely on measuring the small difference in signal from sensors only a small distance apart.

7.6.5 Magnetic gradiometers

The sensing elements of fluxgate, proton and optically pumped magnetometers can be used in pairs to measure either horizontal or vertical magnetic field gradients. *Magnetic gradiometers* are differential magnetometers in which the spacing between the sensors is fixed and small with respect to the distance of the causative body whose magnetic field gradient is to be measured. Magnetic gradients can be measured, albeit less conveniently, with a magnetometer by taking two successive measurements at close vertical or horizontal spacings. Magnetic gradiometers are employed in surveys of shallow magnetic features as the gradient anomalies tend to resolve complex anomalies into their individual components, which can be used in the determination of the location, shape and depth of the causative bodies. The method has the further advantages that regional and temporal variations in the geomagnetic field are automatically removed. Marine and airborne versions of magnetometers and gradiometers are discussed by Wold and Cooper (1989) and Hood and Teskey (1989), respectively.

7.7 Ground magnetic surveys

Ground magnetic surveys are usually performed over relatively small areas on a previously defined target. Consequently, station spacing is commonly of the order of 10–100 m, although smaller spacings may be employed where magnetic gradients are high. Readings should not be taken in the vicinity of metallic objects such as railway lines, cars, roads, fencing, houses, etc, which might perturb the local magnetic field. For similar reasons, operators of magnetometers should not carry metallic objects.

Base station readings are not necessary for monitoring instrumental drift as fluxgate and proton magnetometers do not drift, but are important in monitoring diurnal variations (see Section 7.9).

Since modern magnetic instruments require no precise levelling, a magnetic survey on land invariably proceeds much more rapidly than a gravity survey.

7.8 Aeromagnetic and marine surveys

The vast majority of magnetic surveys are carried out in the air, with the sensor towed in a housing known as a 'bird' to remove the instrument from the magnetic effects of the aircraft or fixed in a 'stinger' in the tail of the aircraft, in which case inboard coil installations compensate for the aircraft's magnetic field.

Aeromagnetic surveying is rapid and cost-effective, typically costing some 40% less per line kilometre than a ground survey. Vast areas can be surveyed rapidly without the cost of sending a field party into the survey area and data can be obtained from areas inaccessible to ground survey.

The most difficult problem in airborne surveys used to be position fixing. Nowadays, however, the availability of GPS obviates the positioning problem.

Marine magnetic surveying techniques are similar to those of airborne surveying. The sensor is towed in a 'fish' at least two ships' lengths behind the vessel to remove its magnetic effects. Marine surveying is obviously slower than aeromagnetic surveying, but is frequently carried out in conjunction with several other geophysical methods, such as gravity surveying and continuous seismic profiling, which cannot be employed in the air.

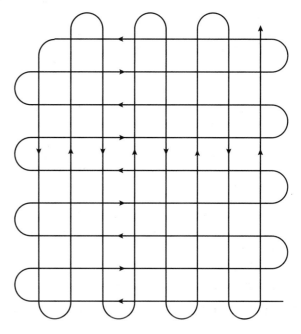

Fig. 7.12 A typical flight plan for an aeromagnetic survey.

7.9 Reduction of magnetic observations

The reduction of magnetic data is necessary to remove all causes of magnetic variation from the observations other than those arising from the magnetic effects of the subsurface.

7.9.1 Diurnal variation correction

The effects of diurnal variation may be removed in several ways. On land a method similar to gravimeter drift monitoring may be employed in which the magnetometer is read at a fixed base station periodically throughout the day. The differences observed in base readings are then distributed among the readings at stations occupied during the day according to the time of observation. It should be remembered that base readings taken during a gravity survey are made to correct for both the drift of the gravimeter and tidal effects; magnetometers do not drift and base readings are taken solely to correct for temporal variation in the measured field. Such a procedure is inefficient as the instrument has to be returned periodically to a base location and is not practical in marine or airborne surveys. These problems may be overcome by use of a base magnetometer, a

continuous-reading instrument which records magnetic variations at a fixed location within or close to the survey area. This method is preferable on land as the survey proceeds faster and the diurnal variations are fully charted. Where the survey is of regional extent the records of a magnetic observatory may be used. Such observatories continuously record changes in all the geomagnetic elements. However, diurnal variations differ quite markedly from place to place and so the observatory used should not be more than about 100 km from the survey area.

Diurnal variation during an aeromagnetic survey may alternatively be assessed by arranging numerous crossover points in the survey plan (Fig. 7.12). Analysis of the differences in readings at each crossover, representing the field change over a series of different time periods, allows the whole survey to be corrected for diurnal variation by a process of network adjustment, without the necessity of a base instrument.

Diurnal variations, however recorded, must be examined carefully. If large, high-frequency variations are apparent, resulting from a magnetic storm, the survey results should be discarded.

7.9.2 Geomagnetic correction

The magnetic equivalent of the latitude correction in gravity surveying is the *geomagnetic correction* which removes the effect of a geomagnetic reference field from the survey data. The most rigorous method of geomagnetic correction is the use of the IGRF (Section 7.4), which expresses the undisturbed geomagnetic field in terms of a large number of harmonics and includes temporal terms to correct for secular variation. The complexity of the IGRF requires the calculation of corrections by computer. It must be realized, however, that the IGRF is imperfect as the harmonics employed are based on observations at relatively few, scattered, magnetic observatories. The IGRF is also predictive in that it extrapolates forwards the spherical harmonics derived from observatory data. Consequently, the IGRF in areas remote from observatories can be substantially in error.

Over the area of a magnetic survey the geomagnetic reference field may be approximated by a uniform gradient defined in terms of latitudinal and longitudinal gradient components. For example, the geomagnetic field over the British Isles is approximated by the following gradient components: $2.13\,\text{nT}\,\text{km}^{-1}\,\text{N}$; $0.26\,\text{nT}\,\text{km}^{-1}$ W; these vary with time. For any survey area the relevant gradient values may be assessed from magnetic maps covering a much larger region.

The appropriate regional gradients may also be obtained by employing a single dipole approximation of the Earth's field and using the well-known equations for the magnetic field of a dipole to derive local field gradients:

$$Z = \frac{\mu_0}{4\pi}\frac{2M}{R^3}\cos\theta, \qquad H = \frac{\mu_0}{4\pi}\frac{M}{R^3}\sin\theta \qquad (7.12)$$

$$\frac{\partial Z}{\partial \theta} = -2H, \qquad \frac{\partial H}{\partial \theta} = \frac{Z}{2} \qquad (7.13)$$

where Z and H are the vertical and horizontal field components, θ the colatitude in radians, R the radius of the Earth, M the magnetic moment of the Earth and $\partial Z/\partial\theta$ and $\partial H/\partial\theta$ the rate of change of Z and H with colatitude, respectively.

An alternative method of removing the regional gradient over a relatively small survey area is by use of trend analysis. A trend line (for profile data) or trend surface (for areal data) is fitted to the observations using the least squares criterion, and subsequently subtracted from the observed data to leave the local anomalies as positive and negative residuals (Fig. 7.13).

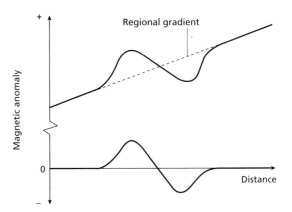

Fig. 7.13 The removal of a regional gradient from a magnetic field by trend analysis. The regional field is approximated by a linear trend.

7.9.3 Elevation and terrain corrections

The vertical gradient of the geomagnetic field is only some $0.03\,\text{nT}\,\text{m}^{-1}$ at the poles and $-0.015\,\text{nT}\,\text{m}^{-1}$ at the equator, so an *elevation correction* is not usually applied. The influence of topography can be significant in ground magnetic surveys but is not completely predictable as it depends upon the magnetic properties of the topographic features. Therefore, in magnetic surveying *terrain corrections* are rarely applied.

Having applied diurnal and geomagnetic corrections, all remaining magnetic field variations should be caused solely by spatial variations in the magnetic properties of the subsurface and are referred to as magnetic anomalies.

7.10 Interpretation of magnetic anomalies

7.10.1 Introduction

The interpretation of magnetic anomalies is similar in its procedures and limitations to gravity interpretation as both techniques utilize natural potential fields based on inverse square laws of attraction. There are several differences, however, which increase the complexity of magnetic interpretation.

Whereas the gravity anomaly of a causative body is entirely positive or negative, depending on whether the body is more or less dense than its surroundings, the magnetic anomaly of a finite body invariably contains positive and negative elements arising from the dipolar nature of magnetism (Fig. 7.14). Moreover, whereas

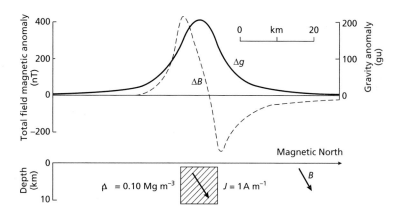

Fig. 7.14 Gravity (Δg) and magnetic (ΔB) anomalies over the same two-dimensional body.

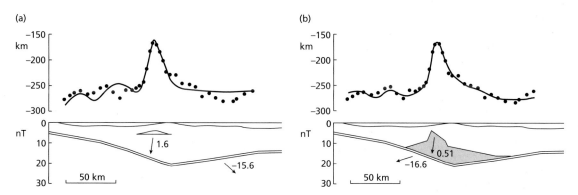

Fig. 7.15 An example of ambiguity in magnetic interpretation. The arrows correspond to the directions of magnetization vectors, whose magnitude is given in $A\,m^{-1}$. (After Westbrook 1975.)

density is a scalar, intensity of magnetization is a vector, and the direction of magnetization in a body closely controls the shape of its magnetic anomaly. Thus bodies of identical shape can give rise to very different magnetic anomalies. For the above reasons magnetic anomalies are often much less closely related to the shape of the causative body than are gravity anomalies.

The intensity of magnetization of a rock is largely dependent upon the amount, size, shape and distribution of its contained ferrimagnetic minerals and these represent only a small proportion of its constituents. By contrast, density is a bulk property. Intensity of magnetization can vary by a factor of 10^6 between different rock types, and is thus considerably more variable than density, where the range is commonly 1.50–$3.50\,Mg\,m^{-3}$.

Magnetic anomalies are independent of the distance units employed. For example, the same magnitude anomaly is produced by, say, a 3 m cube (on a metre scale) as a 3 km cube (on a kilometre scale) with the same

magnetic properties. The same is not true of gravity anomalies.

The problem of ambiguity in magnetic interpretation is the same as for gravity, that is, the same inverse problem is encountered. Thus, just as with gravity, all external controls on the nature and form of the causative body must be employed to reduce the ambiguity. An example of this problem is illustrated in Fig. 7.15, which shows two possible interpretations of a magnetic profile across the Barbados Ridge in the eastern Caribbean. In both cases the regional variations are attributed to the variation in depth of a 1 km thick oceanic crustal layer 2. The high-amplitude central anomaly, however, can be explained by either the presence of a detached sliver of oceanic crust (Fig. 7.15(a)) or a rise of metamorphosed sediments at depth (Fig. 7.15(b)).

Much qualitative information may be derived from a magnetic contour map. This applies especially to aeromagnetic maps which often provide major clues as to the

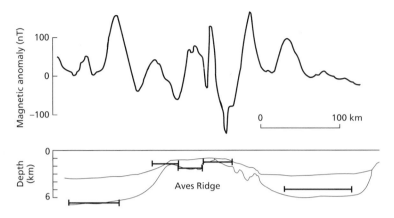

Fig. 7.16 Magnetic anomalies over the Aves Ridge, eastern Caribbean. Lower diagram illustrates bathymetry and basement/sediment interface. Horizontal bars indicate depth estimates of the magnetic basement derived by spectral analysis of the magnetic data.

geology and structure of a broad region from an assessment of the shapes and trends of anomalies. Sediment-covered areas with relatively deep basement are typically represented by smooth magnetic contours reflecting basement structures and magnetization contrasts. Igneous and metamorphic terrains generate far more complex magnetic anomalies, and the effects of deep geological features may be obscured by short-wavelength anomalies of near-surface origin. In most types of terrain an aeromagnetic map can be a useful aid to reconnaissance geological mapping. Such qualitative interpretations may be greatly facilitated by the use of digital image processing techniques (see Section 6.8.6).

In carrying out quantitative interpretation of magnetic anomalies, both direct and indirect methods may be employed, but the former are much more limited than for gravity interpretation and no equivalent general equations exist for total field anomalies.

7.10.2 Direct interpretation

Limiting depth is the most important parameter derived by direct interpretation, and this may be deduced from magnetic anomalies by making use of their property of decaying rapidly with distance from source. Magnetic anomalies caused by shallow structures are more dominated by short-wavelength components than those resulting from deeper sources. This effect may be quantified by computing the power spectrum of the anomaly as it can be shown, for certain types of source body, that the log-power spectrum has a linear gradient whose magnitude is dependent upon the depth of the source (Spector & Grant 1970). Such techniques of spectral analysis provide rapid depth estimates from regularly-spaced digital field data; no geomagnetic or diurnal corrections are

necessary as these remove only low-wavenumber components and do not affect the depth estimates which are controlled by the high-wavenumber components of the observed field. Figure 7.16 shows a magnetic profile across the Aves Ridge in the eastern Caribbean. In this region the configuration of the sediment/basement interface is reasonably well known from both seismic reflection and refraction surveys. The magnetic anomalies clearly show their shortest wavelength over areas of relatively shallow basement, and this observation is quantified by the power spectral depth estimates (horizontal bars) which show excellent correlation with the known basement relief.

A more complex, but more rigorous method of determining the depth to magnetic sources derives from a technique known as *Euler deconvolution* (Reid *et al.* 1990). Euler's homogeneity relation can be written:

$$(x - x_0)\frac{\partial T}{\partial x} + (y - y_0)\frac{\partial T}{\partial y} + (z - z_0)\frac{\partial T}{\partial z} = N(B - T)$$

(7.14)

where (x_0, y_0, z_0) is the location of a magnetic source, whose total field magnetic anomaly at the point (x, y, z) is T and B is the regional field. N is a measure of the rate of change of a field with distance and assumes different values for different types of magnetic source. Equation (7.14) is solved by calculating or measuring the anomaly gradients for various areas of the anomaly and selecting a value for N. This method produces more rigorous depth estimates than other methods, but is considerably more difficult to implement. An example of an Euler deconvolution is shown in Fig. 7.17. The aeromagnetic field shown in Fig. 7.17(a) has the solutions shown in Fig. 17(b–d) for structural indices (N) of 0.0, 0.5 and 0.6

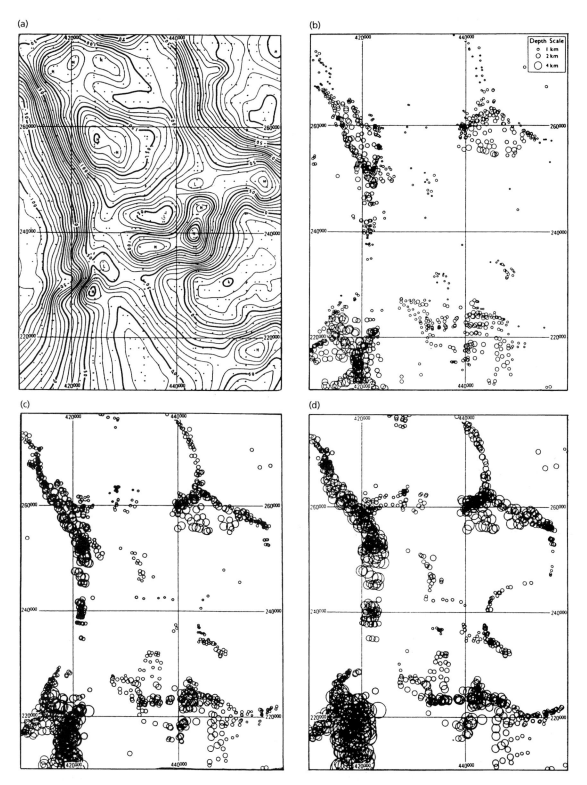

Fig. 7.17 (a) Observed aeromagnetic anomaly of a region in the English Midlands. Contour interval 10 nT. (b–d) Euler deconvolutions for structural indices 0.0 (b), 0.5 (c) and 1.0 (d). Source depth is indicated by the size of the circles. (e) Geological interpretation (*overleaf*). Grid squares are 10 km × 10 km in size. (After Reid *et al*. 1990.)

(e)

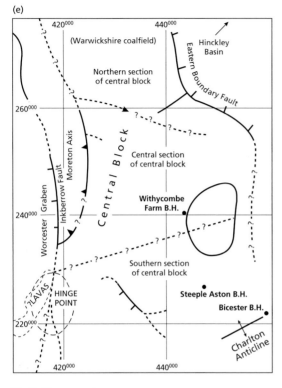

Fig. 7.17 *Continued*

respectively. The boundaries implied by the solutions have been used to construct the interpretation shown in Fig. 7.17(e).

7.10.3 Indirect interpretation

Indirect interpretation of magnetic anomalies is similar to gravity interpretation in that an attempt is made to match the observed anomaly with that calculated for a model by iterative adjustments to the model. Simple magnetic anomalies may be simulated by a single dipole. Such an approximation to the magnetization of a real geological body is often valid for highly magnetic ore bodies whose direction of magnetization tends to align with their long dimension (Fig. 7.18). In such cases the anomaly is calculated by summing the effects of both poles at the observation points, employing equations (7.10), (7.11) and (7.9). More complicated magnetic bodies, however, require a different approach.

The magnetic anomaly of most regularly-shaped bodies can be calculated by building up the bodies from a series of dipoles parallel to the magnetization direction

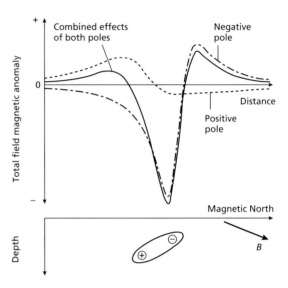

Fig. 7.18 The total field magnetic anomaly of an elongate body approximated by a dipole.

(Fig. 7.19). The poles of the magnets are negative on the surface of the body where the magnetization vector enters the body and positive where it leaves the body. Thus any uniformly-magnetized body can be represented by a set of magnetic poles distributed over its surface. Consider one of these elementary magnets of length l and cross-sectional area δA in a body with intensity of magnetization J and magnetic moment M. From equation (7.5)

$$M = J\delta Al \qquad (7.15)$$

If the pole strength of the magnet is m, from equation (7.4) $m = M/l$, and substituting in equation (7.15)

$$m = J\delta A \qquad (7.16)$$

If $\delta A'$ is the area of the end of the magnet and θ the angle between the magnetization vector and a direction normal to the end face

$$\delta A = \delta A' \cos \theta$$

Substituting in equation (7.16)

$$m = J\delta A' \cos \theta$$

thus

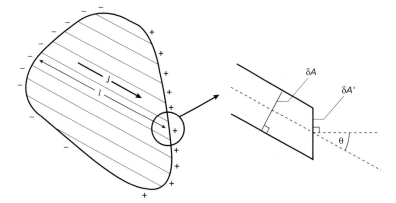

Fig. 7.19 The representation of the magnetic effects of an irregularly-shaped body in terms of a number of elements parallel to the magnetization direction. Inset shows in detail the end of one such element.

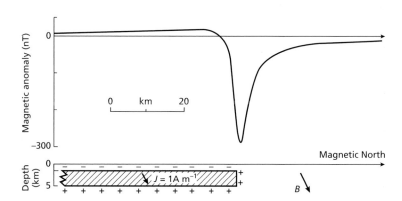

Fig. 7.20 The total field magnetic anomaly of a faulted sill.

pole strength per unit area $= J \cos \theta$ (7.17)

A consequence of the distribution of an equal number of positive and negative poles over the surface of a magnetic body is that an infinite horizontal layer produces no magnetic anomaly since the effects of the poles on the upper and lower surfaces are self-cancelling. Consequently, magnetic anomalies are not produced by continuous sills or lava flows. Where, however, the horizontal structure is truncated, the vertical edge will produce a magnetic anomaly (Fig. 7.20).

The magnetic anomaly of a body of regular shape is calculated by determining the pole distribution over the surface of the body using equation (7.17). Each small element of the surface is then considered and its vertical and horizontal component anomalies are calculated at each observation point using equations (7.10) and (7.11). The effects of all such elements are summed (integrated) to produce the vertical and horizontal anomalies for the whole body and the total field anomaly is calculated using equation (7.9). The integration can be per-

formed analytically for bodies of regular shape, while irregularly-shaped bodies may be split into regular shapes and the integration performed numerically.

In two-dimensional modelling, an approach similar to gravity interpretation can be adopted (see Section 6.10.4) in which the cross-sectional form of the body is approximated by a polygonal outline. The anomaly of the polygon is then computed by adding or subtracting the anomalies of semi-infinite slabs with sloping edges corresponding to the sides of the polygon (Fig. 7.21). In the magnetic case, the horizontal ΔH, vertical ΔZ and total field ΔB anomalies (nT) of the slab shown in Fig. 7.21 are given by (Talwani *et al.* 1965)

$$\Delta Z = 200 \sin \theta [J_x \{ \sin \theta \log_e(r_2/r_1) + \phi \cos \theta \}$$
$$+ J_z \{ \cos \theta \log_e(r_2/r_1) - \phi \sin \theta \}] \quad (7.18a)$$

$$\Delta H = 200 \sin \theta [J_x \{ \phi \sin \theta - \cos \theta \log_e(r_2/r_1) \}$$
$$+ J_z \{ \phi \cos \theta + \sin \theta \log_e(r_2/r_1) \}] \sin \alpha \quad (7.18b)$$

$$\Delta B = \Delta Z \sin I + \Delta H \cos I \quad (7.18c)$$

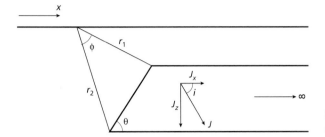

Fig. 7.21 Parameters used in defining the magnetic anomaly of a semi-infinite slab with a sloping edge.

where angles are expressed in radians, J_x ($= J\cos i$) and J_z ($= J\sin i$) are the horizontal and vertical components of the magnetization J, α is the horizontal angle between the direction of the profile and magnetic north, and I is the inclination of the geomagnetic field. Examples of this technique have been presented in Fig. 7.15. An important difference from gravity interpretation is the increased stringency with which the two-dimensional approximation should be applied. It can be shown that two-dimensional magnetic interpretation is much more sensitive to errors associated with variation along strike than is the case with gravity interpretation; the length–width ratio of a magnetic anomaly should be at least 10:1 for a two-dimensional approximation to be valid, in contrast to gravity interpretation where a 2:1 length–width ratio is sufficient to validate two-dimensional interpretation.

Three-dimensional modelling of magnetic anomalies is complex. Probably the most convenient methods are to approximate the causative body by a cluster of right rectangular prisms or by a series of horizontal slices of polygonal outline.

Because of the dipolar nature of magnetic anomalies, trial and error methods of indirect interpretation are difficult to perform manually since anomaly shape is not closely related to the geometry of the causative body. Consequently, the automatic methods of interpretation described in Section 6.10.3 are widely employed.

The continuation and filtering operations used in gravity interpretation and described in Section 6.11 are equally applicable to magnetic fields. A further processing operation that may be applied to magnetic anomalies is known as *reduction to the pole*, and involves the conversion of the anomalies into their equivalent form at the north magnetic pole (Baranov & Naudy 1964). This process usually simplifies the magnetic anomalies as the ambient field is then vertical and bodies with magnetizations which are solely induced produce anomalies that are axisymmetric. The existence of remanent magneti-

zation, however, commonly prevents reduction to the pole from producing the desired simplification in the resultant pattern of magnetic anomalies.

7.11 Potential field transformations

The formulae for the gravitational potential caused by a point mass and the magnetic potential due to an isolated pole were presented in equations (6.3) and (7.3). A consequence of the similar laws of attraction governing gravitating and magnetic bodies is that these two equations have the variable of inverse distance ($1/r$) in common. Elimination of this term between the two formulae provides a relationship between the gravitational and magnetic potentials known as *Poisson's equation*. In reality the relationship is more complex than implied by equations (6.3) and (7.3) as isolated magnetic poles do not exist. However, the validity of the relationship between the two potential fields remains. Since gravity or magnetic fields can be determined by differentiation of the relevant potential in the required direction, Poisson's equation provides a method of transforming magnetic fields into gravitational fields and *vice versa* for bodies in which the ratio of intensity of magnetization to density remains constant. Such transformed fields are known as *pseudogravitational* and *pseudomagnetic* fields (Garland 1951).

One application of this technique is the transformation of magnetic anomalies into pseudogravity anomalies for the purposes of indirect interpretation, as the latter are significantly easier to interpret than their magnetic counterpart. The method is even more powerful when the pseudofield is compared with a corresponding measured field. For example, the comparison of gravity anomalies with the pseudogravity anomalies derived from magnetic anomalies over the same area can show whether the same geological bodies are the cause of the two types of anomaly. Performing the transformation for

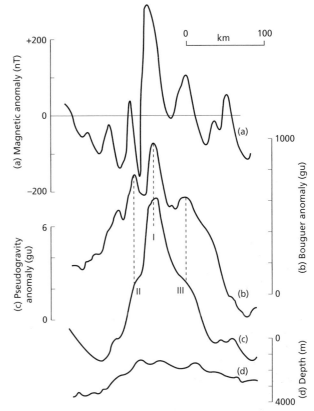

Fig. 7.22 (a) Observed magnetic anomalies over the Aves Ridge, eastern Caribbean. (b) Bouguer gravity anomalies with long-wavelength regional field removed. (c) Pseudo-gravity anomalies computed for induced magnetization and a density : magnetization ratio of unity. (d) Bathymetry.

different orientations of the magnetization vector provides an estimate of the true vector orientation since this will produce a pseudogravity field which most closely approximates the observed gravity field. The relative amplitudes of these two fields then provide a measure of the ratio of intensity of magnetization to density (Ates & Kearey 1995). These potential field transformations provide an elegant means of comparing gravity and magnetic anomalies over the same area and sometimes allow greater information to be derived about their causative bodies than would be possible if the techniques were treated in isolation. A computer program which performs pseudofield transformations is given in Gilbert and Galdeano (1985).

Figures 7.22(a) and (b) show magnetic and residual gravity anomaly profiles across the Aves Ridge, a submarine prominence in the eastern Caribbean which runs parallel to the island arc of the Lesser Antilles. The pseudogravity profile calculated from the magnetic profile assuming induced magnetization is presented in Fig. 7.22(c). It is readily apparent that the main pseudogra-

vity peak correlates with peak I on the gravity profile and that peaks II and III correlate with much weaker features on the pseudofield profile. The data thus suggest that the density features responsible for the gravity maxima are also magnetic, with the causative body of the central peak having a significantly greater susceptibility than the flanking bodies.

Figure 7.23 shows how a variety of processing methods can be used on a synthetic magnetic anomaly map and Fig. 7.24 shows their application to real data.

7.12 Applications of magnetic surveying

Magnetic surveying is a rapid and cost-effective technique and represents one of the most widely-used geophysical methods in terms of line length surveyed (Paterson & Reeves 1985).

Magnetic surveys are used extensively in the search for metalliferous mineral deposits, a task accomplished rapidly and economically by airborne methods.

Magnetic surveys are capable of locating massive sulphide deposits (Fig. 7.25), especially when used in conjunction with electromagnetic methods (see Section 9.12). However, the principal target of magnetic surveying is iron ore. The ratio of magnetite to haematite must be high for the ore to produce significant anomalies, as haematite is commonly non-magnetic (see Section 7.2). Figure 7.26 shows total field magnetic anomalies from an airborne survey of the Northern Middleback Range,

South Australia, in which it is seen that the haematitic ore bodies are not associated with the major anomalies. Figure 7.27 shows the results from an aeromagnetic survey of part of the Eyre Peninsula of South Australia which reveal the presence of a large anomaly elongated east–west. Subsequent ground traverses were performed over this anomaly using both magnetic and gravity methods (Fig. 7.28) and it was found that the magnetic and gravity profiles exhibit coincident highs. Subsequent drilling on these highs revealed the presence of a magnetite-bearing ore body at shallow depth with an iron content of about 30%.

Gunn (1998) has reported on the location of prospective areas for hydrocarbon deposits in Australia by aeromagnetic surveying, although it is probable that this application is only possible in quite specific environments.

In geotechnical and archaeological investigations, magnetic surveys may be used to delineate zones of faulting in bedrock and to locate buried metallic, man-made features such as pipelines, old mine workings and buildings. Figure 7.29 shows a total magnetic field contour map of the site of a proposed apartment block in Bristol, England. The area had been exploited for coal in the past and stability problems would arise from the presence of old shafts and buried workings (Clark 1986). Lined shafts of up to 2 m diameter were subsequently found beneath anomalies A and D, while other isolated anomalies such as B and C were known, or suspected, to be associated with buried metallic objects.

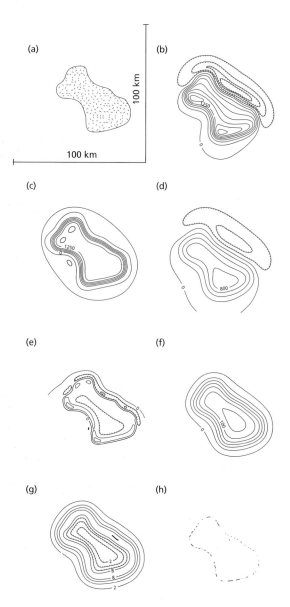

Fig. 7.23 The processing of aeromagnetic data. North direction is from bottom to top. (a) Source body with vertical sides, 2 km thick and a magnetization of 10 A m^{-1}, inclination 60° and declination 20°. (b) Total field magnetic anomaly of the body with induced magnetization measured on a horizontal surface 4 km above the body. Contour interval 250 nT. (c) Reduction to the pole of anomaly shown in (b). Contour interval 250 nT. (d) Anomaly shown in (b) upward continued 5 km above the measurement surface. Contour interval 200 nT. (e) Second vertical derivative of the anomaly shown in (b). Contour interval 50 nT km^{-2}. (f) Pseudogravity transform of anomaly shown in (b) assuming an intensity of magnetization of 1 A m^{-1} and a density contrast of 0.1 Mg m^{-3}. Contour interval 200 gu. (g) Magnitude of maximum horizontal gradient of the pseudogravity transform shown in (f). Contour interval 20 gu km^{-1}. (h) Locations of maxima of data shown in (g). Note correspondence with the actual edges of the source shown in (a). (Redrawn from Blakely & Connard 1989.)

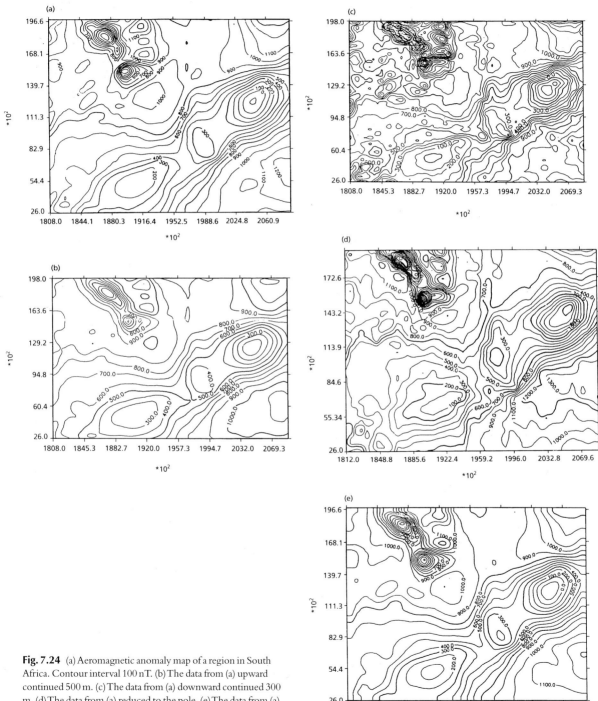

Fig. 7.24 (a) Aeromagnetic anomaly map of a region in South Africa. Contour interval 100 nT. (b) The data from (a) upward continued 500 m. (c) The data from (a) downward continued 300 m. (d) The data from (a) reduced to the pole. (e) The data from (a) low-pass filtered. Scales are in units of metres. (After Cooper 1997.)

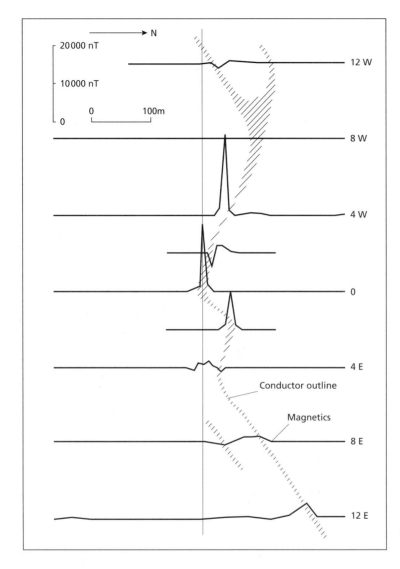

Fig. 7.25 Vertical field ground magnetic anomaly profiles over a massive sulphide ore body in Quebec, Canada. The shaded area represents the location of the ore body inferred from electromagnetic measurements. (After White 1966.)

In academic studies, magnetic surveys can be used in regional investigations of large-scale crustal features, although the sources of major magnetic anomalies tend to be restricted to rocks of basic or ultrabasic composition. Moreover, magnetic surveying is of limited use in the study of the deeper geology of the continental crust because the Curie isotherm for common ferrimagnetic minerals lies at a depth of about 20 km and the sources of major anomalies are consequently restricted to the upper part of the continental crust.

Although the contribution of magnetic surveying to knowledge of continental geology has been modest, magnetic surveying in oceanic areas has had a profound influence on the development of plate tectonic theory

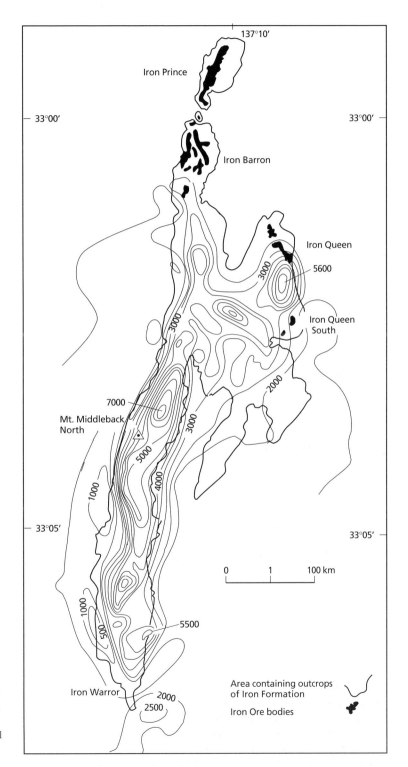

Fig. 7.26 Aeromagnetic anomalies over the Northern Middleback Range, South Australia. The iron ore bodies arc of haematite composition. Contour interval 500 nT. (After Webb 1966.)

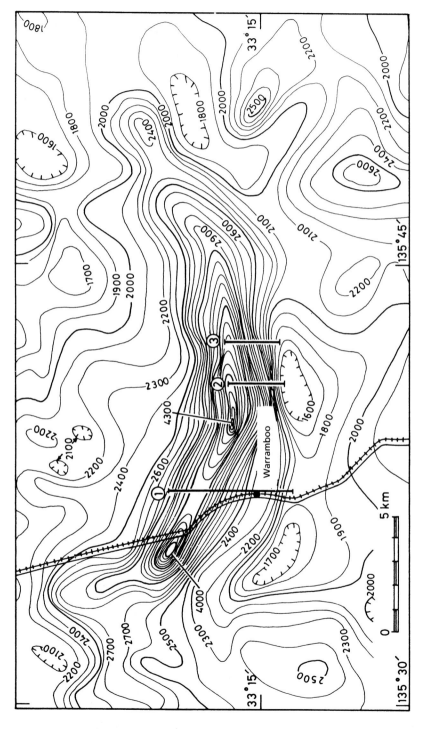

Fig. 7.27 High-level aeromagnetic anomalies over part of the Eyre Peninsula, South Australia. Contour interval 100 nT. (After Webb 1966.)

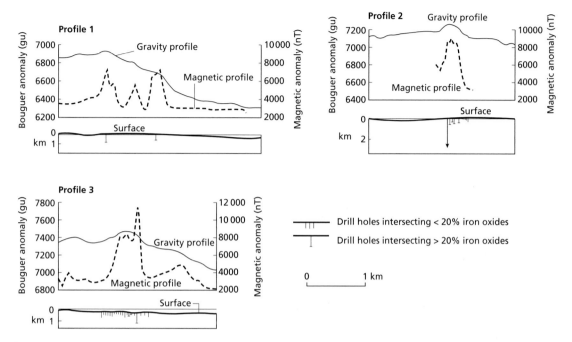

Fig. 7.28 Gravity and magnetic ground profiles over part of the Eyre Peninsula, South Australia, at the locations shown in Fig. 7.27 (After Webb 1966.)

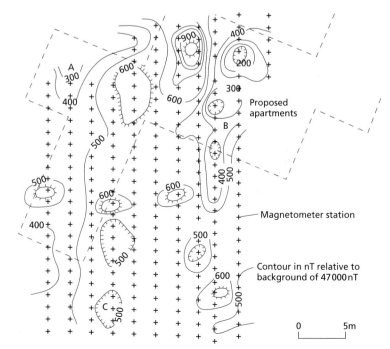

Fig. 7.29 Magnetic anomaly contour map of a site in Bristol, England. Contour interval 100 nT. (After Hooper & McDowell 1977.)

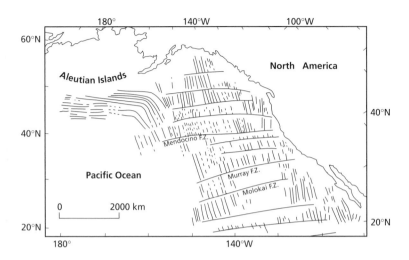

Fig. 7.30 Pattern of linear magnetic anomalies and major fracture zones in the northeast Pacific Ocean.

(Kearey & Vine 1996) and on views of the formation of oceanic lithosphere. Early magnetic surveying at sea showed that the oceanic crust is characterized by a pattern of linear magnetic anomalies (Fig. 7.30) attributable to strips of oceanic crust alternately magnetized in a normal and reverse direction (Mason & Raff 1961). The bilateral symmetry of these linear magnetic anomalies about oceanic ridges and rises (Vine & Matthews 1963) led directly to the theory of sea floor spreading and the establishment of a time scale for polarity transitions of the geomagnetic field (Heirtzler *et al.* 1968). Consequently, oceanic crust can be dated on the basis of the pattern of magnetic polarity transitions preserved in it.

Transform faults disrupt the pattern of linear magnetic anomalies (see Fig. 7.30) and their distribution can therefore be mapped magnetically. Since these faults lie along arcs of small circles to the prevailing pole of rotation at the time of transform fault movement, individual regimes of spreading during the evolution of an ocean basin can be identified by detailed magnetic surveying.

Such studies have been carried out in all the major oceans and show the evolution of an ocean basin to be a complex process involving several discrete phases of spreading, each with a distinct pole of rotation.

Magnetic surveying is a very useful aid to geological mapping. Over extensive regions with a thick sedimentary cover, structural features may be revealed if magnetic horizons such as ferruginous sandstones and shales, tuffs and lava flows are present within the sedimentary sequence. In the absence of magnetic sediments, magnetic survey data can provide information on the nature and form of the crystalline basement. Both cases are applicable to petroleum exploration in the location of structural traps within sediments or features of basement topography which might influence the overlying sedimentary sequence. The magnetic method may also be used to assist a programme of reconnaissance geological mapping based on widely-spaced grid samples, since aeromagnetic anomalies can be employed to delineate geological boundaries between sampling points.

Problems

1. Discuss the advantages and disadvantages of aeromagnetic surveying.
2. How and why do the methods of reduction of gravity and magnetic data differ?

3. Compare and contrast the techniques of interpretation of gravity and magnetic anomalies.
4. Assuming the magnetic moment of the Earth is $8 \times 10^{22}\,\mathrm{A\,m^2}$, its radius 6370 km and that its

Continued

magnetic field conforms to an axial dipole model, calculate the geomagnetic elements at 60°N and 75°S. Calculate also the total field magnetic gradients in $nT\,km^{-1}\,N$ at these latitudes.

5. Using equations (7.18a,b,c), derive expressions for the horizontal, vertical and total field magnetic anomalies of a vertical dyke of infinite depth striking at an angle α to magnetic north.

Given that geomagnetic inclination I is related to latitude θ by $\tan I = 2\tan\theta$, use these formulae to calculate the magnetic anomalies of east–west striking dykes of width 40 m, depth 20 m and intensity of magnetization $2\,A\,m^{-1}$, at a latitude of 45°, in the following cases:

(a) In the northern hemisphere with induced magnetization.

(b) In the northern hemisphere with reversed magnetization.

(c) In the southern hemisphere with normal magnetization.

(d) In the southern hemisphere with reversed magnetization.

How would the anomalies change if the width and depth were increased to 400 m and 200 m, respectively?

6. (a) Calculate the vertical, horizontal and total field magnetic anomaly profiles across a dipole which strikes in the direction of the magnetic meridian and dips to the south at 30° with the negative pole at the northern end 5 m beneath the surface. The length of the dipole is 50 m and the strength of each pole is 300 A m.

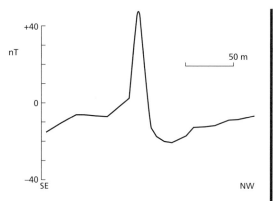

Fig. 7.31 Total field magnetic profile across buried volcanic rocks south of Bristol, England. (After Kearey & Allison 1980.)

The local geomagnetic field dips to the north at 70°.

(b) What is the effect on the profiles if the dipole strikes 25°E of the magnetic meridian?

(c) If the anomalies calculated in (a) actually originate from a cylinder whose magnetic moment is the same as the dipole and whose diameter is 10 m, calculate the intensity of magnetization of the cylinder.

(d) Fig. 7.31 shows a total field magnetic anomaly profile across buried volcanic rocks to the south of Bristol, England. Does the profile constructed in (a) represent a reasonable simulation of this anomaly? If so, calculate the dimensions and intensity of magnetization of a possible magnetic source. What other information would be needed to provide a more detailed interpretation of the anomaly?

Further reading

Arnaud Gerkens, J.C. d' (1989) *Foundations of Exploration Geophysics*. Elsevier, Amsterdam.

Baranov, W. (1975) *Potential Fields and Their Transformation in Applied Geophysics*. Gebrüder Borntraeger, Berlin.

Bott, M.H.P. (1973) Inverse methods in the interpretation of magnetic and gravity anomalies. *In*: Alder, B., Fernbach, S. & Bolt, B.A. (eds), *Methods in Computational Physics*, **13**, 133–62.

Garland, G.D. (1951) Combined analysis of gravity and magnetic anomalies *Geophysics*, **16**, 51–62.

Gibson, R.I. & Millegan, P.S. (eds) (1998) *Geologic Applications of Gravity and Magnetics: Case Histories*. SEG Reference Series 8 & AAPG Studies in Geology 43, Tulsa.

Gunn, P.J. (1975) Linear transformations of gravity and magnetic fields. *Geophys. Prosp.*, **23**, 300–12.

Kanasewich, E.R. & Agarwal, R.G. (1970) Analysis of combined gravity and magnetic fields in wave number domain. *J. Geophys. Res.*, **75**, 5702–12.

Nettleton, L.L. (1971) *Elementary Gravity and Magnetics for Geologists and Seismologists*. Monograph Series No. 1. Society of Exploration Geophysicists, Tulsa.

Nettleton, L.L. (1976) *Gravity and Magnetics in Oil Exploration.* McGraw-Hill, New York.

Sharma, P. (1976) *Geophysical Methods in Geology.* Elsevier, Amsterdam.

Stacey, F.D. & Banerjee, S.K. (1974) *The Physical Principles of Rock Magnetism.* Elsevier, Amsterdam.

Tarling, D.H. (1983) *Palaeomagnetism.* Chapman & Hall, London.

Vacquier, V., Steenland, N.C., Henderson, R.G. & Zeitz, I. (1951) Interpretation of aeromagnetic maps. *Geol. Soc. Am. Mem.,* **47**.

8 Electrical surveying

8.1 Introduction

There are many methods of electrical surveying. Some make use of fields within the Earth while others require the introduction of artificially-generated currents into the ground. The *resistivity* method is used in the study of horizontal and vertical discontinuities in the electrical properties of the ground, and also in the detection of three-dimensional bodies of anomalous electrical conductivity. It is routinely used in engineering and hydrogeological investigations to investigate the shallow subsurface geology. The *induced polarization* method makes use of the capacitive action of the subsurface to locate zones where conductive minerals are dissemin-ated within their host rocks. The *self-potential* method makes use of natural currents flowing in the ground that are generated by electrochemical processes to locate shallow bodies of anomalous conductivity.

Electrical methods utilize direct currents or low-frequency alternating currents to investigate the electrical properties of the subsurface, in contrast to the electromagnetic methods discussed in the next chapter that use alternating electromagnetic fields of higher frequency to this end.

8.2 Resistivity method

8.2.1 Introduction

In the resistivity method, artificially-generated electric currents are introduced into the ground and the resulting potential differences are measured at the surface. Deviations from the pattern of potential differences expected from homogeneous ground provide information on the form and electrical properties of subsurface inhomogeneities.

8.2.2 Resistivities of rocks and minerals

The *resistivity* of a material is defined as the resistance in ohms between the opposite faces of a unit cube of the material. For a conducting cylinder of resistance δR, length δL and cross-sectional area δA (Fig. 8.1) the resistivity ρ is given by

$$\rho = \frac{\delta R \delta A}{\delta L} \tag{8.1}$$

The SI unit of resistivity is the ohm-metre (ohm m) and the reciprocal of resistivity is termed *conductivity* (units: siemens (S) per metre; $1\,\mathrm{S\,m^{-1}} = 1\,\mathrm{ohm^{-1}\,m^{-1}}$; the term 'mho' for inverse ohm is sometimes encountered).

Resistivity is one of the most variable of physical properties. Certain minerals such as native metals and graphite conduct electricity via the passage of electrons. Most rock-forming minerals are, however, insulators, and electrical current is carried through a rock mainly by the passage of ions in pore waters. Thus, most rocks conduct electricity by electrolytic rather than electronic processes. It follows that porosity is the major control of

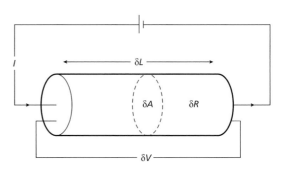

Fig. 8.1 The parameters used in defining resistivity.

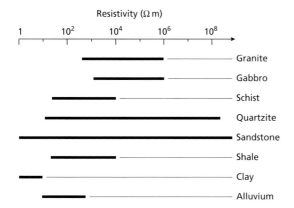

Fig. 8.2 The approximate range of resistivity values of common rock types.

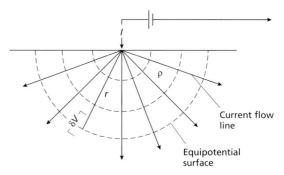

Fig. 8.3 Current flow from a single surface electrode.

the resistivity of rocks, and that resistivity generally increases as porosity decreases. However, even crystalline rocks with negligible intergranular porosity are conductive along cracks and fissures. Figure 8.2 shows the range of resistivities expected for common rock types. It is apparent that there is considerable overlap between different rock types and, consequently, identification of a rock type is not possible solely on the basis of resistivity data. Strictly, equation (8.1) refers to electronic conduction but it may still be used to describe the effective resistivity of a rock; that is, the resistivity of the rock and its pore water. The effective resistivity can also be expressed in terms of the resistivity and volume of the pore water present according to an empirical formula given by Archie (1942)

$$\rho = a\phi^{-b} f^{-c} \rho_w \tag{8.2}$$

where ϕ is the porosity, f the fraction of pores containing water of resistivity ρ_w and a, b and c are empirical constants. ρ_w can vary considerably according to the quantities and conductivities of dissolved materials.

8.2.3 Current flow in the ground

Consider the element of homogeneous material shown in Fig. 8.1. A current I is passed through the cylinder causing a potential drop $-\delta V$ between the ends of the element.

Ohm's law relates the current, potential difference and resistance such that $-\delta V = \delta R I$, and from equation (8.1) $\delta R = \rho \, \delta L / \delta A$. Substituting

$$\frac{\delta V}{\delta L} = -\frac{\rho I}{\delta A} = -\rho i \tag{8.3}$$

$\delta V/\delta L$ represents the potential gradient through the element in volt$\,$m^{-1} and i the current density in A$\,$m^{-2}. In general the current density in any direction within a material is given by the negative partial derivative of the potential in that direction divided by the resistivity.

Now consider a single current electrode on the surface of a medium of uniform resistivity ρ (Fig. 8.3). The circuit is completed by a current sink at a large distance from the electrode. Current flows radially away from the electrode so that the current distribution is uniform over hemispherical shells centred on the source. At a distance r from the electrode the shell has a surface area of $2\pi r^2$, so the current density i is given by

$$i = \frac{I}{2\pi r^2} \tag{8.4}$$

From equation (8.3), the potential gradient associated with this current density is

$$\frac{\partial V}{\partial r} = -\rho i = -\frac{\rho I}{2\pi r^2} \tag{8.5}$$

The potential V_r at distance r is then obtained by integration

$$V_r = \int \partial V = -\int \frac{\rho I \partial r}{2\pi r^2} = \frac{\rho I}{2\pi r} \tag{8.6}$$

The constant of integration is zero since $V_r = 0$ when $r = \infty$.

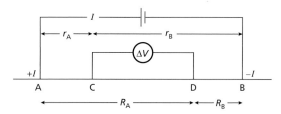

Fig. 8.4 The generalized form of the electrode configuration used in resistivity measurements.

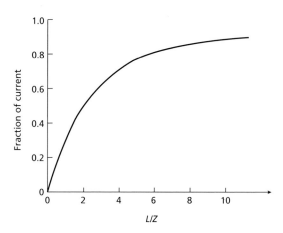

Fig. 8.5 The fraction of current penetrating below a depth Z for a current electrode separation L. (After Telford *et al.* 1990.)

Equation (8.6) allows the calculation of the potential at any point on or below the surface of a homogeneous half-space. The hemispherical shells in Fig. 8.3 mark surfaces of constant voltage and are termed *equipotential surfaces*.

Now consider the case where the current sink is a finite distance from the source (Fig. 8.4). The potential V_C at an internal electrode C is the sum of the potential contributions V_A and V_B from the current source at A and the sink at B

$$V_C = V_A + V_B$$

From equation (8.6)

$$V_C = \frac{\rho I}{2\pi}\left(\frac{1}{r_A} - \frac{1}{r_B}\right) \tag{8.7}$$

Similarly

$$V_D = \frac{\rho I}{2\pi}\left(\frac{1}{R_A} - \frac{1}{R_B}\right) \tag{8.8}$$

Absolute potentials are difficult to monitor so the potential difference ΔV between electrodes C and D is measured

$$\Delta V = V_C - V_D = \frac{\rho I}{2\pi}\left\{\left(\frac{1}{r_A} - \frac{1}{r_B}\right) - \left(\frac{1}{R_A} - \frac{1}{R_B}\right)\right\}$$

Thus

$$\rho = \frac{2\pi \Delta V}{I\left\{\left(\dfrac{1}{r_A} - \dfrac{1}{r_B}\right) - \left(\dfrac{1}{R_A} - \dfrac{1}{R_B}\right)\right\}} \tag{8.9}$$

Where the ground is uniform, the resistivity calculated from equation (8.9) should be constant and inde-

pendent of both electrode spacing and surface location. When subsurface inhomogeneities exist, however, the resistivity will vary with the relative positions of the electrodes. Any computed value is then known as the *apparent resistivity* ρ_a and will be a function of the form of the inhomogeneity. Equation (8.9) is the basic equation for calculating the apparent resistivity for any electrode configuration.

In homogeneous ground the depth of current penetration increases as the separation of the current electrodes is increased, and Fig. 8.5 shows the proportion of current flowing beneath a given depth Z as the ratio of electrode separation L to depth increases. When $L = Z$ about 30% of the current flows below Z and when $L = 2Z$ about 50% of the current flows below Z. The current electrode separation must be chosen so that the ground is energized to the required depth, and should be at least equal to this depth. This places practical limits on the depths of penetration attainable by normal resistivity methods due to the difficulty in laying long lengths of cable and the generation of sufficient power. Depths of penetration of about 1 km are the limit for normal equipment.

Two main types of procedure are employed in resistivity surveys.

Vertical electrical sounding (VES), also known as 'electrical drilling' or 'expanding probe', is used mainly in the study of horizontal or near-horizontal interfaces. The current and potential electrodes are maintained at the same relative spacing and the whole spread is progressively expanded about a fixed central point. Consequently, readings are taken as the current reaches progressively greater depths. The technique is extensively used in geotechnical surveys to determine overburden thickness and also in hydrogeology to define horizontal zones of porous strata.

Constant separation traversing (CST), also known as 'electrical profiling', is used to determine lateral variations of resistivity. The current and potential electrodes are maintained at a fixed separation and progressively moved along a profile. This method is employed in mineral prospecting to locate faults or shear zones and to detect localized bodies of anomalous conductivity. It is also used in geotechnical surveys to determine variations in bedrock depth and the presence of steep discontinuities. Results from a series of CST traverses with a fixed electrode spacing can be employed in the production of resistivity contour maps.

8.2.4 Electrode spreads

Many configurations of electrodes have been designed (Habberjam 1979) and, although several are occasionally employed in specialized surveys, only two are in common use. The *Wenner configuration* is the simpler in that current and potential electrodes are maintained at an equal spacing a (Fig. 8.6). Substitution of this condition into equation (8.9) yields

$$\rho_a = 2\pi a \frac{\Delta V}{I} \qquad (8.10)$$

During VES the spacing a is gradually increased about a fixed central point and in CST the whole spread is moved along a profile with a fixed value of a. The efficiency of performing vertical electrical sounding can be greatly increased by making use of a multicore cable to which a number of electrodes are permanently attached at standard separations (Barker 1981). A sounding can then be rapidly accomplished by switching between different sets of four electrodes. Such a system has the additional advantage that, by measuring ground resistances at two electrode array positions, the effects of near-surface lateral resistivity variations can be substantially reduced.

In surveying with the Wenner configuration all four electrodes need to be moved between successive readings. This labour is partially overcome by the use of the *Schlumberger configuration* (Fig. 8.6) in which the inner, potential electrodes have a spacing 2l which is a small

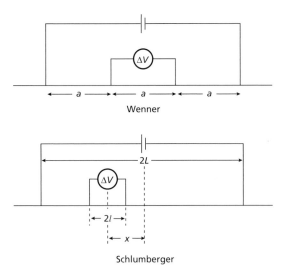

Fig. 8.6 The Wenner and Schlumberger electrode configurations.

proportion of that of the outer, current electrodes (2L). In CST surveys with the Schlumberger configuration several lateral movements of the potential electrodes may be accommodated without the necessity of moving the current electrodes. In VES surveys the potential electrodes remain fixed and the current electrodes are expanded symmetrically about the centre of the spread. With very large values of L it may, however, be necessary to increase l also in order to maintain a measurable potential.

For the Schlumberger configuration

$$\rho_a = \frac{\pi}{2l} \frac{(L^2 - x^2)^2}{(L^2 + x^2)} \frac{\Delta V}{I} \qquad (8.11)$$

where x is the separation of the mid-points of the potential and current electrodes. When used symmetrically, x = 0, so

$$\rho_a = \frac{\pi L^2}{2l} \frac{\Delta V}{I} \qquad (8.12)$$

8.2.5 Resistivity surveying equipment

Resistivity survey instruments are designed to measure the resistance of the ground, that is, the ratio $(\Delta V/I)$ in equations (8.10), (8.11) and (8.12), to a very high accuracy. They must be capable of reading to the very low

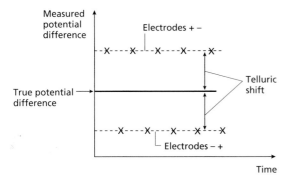

Fig. 8.7 The use of alternating current to remove the effects of telluric currents during a resistivity measurement. Summing the measured potential difference over several cycles provides the true potential difference.

levels of resistance commonly encountered in resistivity surveying. Apparent resistivity values are computed from the resistance measurements using the formula relevant to the electrode configuration in use.

Most resistivity meters employ low-frequency alternating current rather than direct current, for two main reasons. Firstly, if direct current were employed there would eventually be a build-up of anions around the negative electrode and cations around the positive electrode; that is, electrolytic polarization would occur, and this would inhibit the arrival of further ions at the electrodes. Periodic reversal of the current prevents such an accumulation of ions and thus overcomes electrolytic polarization. Secondly, the use of alternating current overcomes the effects of telluric currents (see Chapter 9), which are natural electric currents in the ground that flow parallel to the Earth's surface and cause regional potential gradients. The use of alternating current nullifies their effects since at each current reversal the telluric currents alternately increase or decrease the measured potential difference by equal amounts. Summing the results over several cycles thus removes telluric effects (Fig. 8.7). The frequency of the alternating current used in resistivity surveying depends upon the required depth of penetration (see equation (9.2)). For penetration of the order of 10 m, a frequency of 100 Hz is suitable, and this is decreased to less than 10 Hz for depths of investigation of about 100 m. For very deep ground penetration direct currents must be used, and more complex measures adopted to overcome electrolytic polarization and telluric current effects. Many modern instruments make use of a square wave current input to overcome the polarization.

Resistivity meters are designed to measure potential differences when no current is flowing. Such a null method is used to overcome the effects of contact resistance of the electrodes with the ground. The potential between the potential electrodes is balanced by the potential tapped from a variable resistance. No current then flows in the resistivity circuit so that contact resistance will not register, and the variable resistance reading represents the true resistance of the ground (equal to the ratio $\Delta V/I$ in the relevant equations).

Previous generations of resistivity meters required the nulling of a displayed voltage by manual manipulation of a resistor bank. Modern instruments have microprocessor-controlled electronic circuitry which accomplishes this operation internally and, moreover, performs checks on the circuitry before display of the result.

Resistivity surveying for shallow penetration can be made more efficient by the use of spike electrodes which are mounted on small wheels and towed along a profile by the operator. Improvements in instrument technology have also led to the development of electrodes in the form of antennae which are capacitively coupled to the ground (Panissod *et al.* 1998), so that there is no need for spike electrodes to be placed in the ground and a CST may be accomplished by an operator towing the array at a walking pace by foot or vehicle. Measurements can be taken automatically and are no longer restricted to areas where electrodes can be in-serted, such as road metal, ice, permafrost, etc. Such a system allows the collection, by a single operator, of 500% more data in the same time as a conventional instrument with a crew of two. However, the limitations of the physical dimensions of such equipment considerably restricts penetration.

8.2.6 Interpretation of resistivity data

Electrical surveys are among the most difficult of all the geophysical methods to interpret quantitatively because of the complex theoretical basis of the technique. In resistivity interpretation, mathematical analysis is most highly developed for VES, less well for CST over two-dimensional structures and least well for CST over three-dimensional bodies. The resistivity method utilizes a potential field and consequently suffers from similar ambiguity problems to the gravitational and magnetic methods.

Since a potential field is involved, the apparent resistivity signature of any structure should be computed by

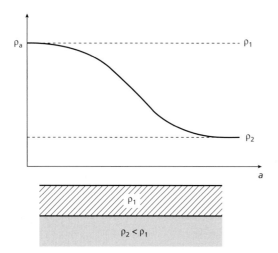

Fig. 8.8 The variation of apparent resistivity ρ_a with electrode separation a over a single horizontal interface between media with resistivities ρ_1 and ρ_2.

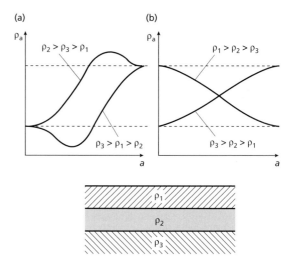

Fig. 8.9 The variation of apparent resistivity ρ_a, with electrode separation a over three horizontal layers.

solution of Laplace's equation (Section 6.11) and insertion of the boundary conditions for the particular structure under consideration, or by integrating it directly. In practice such solutions are invariably complex. Consequently, a simplified approach is initially adopted here in which electric fields are assumed to act in a manner similar to light. It should be remembered, however, that such an optical analogue is not strictly valid in all cases.

8.2.7 Vertical electrical sounding interpretation

Consider a Wenner electrode spread above a single horizontal interface between media with resistivities ρ_1 (upper) and ρ_2 (lower) with $\rho_1 > \rho_2$ (Fig. 8.8). On passing through the interface the current flow lines are deflected towards the interface in a fashion similar to refracted seismic waves (Chapter 3) since the less resistive lower layer provides a more attractive path for the current. When the electrode separation is small, most of the current flows in the upper layer with the consequence that the apparent resistivity tends towards ρ_1. As the electrode separation is gradually increased, more and more current flows within the lower layer and the apparent resistivity then approaches ρ_2. A similar situation obtains when $\rho_2 > \rho_1$, although in this case the apparent resistivity approaches ρ_2 more gradually as the more resistive lower layer is a less attractive path for the current.

Where three horizontal layers are present the apparent resistivity curves are more complex (Fig. 8.9). Although the apparent resistivity approaches ρ_1 and ρ_3 for small and

large electrode spacings, the presence of the intermediate layer causes a deflection of the apparent resistivity curve at intermediate spacings. If the resistivity of the intermediate layer is greater or less than the resistivities of the upper and lower layers the apparent resistivity curve is either bell-shaped or basin-shaped (Fig. 8.9(a)). A middle layer with a resistivity intermediate between ρ_1 and ρ_3 produces apparent resistivity curves characterized by a progressive increase or decrease in resistivity as a function of electrode spacing (Fig. 8.9(b)). The presence of four or more layers further increases the complexity of apparent resistivity curves.

Simple examination of the way in which apparent resistivity varies with electrode spacing may thus provide estimates of the resistivities of the upper and lower layers and indicate the relative resistivities of any intermediate layers. In order to compute layer thicknesses it is necessary to be able to calculate the apparent resistivity of a layered structure. The first computation of this type was performed by Hummel in the 1930s using an optical analogue to calculate the apparent resistivity signature of a simple two-layered model.

Referring to Fig. 8.10, current I is introduced into the ground at point C_0 above a single interface at depth z between an upper medium 1 of resistivity ρ_1 and a lower medium 2 of resistivity ρ_2. The two parallel interfaces between media 1 and 2 and between medium 1 and the air produce an infinite series of images of the source, located above and below the surface. Thus C_1 is the image of C_0 in the medium 1/2 interface at depth $2z$,

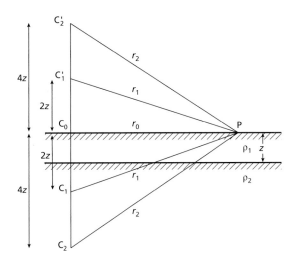

Fig. 8.10 Parameters used in the calculation of the potential due to a single surface electrode above a single horizontal interface using the method of images.

Table 8.1 Distribution and intensity of electrical sources due to a single horizontal interface.

Source	Intensity	Depth/height	Distance
C_0	I	0	r_0
C_1	kI	$2z$	r_1
C_1'	kI	$2z$	r_1
C_2	k^2I	$4z$	r_2
C_2'	k^2I	$4z$	r_2
etc			

C_2 is the image of C_1' in the medium 1/air interface at height $2z$, C_2 is the image of C_1' in the medium 1/2 interface at depth $4z$, etc. Each image in the medium 1/2 interface is reduced in intensity by a factor k, the reflection coefficient of the interface. (There is no reduction in intensity of images in the medium 1/air interface, as its reflection coefficient is unity.) A consequence of the progressive reduction in intensity is that only a few images have to be considered in arriving at a reasonable estimate of the potential at point P. Table 8.1 summarizes this argument.

The potential V_P at point P is the sum of the contributions of all sources. Employing equation (8.6)

$$V_P = \frac{I\rho_1}{2\pi r_0} + \frac{2k I\rho_1}{2\pi r_1} + \frac{2k^2 I\rho_1}{2\pi r_2} + \ldots + \frac{2k^i \rho_1}{2\pi r_i} + \ldots$$

Thus

$$V_P = \frac{I\rho_1}{2\pi}\left(\frac{1}{r_0} + 2\sum_{n=1}^{\infty}\frac{k^n}{r_n}\right) \tag{8.13}$$

where

$$r_n = \sqrt{r_0^2 + (2nz)^2}$$

The first term in the brackets of equation (8.13) refers to the normal potential pertaining if the subsurface were homogeneous, and the second term to the disturbing potential caused by the interface. The series is convergent as the dimming factor, or reflection coefficient, k is less than unity ($k = (\rho_2 - \rho_1)/(\rho_2 + \rho_1)$, cf. Section 3.6.1).

Knowledge of the potential resulting at a single point from a single current electrode allows the computation of the potential difference ΔV between two electrodes, resulting from two current electrodes, by the addition and subtraction of their contribution to the potential at these points. For the Wenner system with spacing a

$$\Delta V = \frac{I\rho_1}{2\pi a}(1 + 4F) \tag{8.14}$$

where

$$F = \sum_{n=1}^{\infty} k^n\left(\frac{1}{\sqrt{1 + 4n^2z^2/a^2}} - \frac{1}{\sqrt{4 + 4n^2z^2/a^2}}\right) \tag{8.15}$$

Relating this to the apparent resistivity ρ_a measured by the Wenner system (equation (8.10))

$$\rho_a = \rho_1(1 + 4F) \tag{8.16}$$

Consequently the apparent resistivity can be computed for a range of electrode spacings.

Similar computations can be performed for multi-layer structures, although the calculations are more easily executed using recurrence formulae and filtering techniques designed for this purpose (see later). Field data can then be compared with graphs (master curves) representing the calculated effects of layered models derived by such methods, a once-important but now little-used technique known as *curve matching*. Figure 8.11 shows an interpretation using a set of master curves for vertical electrical sounding with a Wenner spread over two horizontal layers. The master curves are prepared in

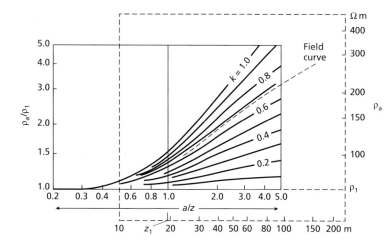

Fig. 8.11 The interpretation of a two-layer apparent resistivity graph by comparison with a set of master curves. The upper layer resistivity ρ_1 is $68\,\Omega\,\text{m}$ and its thickness z_1 is $19.5\,\text{m}$. (After Griffiths & King 1981.)

dimensionless form for a number of values of the reflection coefficient k by dividing the calculated apparent resistivity values ρ_a by the upper layer resistivity ρ_1 (the latter derived from the field curve at electrode spacings approaching zero), and by dividing the electrode spacings a by the upper layer thickness z_1. The curves are plotted on logarithmic paper, which has the effect of producing a more regular appearance as the fluctuations of resistivity then tend to be of similar wavelength over the entire length of the curves. The field curve to be interpreted is plotted on transparent logarithmic paper with the same modulus as the master curves. It is then shifted over the master curves, keeping the coordinate axes parallel, until a reasonable match is obtained with one of the master curves or with an interpolated curve. The point at which $\rho_a/\rho_1 = a/z = 1$ on the master sheet gives the true values of ρ_1 and z_1 on the relevant axes. ρ_2 is obtained from the k-value of the best-fitting curve.

Curve matching is simple for the two-layer case since only a single sheet of master curves is required. When three layers are present much larger sets of curves are required to represent the increased number of possible combinations of resistivities and layer thicknesses. Curve matching is simplified if the master curves are arranged according to curve type (Fig. 8.9), and sets of master curves for both Wenner and Schlumberger electrode configurations are available (Orellana & Mooney 1966, 1972). The number of master curves required for full interpretation of a four-layer field curve is prohibitively large although limited sets have been published.

The interpretation of resistivity curves over multilayered structures may alternatively be performed by *partial curve matching* (Bhattacharya & Patra 1968). The method

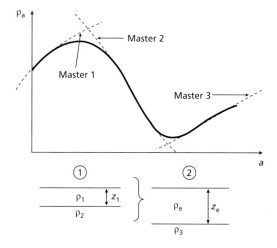

Fig. 8.12 The technique of partial curve matching. A two-layer curve is fitted to the early part of the graph and the resistivities ρ_1 and ρ_2 and thickness z_1 of the upper layer determined. ρ_1, ρ_2 and z_1 are combined into a single equivalent layer of resistivity ρ_e and thickness z_e, which then forms the upper layer in the interpretation of the next segment of the graph with a second two-layer curve.

involves the matching of successive portions of the field curve by a set of two-layer curves. After each segment is fitted the interpreted resistivities and layer thickness are combined by use of auxiliary curves into a single layer with an equivalent thickness z_e and resistivity ρ_e. This equivalent layer then forms the upper layer in the interpretation of the next segment of the field curve with another two-layer curve (Fig. 8.12). Similar techniques are available in which successive use is made of three-layer master curves.

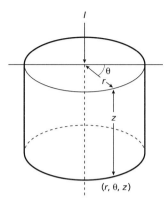

Fig. 8.13 Cylindrical polar coordinates.

The curve-matching methods have been almost completely superseded by more sophisticated interpretational techniques described below. Curve-matching methods might still be used, however, to obtain interpretations in the field in the absence of computing facilities, or to derive an approximate model that is to be used as a starting point for one of the more complex routines.

Equation (8.13) represents the potential at the surface resulting from a single point of current injection over two horizontal layers as predicted by the method of images. In general, however, the potential arising from any number of horizontal layers is derived by solution of Laplace's equation (see Section 6.11). The equation in this case is normally represented in cylindrical coordinates as electrical fields have cylindrical symmetry with respect to the vertical line through the current source (Fig. 8.13). The solution and application of the relevant boundary conditions are complex (e.g. Koefoed 1979), but show that the potential V at the surface over a series of horizontal layers, the uppermost of which has a resistivity ρ_1, at a distance r from a current source of strength I is given by

$$V = \frac{\rho_1 I}{2\pi} \int_0^\infty K(\lambda) J_0(\lambda r)\, d\lambda \qquad (8.17)$$

λ is the variable of integration. $J_0(\lambda r)$ is a specialized function known as a Bessel function of order zero whose behaviour is known completely. $K(\lambda)$ is known as a kernel function and is controlled by the thicknesses and resistivities of the underlying layers. The kernel function can be built up relatively simply for any number of layers using *recurrence relationships* (Koefoed 1979) which progressively add the effects of successive layers in the se-

quence. A useful additional parameter is the resistivity transform $T(\lambda)$ defined by

$$T_i(\lambda) = \rho_i K_i(\lambda) \qquad (8.18)$$

where $T_i(\lambda)$ is the resistivity transform of the ith layer which has a resistivity ρ_i and a kernel function $K_i(\lambda)$. $T(\lambda)$ can similarly be constructed using recurrence relationships.

By methods analogous to those used to construct equation (8.16), a relationship between the apparent resistivity and resistivity transform can be derived. For example, this relationship for a Wenner spread with electrode spacing a is

$$\rho_a = 2a \int_0^\infty T(\lambda)[J_0(\lambda a) - J_0(2\lambda a)]\, d\lambda \qquad (8.19)$$

The resistivity transform function has the dimensions of resistivity and the variable λ has the dimensions of inverse length. It has been found that if $T(\lambda)$ is plotted as a function of λ^{-1} the relationship is similar to the variation of apparent resistivity with electrode spacing for the same sequence of horizontal layers. Indeed only a simple filtering operation is required to transform the $T(\lambda)$ vs. λ^{-1} relationship (resistivity transform) into the ρ_a vs. a relationship (apparent resistivity function). Such a filter is known as an indirect filter. The inverse operation, that is, the determination of the resistivity transform from the apparent resistivity function, can be performed using a direct filter.

Apparent resistivity curves over multilayered models can be computed relatively easily by determining the resistivity transform from the layer parameters using a recurrence relationship and then filtering the transform to derive the apparent resistivity function. Such a technique is considerably more efficient than the method used in the derivation of equation (8.13).

This method leads to a form of interpretation similar to the indirect interpretation of gravity and magnetic anomalies, in which field data are compared with data calculated for a model whose parameters are varied in order to simulate the field observations. This comparison can be made between either observed and calculated apparent resistivity profiles or the equivalent resistivity transforms, the latter method requiring the derivation of the resistivity transform from the field resistivity data by direct filtering. Such techniques lend themselves well to automatic iterative processes of interpretation in which a computer performs the adjustments necessary to a layered model derived by an approximate interpretation method in order to improve the correspondence between observed and calculated functions.

In addition to this indirect modelling there are also a number of direct methods of interpreting resistivity data which derive the layer parameters directly from the field profiles (e.g. Zohdy 1989). Such methods usually involve the following steps:

1. Determination of the resistivity transform of the field data by direct filtering.

2. Determination of the parameters of the upper layer by fitting the early part of the resistivity transform curve with a synthetic two-layer curve.

3. Subtraction of the effects of the upper layer by reducing all observations to the base of the previously determined layer by the use of a *reduction equation* (the inverse of a recurrence relationship).

Steps **2** and **3** are then repeated so that the parameters of successively deeper layers are determined. Such methods suffer from the drawback that errors increase with depth so that any error made early in the interpretation becomes magnified. The direct interpretation methods consequently employ various techniques to suppress such error magnification.

The indirect and direct methods described above have now largely superseded curve-matching techniques and provide considerably more accurate interpretations.

Interpretation of VES data suffers from non-uniqueness arising from problems known as *equivalence* and *suppression*. The problem of equivalence (see e.g. van Overmeeren 1989) is illustrated by the fact that identical bell-shaped or basin-shaped resistivity curves (Fig. 8.9(a)) can be obtained for different layered models. Identical bell-shaped curves are obtained if the product of the thickness z and resistivity ρ, known as the transverse resistance, of the middle layer remains constant. For basin-shaped curves the equivalence function of the middle layer is z/ρ, known as the longitudinal conductance. The problem of suppression applies to resistivity curves in which apparent resistivity progressively increases or decreases as a function of electrode spacing (Fig. 8.9(b)). In such cases the addition of an extra intermediate layer causes a slight horizontal shift of the curve without altering its overall shape. In the interpretation of relatively noisy field data such an intermediate layer may not be detected.

It is the conventional practice in VES interpretation to make the assumption that layers are horizontal and isotropic. Deviations from these assumptions result in errors in the final interpretation.

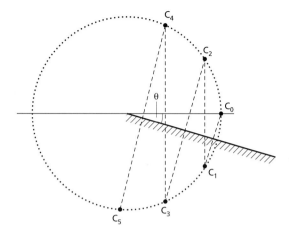

Fig. 8.14 Apparent current sources caused by a dipping interface. The sources C_1–C_5 are successive images of the primary source C_0 in the interface and the surface. The sources lie on a circle centred on the outcrop of the interface, and their number is dependent upon the magnitude of the dip of the interface, θ.

The assumption of isotropy can be incorrect for individual layers. For example, in sediments such as clay or shale the resistivity perpendicular to the layering is usually greater than parallel to the direction of the layering. Anisotropy cannot be detected in subsurface layers during vertical electrical sounding and normally results in too large a thickness being assigned to the layers. Other anisotropic effects are depth-dependent, for example the reduction with depth of the degree of weathering, and the increase with depth of both compaction of sediments and salinity of pore fluids. The presence of a vertical contact, such as a fault, gives rise to lateral inhomogeneity which can greatly affect the interpretation of an electrical sounding in its vicinity.

If the layers are dipping, the basic theory discussed above is invalid. Using the optical analogue, the number of images produced by a dipping interface is finite, the images being arranged around a circle (Fig. 8.14). Because the intensity of the images progressively decreases, only the first few need to be considered in deriving a reasonable estimate of the resulting potential. Consequently, the effect of dip can probably be ignored for inclinations up to about 20°, which provide a sufficient number of images.

Topography can influence electrical surveys as current flow lines tend to follow the ground surface. Equipotential surfaces are thus distorted and anomalous readings can result.

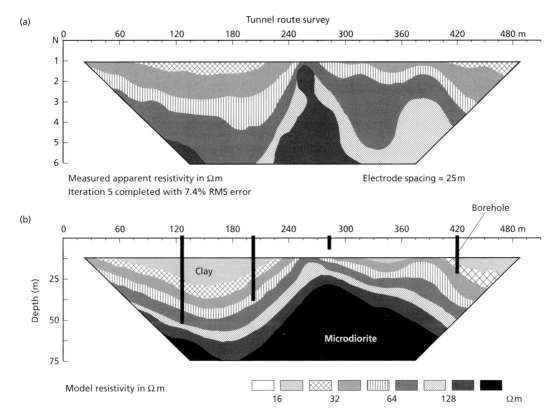

Fig. 8.15 (a) Contoured apparent resistivity pseudosection measured along the route of a proposed tunnel. (b) Electrical image and depths to bedrock determined in four boreholes. (After Barker 1997.)

VES data from several soundings can be presented in the form of a pseudosection (Section 8.3.3) and it is now possible to invert the data into a full, two-dimensional geoelectric model (e.g. Loke & Barker 1995, 1996) rather than a sequence of discrete, unidimensional geoelectric sections. This technique is known as electrical imaging or electrical tomography. An example of electrical imaging illustrating how a pseudosection can be transformed into a geoelectric structure is given in Fig. 8.15. Cross-borehole tomography can also be undertaken (Daily & Owen 1991).

If electrode spreads are arranged in parallel, many 2D pseudosections can be determined that can be combined into a 3D model.

8.2.8 Constant separation traversing interpretation

Constant separation traverses are obtained by moving an electrode spread with fixed electrode separation along a traverse line, the array of electrodes being aligned either in the direction of the traverse (longitudinal traverse) or at right angles to it (transverse traverse). The former technique is more efficient as only a single electrode has to be moved from one end of the spread to the other, and the electrodes reconnected, between adjacent readings.

Figure 8.16(a) shows a transverse traverse across a single vertical contact between two media of resistivities ρ_1 and ρ_2. The apparent resistivity curve varies smoothly from ρ_1 to ρ_2 across the contact.

A longitudinal traverse over a similar structure shows the same variation from ρ_1 to ρ_2 at its extremities, but the intermediate parts of the curve exhibit a number of cusps (Fig. 8.16(b)), which correspond to locations where successive electrodes cross the contact. There will be four cusps on a Wenner profile but two on a Schlumberger profile where only the potential electrodes are mobile.

Figure 8.17 shows the results of transverse and

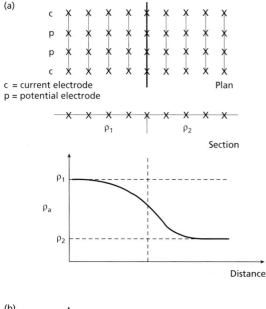

(a)

c = current electrode
p = potential electrode

Plan

Section

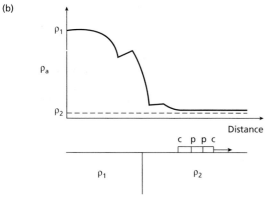

(b)

Distance

Fig. 8.16 (a) A transverse traverse across a single vertical interface. (b) A longitudinal traverse across a single vertical interface employing a configuration in which all four electrodes are mobile. (After Parasnis 1973.)

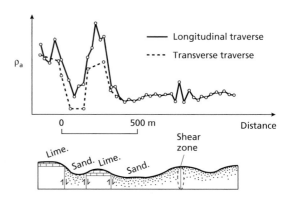

Fig. 8.17 Longitudinal and transverse traverses across a series of faulted strata in Illinois, USA. (After Hubbert 1934.)

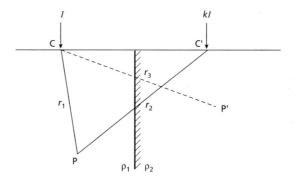

Fig. 8.18 Parameters used in the calculation of the potential due to a single surface current electrode on either side of a single vertical interface.

longitudinal traversing across a series of faulted strata in Illinois, USA. Both sets of results illustrate well the strong resistivity contrasts between the relatively conductive sandstone and relatively resistive limestone.

A vertical discontinuity distorts the direction of current flow and thus the overall distribution of potential in its vicinity. The potential distribution at the surface can be determined by an optical analogue in which the discontinuity is compared with a semitransparent mirror which both reflects and transmits light. Referring to Fig. 8.18, current I is introduced at point C on the surface of a medium of resistivity ρ_1 in the vicinity of a vertical contact with a second medium of resistivity ρ_2.

In the optical analogue, a point P on the same side of the mirror as the source would receive light directly and via a single reflection. In the latter case the light would appear to originate from the image of C in the mirror, C′, and would be decreased in intensity with respect to the source by a factor corresponding to the reflection coefficient. Both the electric source and its image contribute to the potential V_p at P, the latter being decreased in intensity by a factor k, the reflection coefficient. From equation (8.6)

$$V_P = \frac{I\rho_1}{2\pi}\left(\frac{1}{r_1} + \frac{k}{r_2}\right) \tag{8.20}$$

For a point P′ on the other side of the interface from the

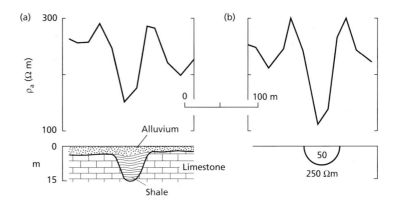

Fig. 8.19 (a) The observed Wenner resistivity profile over a shale-filled sink of known geometry in Kansas, USA. (b) The theoretical profile for a buried hemisphere. (After Cook & Van Nostrand 1954.)

source, the optical analogue indicates that light would be received only after transmission through the mirror, resulting in a reduction in intensity by a factor corresponding to the transmission coefficient. The only contributor to the potential $V_{p'}$ at P′ is the current source reduced in intensity by the factor $(1 − k)$. From equation (8.6)

$$V_{p'} = \frac{I(1-k)\rho_2}{2\pi r_3} \tag{8.21}$$

Equations (8.20) and (8.21) may be used to calculate the measured potential difference for any electrode spread between two points in the vicinity of the interface and thus to construct the form of an apparent resistivity profile produced by longitudinal constant separation traversing. In fact, five separate equations are required, corresponding to the five possible configurations of a four-electrode spread with respect to the discontinuity. The method can also be used to construct apparent resistivity profiles for constant separation traversing over a number of adjacent discontinuities. Albums of master curves are available for single and double vertical contacts (Logn 1954).

Three-dimensional resistivity anomalies may be obtained by contouring apparent resistivity values from a number of CST lines. The detection of a three-dimensional body is usually only possible when its top is close to the surface, and traverses must be made directly over the body or very near to its edges if its anomaly is to be registered.

Three-dimensional anomalies may be interpreted by laboratory modelling. For example, metal cylinders, blocks or sheets may be immersed in water whose resistivity is altered by adding various salts and the model moved beneath a set of stationary electrodes. The shape

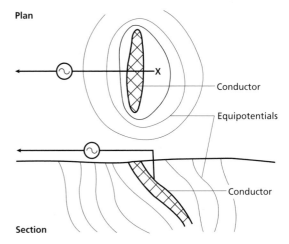

Fig. 8.20 The mise-à-la-masse method.

of the model can then be varied until a reasonable approximation to the field curves is obtained.

The mathematical analysis of apparent resistivity variations over bodies of regular or irregular form is complex but equations are available for simple shapes such as spheres or hemispheres (Fig. 8.19), and it is also possible to compute the resistivity response of two-dimensional bodies with an irregular cross-section (Dey & Morrison 1979).

Three-dimensional anomalies may also be obtained by an extension of the CST technique known as the *mise-à-la-masse method*. This is employed when part of a conductive body, for example an ore body, has been located either at outcrop or by drilling. One current electrode is sited within the body, the other being placed a large distance away on the surface (Fig. 8.20). A pair of potential electrodes is then moved over the surface mapping equipotential lines (lines joining the electrodes when the indicated potential difference is zero). The method provides much more information on the extent, dip, strike and continuity of the body than the normal CST techniques. An example of the delineation of a massive sulphide body by the mise-à-la-masse method is given in Bowker (1991).

8.2.9 Limitations of the resistivity method

Resistivity surveying is an efficient method for delineating shallow layered sequences or vertical discontinuities involving changes of resistivity. It does, however, suffer from a number of limitations:

1. Interpretations are ambiguous. Consequently, independent geophysical and geological controls are necessary to discriminate between valid alternative interpretations of the resistivity data.

2. Interpretation is limited to simple structural configurations. Any deviations from these simple situations may be impossible to interpret.

3. Topography and the effects of near-surface resistivity variations can mask the effects of deeper variations.

4. The depth of penetration of the method is limited by the maximum electrical power that can be introduced into the ground and by the physical difficulties of laying out long lengths of cable. The practical depth limit for most surveys is about 1 km.

8.2.10 Applications of resistivity surveying

Resistivity surveys are usually restricted to relatively small-scale investigations because of the labour involved in physically planting the electrodes prior to each measurement. For this reason resistivity methods are not commonly used in reconnaissance exploration. It is probable, however, that with the increasing availability of non-contacting conductivity measuring devices (see Section 9.7) this restriction will no longer apply.

Resistivity methods are widely used in engineering geological investigations of sites prior to construction (Barker 1997). VES is a very convenient, non-destructive method of determining the depth to rockhead for foundation purposes and also provides information on the degree of saturation of subsurface materials. CST can be used to determine the variation in rockhead depth between soundings and can also indicate the presence of potentially unstable ground conditions. Figure 8.21 shows a CST profile which has revealed the presence of a buried mineshaft from the relatively high resistivity val-

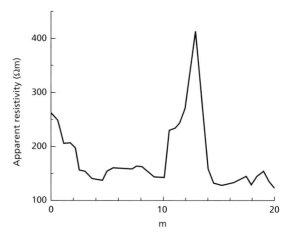

Fig. 8.21 CST resistivity profile across a buried mineshaft. (After Aspinall & Walker 1975.)

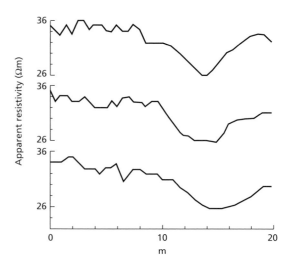

Fig. 8.22 Resistivity profiles across a buried ditch 4 m wide. (After Aspinall & Walker 1975.)

ues associated with its poorly-compacted infill. Similar techniques can be used in archaeological investigations for the location of artefacts with anomalous resistivities. For example, Fig. 8.22 shows CST profiles across an ancient buried ditch.

Probably the most widely-employed use of resistivity surveys is in hydrogeological investigations, as important information can be provided on geological structure, lithologies and subsurface water resources without the large cost of an extensive programme of drilling. The results can determine the locations of the minimum

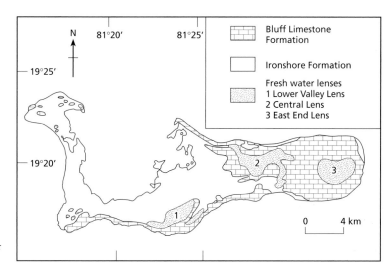

Fig. 8.23 Simplified geology and freshwater lenses of Grand Cayman. (After Bugg & Lloyd 1976.)

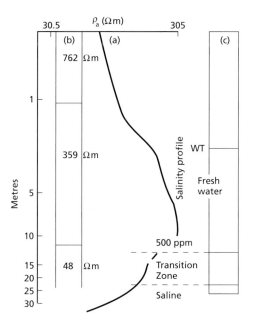

Fig. 8.24 (a) Vertical electrical sounding adjacent to a test borehole in the Central Lens, Grand Cayman. (b) Layered model interpretation of the VES. (c) Interpreted salinity profile. (After Bugg & Lloyd 1976.)

number of exploratory boreholes required for both essential aquifer tests and control of the geophysical interpretation.

The resistivity method was used by Bugg and Lloyd (1976) to delineate freshwater lenses in Grand Cayman Island off the northern Caribbean (Fig. 8.23). Because of its relatively low density, fresh water tends to float on the denser saline water which penetrates the limestone substrate of the island from the sea. Figure 8.24 shows a fluid conductivity profile from a borehole sunk in the Central Lens compared with the results of a VES interpretation from a sounding adjacent to the borehole. It is apparent that fresh water can be distinguished from saline water by its much higher resistivity. The resistivity survey took the form of a series of VES which were interpreted using the sounding by the borehole as control. Contours on the base of the Central Lens, defined from these interpretations, are shown in Fig. 8.25. A similar investigation using resistivity to investigate the intrusion of saline water into a coastal aquifer is given in Gondwe (1991).

Resistivity surveys can also be used to locate and monitor the extent of groundwater pollution. Merkel (1972) described the use of this technique in the delineation of contaminated mine drainage from old coal workings in Pennsylvania, USA. Figure 8.26 shows a geoelectric section across part of the area, constructed from a series of VES, and its geological interpretation which indicates that no pollution is present. Figure 8.27 shows a further geoelectric section from an adjacent area in which acid mine drainage has increased the conductivity of the groundwater, allowing its delineation as a band of low resistivity. Further VES enabled the extent of the pollution to be defined. Since contamination of this type is associated with a significant change in resistivity, periodic measurements at electrodes sited in a borehole penetrating the water table could be used to monitor the onset of pollution and the degree of contamination. Ebraheem *et al.*

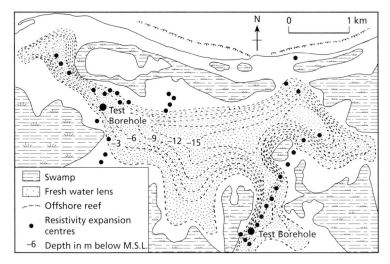

Fig. 8.25 Configuration of base of Central Lens, Grand Gayman. (After Bugg & Lloyd 1976.)

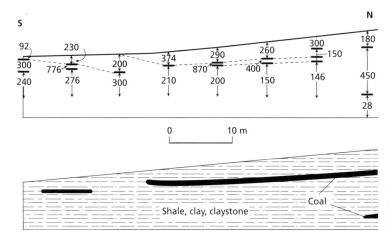

Fig. 8.26 Geoelectric section and geological interpretation of a profile near Kylertown, Pennsylvania. Numbers refer to resistivity in ohm m. (After Merkel 1972.)

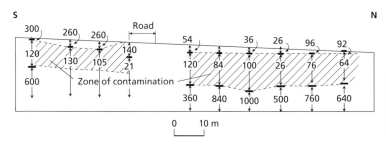

Fig. 8.27 A further geoelectric section from Kylertown, Pennsylvania. Shaded area shows zone of contamination. Numbers refer to resistivity in ohm m. (After Merkel 1972.)

(1990) have also described how the resistivity method can be used to study acid mine drainage and, in a similar environmental context, Carpenter *et al.* (1991) have reported the use of repeated resistivity measurements to monitor the integrity of the cover of a landfill site in Chicago.

8.3 Induced polarization (IP) method

8.3.1 Principles

When using a standard four-electrode resistivity spread in a DC mode, if the current is abruptly switched off, the voltage between the potential electrodes does not drop to zero immediately. After a large initial decrease the voltage suffers a gradual decay and can take many seconds to reach a zero value (Fig. 8.28). A similar phenomenon is observed as the current is switched on. After an initial sudden voltage increase, the voltage increases gradually over a discrete time interval to a steady-state value. The ground thus acts as a capacitor and stores electrical charge, that is, becomes electrically polarized.

If, instead of using a DC source for the measurement of resistivity, a variable low-frequency AC source is used, it is found that the measured apparent resistivity of the subsurface decreases as the frequency is increased. This is because the capacitance of the ground inhibits the passage of direct currents but transmits alternating currents with increasing efficiency as the frequency rises.

The capacitive property of the ground causes both the transient decay of a residual voltage and the variation of apparent resistivity as a function of frequency. The two effects are representations of the same phenomenon in the time and frequency domains, and are linked by Fourier transformation (see Chapter 2). These

two manifestations of the capacitance property of the ground provide two different survey methods for the investigations of the effect.

The measurement of a decaying voltage over a certain time interval is known as *time-domain* IP surveying. Measurement of apparent resistivity at two or more low AC frequencies is known as *frequency-domain* IP surveying.

8.3.2 Mechanisms of induced polarization

Laboratory experiments indicate that electrical energy is stored in rocks mainly by electrochemical processes. This is achieved in two ways.

The passage of current through a rock as a result of an externally imposed voltage is accomplished mainly by electrolytic flow in the pore fluid. Most of the rock-forming minerals have a net negative charge on their outer surfaces in contact with the pore fluid and attract positive ions onto this surface (Fig. 8.29(a)). The concentration of positive ions extends about $100\,\mu$m into the pore fluid, and if this distance is of the same order as the diameter of the pore throats, the movement of ions in the fluid resulting from the impressed voltage is inhibited. Negative and positive ions thus build up on either side of the blockage and, on removal of the impressed voltage, return to their original locations over a finite period of time causing a gradually decaying voltage.

This effect is known as *membrane polarization* or *electrolytic polarization*. It is most pronounced in the presence of clay minerals where the pores are particularly small. The effect decreases with increasing salinity of the pore fluid.

When metallic minerals are present in a rock, an alternative, electronic path is available for current flow. Figure 8.29(b) shows a rock in which a metallic mineral

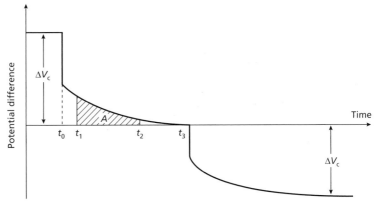

Fig. 8.28 The phenomenon of induced polarization. At time t_0 the current is switched off and the measured potential difference, after an initial large drop from the steady-state value ΔV_c, decays gradually to zero. A similar sequence occurs when the current is switched on at time t_3. A represents the area under the decay curve for the time increment t_1–t_2.

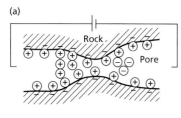 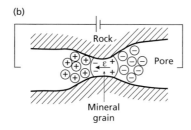

Fig. 8.29 Mechanisms of induced polarization: (a) membrane polarization and (b) electrode polarization.

grain blocks a pore, When a voltage is applied to either side of the pore space, positive and negative charges are imposed on opposite sides of the grain. Negative and positive ions then accumulate on either side of the grain which are attempting either to release electrons to the grain or to accept electrons conducted through the grain. The rate at which the electrons are conducted is slower than the rate of electron exchange with the ions. Consequently, ions accumulate on either side of the grain and cause a build-up of charge. When the impressed voltage is removed the ions slowly diffuse back to their original locations and cause a transitory decaying voltage.

This effect is known as *electrode polarization* or *overvoltage*. All minerals which are good conductors (e.g. metallic sulphides and oxides, graphite) contribute to this effect. The magnitude of the electrode polarization effect depends upon both the magnitude of the impressed voltage and the mineral concentration. It is most pronounced when the mineral is disseminated throughout the host rock as the surface area available for ionic–electronic interchange is then at a maximum. The effect decreases with increasing porosity as more alternative paths become available for the more efficient ionic conduction.

In prospecting for metallic ores, interest is obviously in the electrode polarization (overvoltage) effect. Membrane polarization, however, is indistinguishable from this effect during IP measurements. Membrane polarization consequently reduces the effectiveness of IP surveys and causes geological 'noise' which may be equivalent in magnitude to the overvoltage effect of a rock with up to 2% metallic minerals.

8.3.3 Induced polarization measurements

Time–domain IP measurements involve the monitoring of the decaying voltage after the current is switched off. The most commonly measured parameter is the *chargeability M*, defined as the area *A* beneath the decay curve over a certain time interval (t_1–t_2) normalized by the steady-state potential difference ΔV_c (Fig. 8.28)

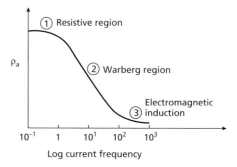

Fig. 8.30 The relationship between apparent resistivity and log measuring current frequency.

$$M = \frac{A}{\Delta V_c} = \frac{1}{\Delta V_c} \int_{t_1}^{t_2} v(t)\, dt \qquad (8.22)$$

Chargeability is measured over a specific time interval shortly after the polarizing current is cut off (Fig. 8.28) The area *A* is determined within the measuring apparatus by analogue integration. Different minerals are distinguished by characteristic chargeabilities, for example pyrite has $M = 13.4$ ms over an interval of 1 s, and magnetite 2.2 ms over the same interval. Figure 8.28 also shows that current polarity is reversed between successive measurements in order to destroy any remanent polarization.

Frequency-domain techniques involve the measurement of apparent resistivity at two or more AC frequencies. Figure 8.30 shows the relationship between apparent resistivity and log current frequency. Three distinct regions are apparent: region 1 is in low frequencies where resistivity is independent of frequency; region 2 is the Warberg region where resistivity is a linear function of log frequency; region 3 is the region of electromagnetic induction (Chapter 9) where current flow is by induction rather than simple conduction. Since the relationship illustrated in Fig. 8.30 varies with rock type and mineral concentration, IP measurements are usually made at frequencies at, or below, 10 Hz to remain in the non-inductive regions.

Two measurements are commonly made. The *percentage frequency effect* (PFE) is defined as

$$\text{PFE} = 100\frac{(\rho_{0.1} - \rho_{10})}{\rho_{10}} \tag{8.23}$$

where $\rho_{0.1}$ and ρ_{10} are apparent resistivities at measuring frequencies of 0.1 and 10 Hz. The *metal factor* (MF) is defined as

$$\text{MF} = 2\pi \times 10^5 \frac{(\rho_{0.1} - \rho_{10})}{\rho_{0.1}\rho_{10}} \tag{8.24}$$

This factor normalizes the PFE with respect to the lower frequency resistivity and consequently removes, to a certain extent, the variation of the IP effect with the effective resistivity of the host rock.

A common method of presenting IP measurements is the *pseudosection*, in which readings are plotted so as to reflect the depth of penetration. Figure 8.31 illustrates how a pseudosection is constructed for the double-dipole array geometry illustrated in Fig. 8.33. Measured values are plotted at the intersections of lines sloping at 45° from the centres of the potential and current electrode pairs. Values are thus plotted at depths which reflect the increasing depth of penetration as the length of the dipole separation increases. The values are then contoured. VES resistivity data can also be presented in this way with the plotted depth proportional to the current electrode separation. Pseudosections give only a crude representation of the IP response distribution at depth: for example, the apparent dip of the anomalous body is not always the same as the true dip. An example of this method of presentation is shown in Fig. 8.32.

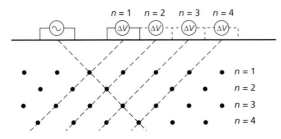

Fig. 8.31 The presentation of double-dipole IP results on a pseudosection. *n* represents the relative spacing between the current and potential electrode pairs.

8.3.4 Field operations

IP equipment is similar to resistivity apparatus but uses a current about 10 times that of a resistivity spread; it is also rather more bulky and elaborate. Theoretically, any standard electrode spread may be employed but in practice the double-dipole, pole–dipole and Schlumberger configurations (Fig. 8.33) are the most effective. Electrode spacings may vary from 3 to 300 m with the larger spacings used in reconnaissance surveys. To reduce the labour of moving current electrodes and generator, several pairs of current electrodes may be used, all connected via a switching device to the generator. Traverses are made over the area of interest plotting the IP reading at the mid-point of the electrode array (marked by crosses in Fig. 8.33).

Noise in an IP survey can result from several phenomena. Telluric currents cause similar anomalous effects to those encountered in resistivity measurements. Noise also results from the general IP effect of barren rocks caused by membrane polarization. Noise generated by the measuring equipment results from electromagnetic coupling between adjacent wires. Such effects are common when alternating current is used since currents can be induced to flow in adjacent conductors. Consequently, cables should be at least 10 m apart and if they must cross they should do so at right angles to minimize electromagnetic induction effects.

8.3.5 Interpretation of induced polarization data

Quantitative interpretation is considerably more complex than for the resistivity method. The IP response has been computed analytically for simple features such as spheres, ellipsoids, dykes, vertical contacts and horizontal layers, enabling indirect interpretation (numerical modelling) techniques to be used.

Laboratory modelling can also be employed in indirect interpretation to simulate an observed IP anomaly. For example, apparent resistivities may be measured for various shapes and resistivities of a gelatine–copper sulphate body immersed in water.

Much IP interpretation is, however, only qualitative. Simple parameters of the anomalies, such as sharpness, symmetry, amplitude and spatial distribution may be used to estimate the location, lateral extent, dip and depth of the anomalous zone.

The IP method suffers from the same disadvantages as resistivity surveying (see Section 8.2.9). Further, the sources of significant IP anomalies are often not of

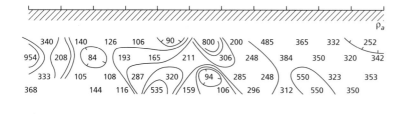

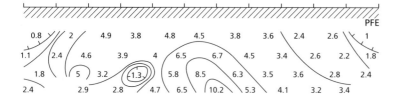

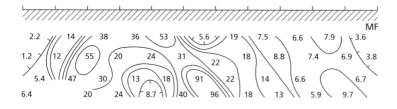

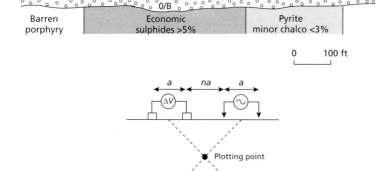

Fig. 8.32 Pseudosections of apparent resistivity (ρ_a), percentage frequency effect (PFE) and metal factor parameter (MF) for a double-dipole IP traverse across a zone of massive sulphides whose shape is known from subsequent test drilling. Current and potential electrode spacing *a* was 100 feet (30.5 m). Frequencies used for the IP measurements were 0.31 and 5.0 Hz. (After Fountain 1972.)

economic importance, for example water-filled shear zones and graphite-bearing sediments can both generate strong IP effects. Field operations are slow and the method is consequently far more expensive than most other ground geophysical techniques, survey costs being comparable with those of a gravity investigation.

8.3.6 Applications of induced polarization surveying

In spite of its drawbacks, the IP method is extensively used in base metal exploration as it has a high success rate in locating low-grade ore deposits such as disseminated sulphides (e.g. Langore *et al.* 1989). These have a strong IP effect but are non-conducting and therefore are not readily detectable by the electromagnetic methods discussed in Chapter 9. IP is by far the most effective geophysical method that can be used in the search for such targets.

Figure 8.34 shows the chargeability profile for a time-domain IP survey using a pole–dipole array across the Gortdrum copper–silver ore body in Ireland. Although the deposit is of low grade, containing less than 2% conducting minerals, the chargeability anomaly is well defined and centred over the ore body. In contrast, the

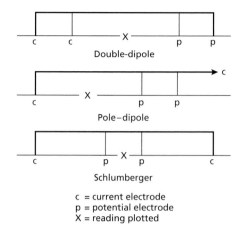

Fig. 8.33 Electrode configurations used in induced polarization measurements.

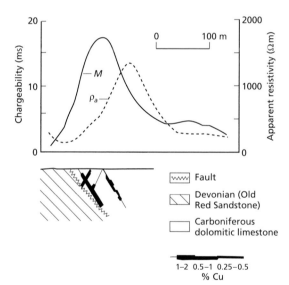

Fig. 8.34 Time-domain IP profile using a pole–dipole array over the Gortdrum copper–silver body, Ireland. (After Seigel 1967.)

clearly show the presence of the mineralization, allow its limits to be determined and provide estimates of the depth to its upper surface.

8.4 Self-potential (SP) method

8.4.1 Introduction

The self-potential (or spontaneous polarization) method is based on the surface measurement of natural potential differences resulting from electrochemical reactions in the subsurface. Typical SP anomalies may have an amplitude of several hundred millivolts with respect to barren ground. They invariably exhibit a central negative anomaly and are stable over long periods of time. They are usually associated with deposits of metallic sulphides (Corry 1985), magnetite or graphite.

8.4.2 Mechanism of self-potential

Field studies indicate that for a self-potential anomaly to occur its causative body must lie partially in a zone of oxidation. A widely-accepted mechanism of self-potential (Sato & Mooney 1960; for a more recent analysis see Kilty 1984) requires the causative body to straddle the water table (Fig. 8.36). Below the water table electrolytes in the pore fluids undergo oxidation and release electrons which are conducted upwards through the ore body. At the top of the body the released electrons cause reduction of the electrolytes. A circuit thus exists in which current is carried electrolytically in the pore fluids and electronically in the body so that the top of the body acts as a negative terminal. This explains the negative SP anomalies that are invariably observed and, also, their stability as the ore body itself undergoes no chemical reactions and merely serves to transport electrons from depth. As a result of the subsurface currents, potential differences are produced at the surface.

8.4.3 Self-potential equipment and survey procedure

Field equipment consists simply of a pair of electrodes connected via a high-impedance millivoltmeter. The electrodes must be non-polarizing as simple metal spikes would generate their own SP effects. Non-polarizing electrodes consist of a metal immersed in a saturated solution of its own salt, such as copper in copper sulphate. The salt is contained in a porous pot which allows slow leakage of the solution into the ground.

corresponding apparent resistivity profile reflects the large resistivity contrast between the Old Red Sandstone and dolomitic limestone but gives no indication of the presence of the mineralization.

A further example of an IP survey is illustrated in Fig. 8.35 which shows a traverse over a copper porphyry body in British Columbia, Canada. IP and resistivity traverses were made at three different electrode spacings of a pole–dipole array. The CST results exhibit little variation over the body, but the IP (chargeability) profiles

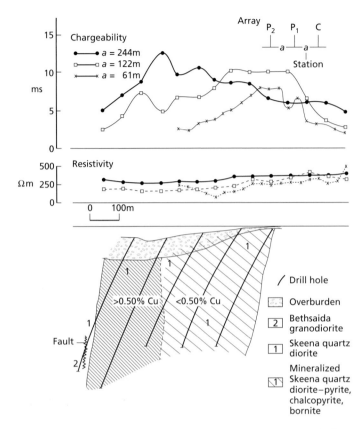

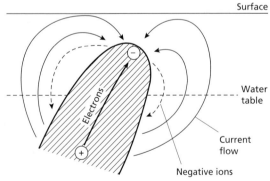

Fig. 8.35 Time domain induced polarisation and resistivity profiles over a copper porphyry body in British Columbia, Canada. (After Seigel 1967.)

Fig. 8.36 The mechanism of self-potential anomalies. (After Sato & Mooney 1960.)

Station spacing is generally less than 30 m. Traverses may be performed by leapfrogging successive electrodes or, more commonly, by fixing one electrode in barren ground and moving the other over the survey area.

8.4.4 Interpretation of self-potential anomalies

The interpretation of SP anomalies is similar to mag-

netic interpretation because dipole fields are involved in both cases. It is thus possible to calculate the potential distributions around polarized bodies of simple shape such as spheres, ellipsoids and inclined sheets (Sundararajan *et al.* 1998) by making assumptions about the distribution of charge over their surfaces.

Most interpretation, however, is qualitative. The anomaly minimum is assumed to occur directly over the anomalous body, although it may be displaced downhill in areas of steep topography. The anomaly half-width provides a rough estimate of depth. The symmetry or asymmetry of the anomaly provides information on the attitude of the body, the steep slope and positive tail of the anomaly lying on the downdip side.

The type of overburden can have a pronounced effect on the presence or absence of SP anomalies. Sand has little effect but a clay cover can mask the SP anomaly of an underlying body.

The SP method is only of minor importance in exploration. This is because quantitative interpretation is difficult and the depth of penetration is limited to about 30 m. It is, however, a rapid and cheap method requiring only simple field equipment. Consequently, it can be useful in rapid ground reconnaissance for base metal de-

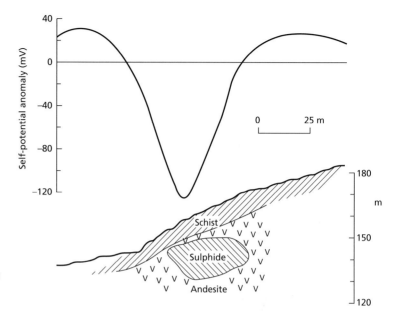

Fig. 8.37 The SP anomaly over a sulphide ore body at Sariyer, Turkey. (After Yüngül 1954.)

posits when used in conjunction with magnetic, electromagnetic and geochemical techniques. It has also been used in hydrogeological investigations (e.g. Fournier 1989), geothermal prospecting (Apostolopoulos *et al.* 1997) and the detection of air-filled drainage galleries (Ogilvy *et al.* 1991).

Figure 8.37 shows the SP profile over a sulphide ore body in Turkey which contains copper concentrations of up to 14%. The SP anomaly is negative and has an amplitude of some 140 mV. The steep topography has displaced the anomaly minimum downhill from the true location of the ore body.

Problems

1. Using the method of electrical images, derive the relationship between apparent resistivity, electrode spacing, layer thicknesses and resistivities for a VES performed with a Schlumberger spread over a single horizontal interface between media with resistivities ρ_1 and ρ_2.

2. At locations A, B, C and D along the gravity profile shown in Fig. 8.38, VES were performed with a Wenner array with the spread laid perpendicular to the profile. It was found that the sounding curves, shown in Fig. 8.39, were similar for locations A and B and for C and D. A borehole close to A penetrated 3 m of drift, 42 m of limestone and bottomed in sandstone. Downhole geophysical surveys (Chapter 11) provided the values of resistivity (ρ_R) and density (ρ_D) shown in the table for the lithologies encountered.

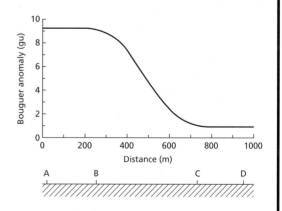

Fig. 8.38 Gravity anomaly profile pertaining to Question 2 showing also the locations of the VES at A, B, C and D.

Unit	ρ_R ($\Omega\,$m)	ρ_D (Mg m^{-3})
Drift	40	2.00
Limestone	2000	2.75
Sandstone	200	2.40

Continued

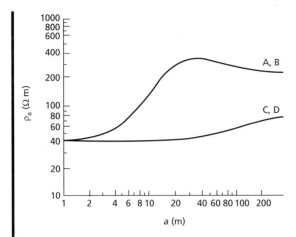

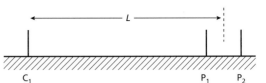

Fig. 8.40 The half-Schlumberger electrode configuration. See Question 4.

Fig. 8.39 Wenner VES sounding data for the locations shown in Fig. 8.38.

A seismic refraction line near to D revealed 15 m of drift, although the nature of the underlying basement could not be assessed from the seismic velocity.

(a) Interpret the geophysical data so as to provide a geological section along the profile.

(b) What further techniques might be used to confirm your interpretation?

(c) If a CST were to be performed along the profile, select, giving reasons, a suitable electrode spacing to map the basement. Sketch the expected form of the CST for both longitudinal and transverse traverses.

3. Calculate the variation in apparent resistivity along a CST profile at right angles to a vertically faulted contact between sandstone and limestone, with apparent resistivities of 50 ohm m and 600 ohm m, respectively, for a Wenner configuration. What would be the effect on the profiles if the contact dipped at a shallower angle?

4. Figure 8.40 shows a half-Schlumberger resistivity array in which the second current electrode is situated at a great distance from the other electrodes. Derive an expression for the apparent resistivity of this array in terms of the electrode spacings and the measured resistance.

The data in the table represent measurements taken with a half-Schlumberger array along a profile across gneissic terrain near Kongsberg, Norway. The potential electrode half-separation was kept constant at 40 m and the current electrode C_1 was fixed at the origin of the profile so

that as L (the current electrode half-separation) increased a CST was built up. R represents the resistance measured by the resistivity apparatus.

L (m)	R (Ω m)
30.2	1244.818
53.8	255.598
80.9	103.812
95.1	73.846
106.0	58.820
120.0	45.502
143.8	31.416
168.4	22.786
179.6	19.993
205.1	15.290
229.3	12.209
244.0	10.785

Calculate the apparent resistivity for each reading and plot a profile illustrating the results.

In this region it is known that the gneiss can be extensively brecciated. Does the CST give any indication of brecciation?

5. The following table represents the results of a frequency-domain IP survey of a Precambrian shield area. A double-dipole array was used with the separation (x) of both the current electrodes and the potential electrodes kept constant at 60 m. n refers to the number of separations between the current and potential electrode pairs and c to the distance of the centre of the array from the origin of the profile, where the results are plotted (Fig. 8.41). Measurements were taken using direct current and an alternating current of 10 Hz. These provided the apparent resistivities ρ_{dc} and ρ_{ac}, respectively.

(a) For each measurement point, calculate the percentage frequency effect (PFE) and metal factor parameter (MF).

(b) For both the PFE and MF plot four profiles for $n = 1, 2, 3$ and 4.

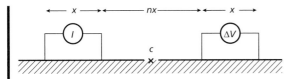

Fig. 8.41 The double-dipole electrode configuration. See Question 5.

(c) Construct and contour pseudosections of the DC apparent resistivity, PFE and MF.

(d) The area is covered by highly-conductive glacial deposits 30–60 m thick. It is possible that massive sulphide mineralization is present within the bedrock. Bearing this information in mind, comment upon and interpret the profiles and pseudosections produced from (b) and (c).

6. Why are the electrical methods of exploration particularly suited to hydrogeological investigations? Describe other geophysical methods which could be used in this context, stating the reasons why they are applicable.

c (m)	ρ_{dc} (Ωm)	ρ_{ac} (Ωm)	ρ_{dc} (Ωm)	ρ_{ac} (Ωm)	ρ_{dc} (Ωm)	ρ_{ac} (Ωm)	ρ_{dc} (Ωm)	ρ_{ac} (Ωm)
	n = 1		n = 2		n = 3		n = 4	
0	49.8	49.6			101.5	100.9		
30			72.8	72.4			99.6	98.5
60	46.0	45.8			86.2	85.2		
90			61.3	60.6			90.0	86.1
120	42.1	41.7			72.8	70.1		
150			55.5	54.4			57.5	53.5
180	44.0	43.5			49.8	46.6		
210			53.6	51.1			47.9	44.0
240	42.1	41.8			44.0	41.4		
270			65.1	64.1			47.9	44.9
300	49.8	49.6			95.8	91.7		
330			82.3	81.3			132.1	129.4
360	51.7	51.3			114.9	114.1		
390			86.2	85.9			164.7	164.0
420	49.8	49.6			120.7	120.1		
450			78.5	78.0			170.4	169.7

Further reading

Bertin, J. (1976) *Experimental and Theoretical Aspects of Induced Polarisation, Vols 1 and 2.* Gebrüder Borntraeger, Berlin.

Fink, J.B., McAlister, E.O. & Wieduwilt, W.G. (eds) (1990) *Induced Polarization. Applications and Case Histories.* Society of Exploration Geophysicists, Tulsa.

Griffiths, D.H. & King, R.F. (1981) *Applied Geophysics for Geologists and Engineers.* Pergamon, Oxford.

Habberjam, G.M. (1979) *Apparent Resistivity and the Use of Square Array Techniques.* Gebrüder Borntraeger, Berlin.

Keller, G.V. & Frischnecht, F.C. (1966) *Electrical Methods in Geophysical Prospecting.* Pergamon, Oxford.

Koefoed, O. (1968) *The Application of the Kernel Function in Interpreting Resistivity Measurements.* Gebrüder Borntraeger, Berlin.

Koefoed, O. (1979) *Geosounding Principles. I – Resistivity Sounding Measurements.* Elsevier, Amsterdam.

Kunetz, G. (1966) *Principles of Direct Current Resistivity Prospecting.* Gebrüder Borntraeger, Berlin.

Marshall, D.J. & Madden, T.R. (1959) Induced polarisation: a study of its causes. *Geophysics*, **24**, 790–816.

Milsom, J. (1989) *Field Geophysics.* Open University Press, Milton Keynes.

Parasnis, D.S. (1973) *Mining Geophysics.* Elsevier, Amsterdam.

Parasnis, D.S. (1996) *Principles of Applied Geophysics*, 5th edn. Chapman & Hall, London.

Parkhomenko, E.I. (1967) *Electrical Properties of Rocks.* Plenum, New York.

Sato, M. & Mooney, H.M. (1960) The electrochemical mechanism of sulphide self potentials. *Geophysics*, **25**, 226–49.

Sumner, J.S. (1976) *Principles of Induced Polarisation for Geophysical Exploration.* Elsevier, Amsterdam.

Telford, W.M., Geldart, L.P. & Sheriff, R.E. (1990) *Applied Geophysics*, 2nd edn. Cambridge University Press, Cambridge.

Ward, S.H. (1987) Electrical methods in geophysical prospecting. In: Samis, C.G. & Henyey, T.L. (eds), *Methods of Experimental Physics, Vol. 24, Part B – Field Measurements*, 265–375. Academic Press, Orlando.

9 Electromagnetic surveying

9.1 Introduction

Electromagnetic (EM) surveying methods make use of the response of the ground to the propagation of electromagnetic fields, which are composed of an alternating electric intensity and magnetizing force. Primary electromagnetic fields may be generated by passing alternating current through a small coil made up of many turns of wire or through a large loop of wire. The response of the ground is the generation of secondary electromagnetic fields and the resultant fields may be detected by the alternating currents that they induce to flow in a receiver coil by the process of electromagnetic induction.

The primary electromagnetic field travels from the transmitter coil to the receiver coil via paths both above and below the surface. Where the subsurface is homogeneous there is no difference between the fields propagated above the surface and through the ground other than a slight reduction in amplitude of the latter with respect to the former. However, in the presence of a conducting body the magnetic component of the electromagnetic field penetrating the ground induces alternating currents, or eddy currents, to flow in the conductor (Fig. 9.1). The eddy currents generate their own secondary electromagnetic field which travels to the receiver. The receiver then responds to the resultant of the arriving primary and secondary fields so that the response differs in both phase and amplitude from the response to the primary field alone. These differences between the transmitted and received electromagnetic fields reveal the presence of the conductor and provide information on its geometry and electrical properties.

The induction of current flow results from the magnetic component of the electromagnetic field. Consequently, there is no need for physical contact of either transmitter or receiver with the ground. Surface EM surveys can thus proceed much more rapidly than electrical surveys, where ground contact is required. More importantly, both transmitter and receiver can be mounted in aircraft or towed behind them. Airborne EM methods are widely used in prospecting for conductive ore bodies (see Section 9.8).

All anomalous bodies with high electrical conductivity (see Section 8.2.2) produce strong secondary electromagnetic fields. Some ore bodies containing minerals that are themselves insulators may produce secondary fields if sufficient quantities of an accessory mineral with a high conductivity are present. For example, electromagnetic anomalies observed over certain sulphide ores are due to the presence of the conducting mineral pyrrhotite distributed throughout the ore body.

9.2 Depth of penetration of electromagnetic fields

The depth of penetration of an electromagnetic field (Spies 1989) depends upon its frequency and the electrical conductivity of the medium through which it is propagating. Electromagnetic fields are attenuated during their passage through the ground, their amplitude decreasing exponentially with depth. The depth of penetration d can be defined as the depth at which the amplitude of the field A_d is decreased by a factor e^{-1} compared with its surface amplitude A_0

$$A_d = A_0 e^{-1} \tag{9.1}$$

In this case

$$d = \frac{503.8}{\sqrt{\sigma f}} \tag{9.2}$$

where d is in metres, the conductivity σ of the ground is in $S\,m^{-1}$ and the frequency f of the field is in Hz.

The depth of penetration thus increases as both the frequency of the electromagnetic field and the conductivity of the ground decrease. Consequently, the

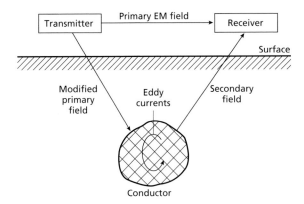

Fig. 9.1 General principle of electromagnetic surveying.

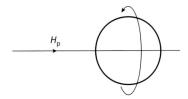

Fig. 9.2 The rotation of a search coil about an axis corresponding to the direction of arriving electromagnetic radiation H_p producing an infinite number of null positions.

frequency used in an EM survey can be tuned to a desired depth range in any particular medium. For example, in relatively dry glacial clays with a conductivity of $5 \times 10^{-4}\,\mathrm{S\,m^{-1}}$, d is about 225 m at a frequency of 10 kHz.

Equation (9.2) represents a theoretical relationship. Empirically, an effective depth of penetration z_e can be defined which represents the maximum depth at which a conductor may lie and still produce a recognizable electromagnetic anomaly

$$z_e \approx \frac{100}{\sqrt{\sigma f}} \tag{9.3}$$

The relationship is approximate as the penetration depends upon such factors as the nature and magnitude of the effects of near-surface variations in conductivity, the geometry of the subsurface conductor and instrumental noise. The frequency dependence of the depth of penetration places constraints on the EM method. Normally, very low frequencies are difficult to generate and measure and the maximum achievable penetration is usually of the order of 500 m.

9.3 Detection of electromagnetic fields

Electromagnetic fields may be mapped in a number of ways, the simplest of which employs a small search coil consisting of several hundred turns of copper wire wound on a circular or rectangular frame typically between 0.5 m and 1 m across. The ends of the coil are connected via an amplifier to earphones. The amplitude of the alternating voltage induced in the coil by an electro-

magnetic field is proportional to the component of the field perpendicular to the plane of the coil. Consequently, the strength of the signal in the earphones is at a maximum when the plane of the coil is at right angles to the direction of the arriving field. Since the ear is more sensitive to sound minima than maxima, the coil is usually turned until a null position is reached. The plane of the coil then lies in the direction of the arriving field.

9.4 Tilt-angle methods

When only a primary electromagnetic field H_p is present at a receiver coil, a null reading is obtained when the plane of the coil lies parallel to the field direction. There are an infinite number of such null positions as the coil is rotated about a horizontal axis in the direction of the field (Fig. 9.2).

In many EM systems the induced secondary field H_s lies in a vertical plane. Since the primary and secondary fields are both alternating, the total field vector describes an ellipse in the vertical plane with time (Fig. 9.3). The resultant field is then said to be *elliptically polarized* in the vertical plane. In this case there is only one null position of the search coil, namely where the plane of the coil coincides with the plane of polarization.

For good conductors it can be shown that the direction of the major axis of the ellipse of polarization corresponds reasonably accurately to that of the resultant of the primary and secondary electromagnetic field directions. The angular deviation of this axis from the horizontal is known as the *tilt-angle* θ of the resultant field (Fig. 9.3). There are a number of EM techniques (known as *tilt-angle* or *dip-angle* methods) which simply measure spatial variations in this angle. The primary field may be generated by a fixed transmitter, which usually consists of a large horizontal or vertical coil, or by a small mobile transmitter. Traverses are made across the survey area normal to the geological strike. At each station the

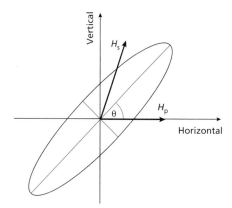

Fig. 9.3 The polarization ellipse and tilt-angle θ. H_p and H_s represent the primary and secondary electromagnetic fields.

search coil is rotated about three orthogonal axes until a null signal is obtained so that the plane of the coil lies in the plane of the polarization ellipse. The tilt-angle may then be determined by rotating the coil about a horizontal axis at right angles to this plane until a further minimum is encountered.

9.4.1 Tilt-angle methods employing local transmitters

In the case of a fixed, vertical transmitter coil, the primary field is horizontal. Eddy currents within a subsurface conductor then induce a magnetic field whose lines of force describe concentric circles around the eddy current source, which is assumed to lie along its upper edge (Fig. 9.4(a)). On the side of the body nearest the transmitter the resultant field dips upwards. The tilt decreases towards the body and dips downwards on the side of the body remote from the transmitter. The body is located directly below the crossover point where the tilt-angle is zero, as here both primary and secondary fields are horizontal. When the fixed transmitter is horizontal the primary field is vertical (Fig. 9.4(b)) and the body is located where the tilt is at a minimum. An example of the use of tilt-angle methods (vertical transmitter) in the location of a massive sulphide body is presented in Fig. 9.5.

If the conductor is near the surface both the amplitude and gradients of the tilt-angle profile are large. These quantities decrease as the depth to the conductor increases and may consequently be used to derive semiquantitative estimates of the conductor depth. A vertical conductor would provide a symmetrical tilt-angle profile with equal gradients on either side of the body. As the inclination of the conductor decreases, the gradients on either side become progressively less similar. The asymmetry of the tilt-angle profile can thus be used to obtain an estimate of the dip of the conductor.

Tilt-angle methods employing fixed transmitters have been largely superseded by survey arrangements in which both transmitter and receiver are mobile and which can provide much more quantitative information on subsurface conductors. However, two tilt-angle methods still in common use are the very low-frequency (VLF) and audio-frequency magnetic field (AFMAG) methods, neither of which requires the erection of a special transmitter.

9.4.2 The VLF method

The source utilized by the VLF method is electromagnetic radiation generated in the low-frequency band of 15–25 kHz by the powerful radio transmitters used in long-range communications and navigational systems. Several stations using this frequency range are available around the world and transmit continuously either an unmodulated carrier wave or a wave with superimposed Morse code. Such signals may be used for surveying up to distances of several thousand kilometres from the transmitter.

At large distances from the source the electromagnetic field is essentially planar and horizontal (Fig. 9.6). The electric component E lies in a vertical plane and the magnetic component H lies at right angles to the direction of propagation in a horizontal plane. A conductor that strikes in the direction of the transmitter is cut by the magnetic vector and the induced eddy currents produce a secondary electromagnetic field. Conductors striking at right angles to the direction of propagation are not cut effectively by the magnetic vector.

The basic VLF receiver is a small hand-held device incorporating two orthogonal aerials which can be tuned to the particular frequencies of the transmitters. The direction of a transmitter is found by rotating the horizontal coil around a vertical axis until a null position is found. Traverses are then performed over the survey area at right angles to this direction. The instrument is rotated about a horizontal axis orthogonal to the traverse and the tilt recorded at the null position. Profiles are similar in form to Fig. 9.4(a), with the conductor lying beneath locations of zero tilt. See Hjelt *et al.* (1985) for a discussion of the interpretation of VLF data and Beamish (1998) for a means of three-dimensional modelling of VLF data.

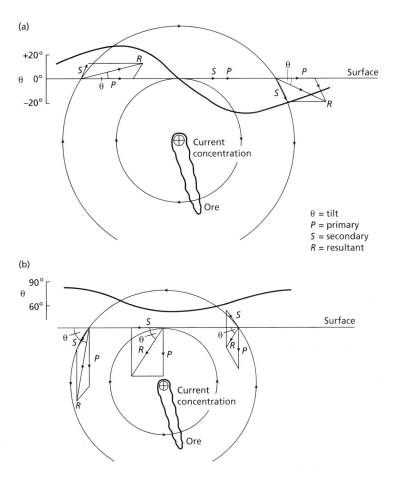

Fig. 9.4 Tilt-angle profiles resulting from (a) vertical and (b) horizontal transmitter loops. (After Parasnis 1973.)

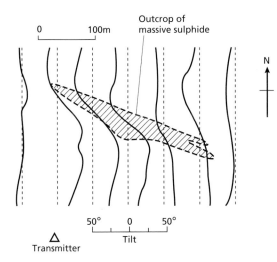

Fig. 9.5 Example of tilt-angle survey using a vertical loop transmitter. (After Parasnis 1973.)

Modern instruments have three coils with their axes at right angles. They can thus detect the signal whatever its direction, and find the null orientation electronically and automatically. Some instruments will measure signals from two or more transmitters simultaneously. In this case transmitters are chosen whose signals arrive in the survey area at very different azimuths.

The VLF method has the advantages that the field equipment is small and light, being conveniently operated by one person, and that there is no need to install a transmitter. However, for a particular survey area, there may be no suitable transmitter providing a magnetic vector across the geological strike. A further disadvantage is that the depth of penetration is somewhat less than that attainable by tilt-angle methods using a local transmitter. The VLF method can be used in airborne EM surveying.

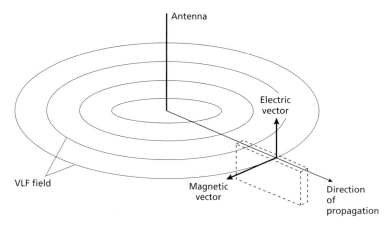

Fig. 9.6 Principle of VLF method. Dashed lines show a tabular conductor striking towards the antenna which is cut by the magnetic vector of the electromagnetic field.

9.4.3 The AFMAG method

The AFMAG method (Labson *et al.* 1985) can similarly be used on land or in the air. The source in this case is the natural electromagnetic fields generated by thunderstorms and known as *sferics*. Sferics propagate around the Earth between the ground surface and the ionosphere. This space constitutes an efficient electromagnetic waveguide and the low attenuation means that thunderstorms anywhere in the world make an effective contribution to the field at any given point. The field also penetrates the subsurface where, in the absence of electrically-conducting bodies, it is practically horizontal. The sferic sources are random so that the signal is generally quite broad-band between 1 and 1000 Hz.

The AFMAG receiver differs from conventional tilt-angle coils since random variations in the direction and intensity of the primary field make the identification of minima impossible with a single coil. The receiver consists of two orthogonal coils each inclined at 45° to the horizontal (Fig. 9.7). In the absence of a secondary field the components of the horizontal primary field perpendicular to the coils are equal and their subtracted output is zero (Fig. 9.7(a)). The presence of a conductor gives rise to a secondary field which causes deflection of the resultant field from the horizontal (Fig. 9.7(b)). The field components orthogonal to the two coils are then unequal, so that the combined output is no longer zero and the presence of a conductor is indicated. The output provides a measure of the tilt.

On land both the azimuths and tilts of the resultant electromagnetic field can be determined by rotating the coils about a vertical axis until a maximum signal is obtained. These are conventionally plotted as dip vectors. In the air, azimuths cannot be determined as the coils are

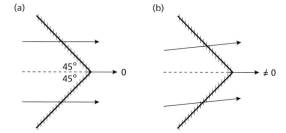

Fig. 9.7 Principle of AFMAG receiver: (a) conductor absent, (b) conductor present.

attached to the aircraft so that their orientation is controlled by the flight direction. Consequently, only perturbations from the horizontal are monitored along the flight lines. The output signal is normally fed into an amplifier tuned to two frequencies of about 140 and 500 Hz. Comparison of the amplitudes of the signals at the two frequencies provides an indication of the conductivity of the anomalous structure as it can be shown that the ratio of low-frequency response to high-frequency response is greater than unity for a good conductor and less than unity for a poor conductor.

The AFMAG method has the advantage that the frequency range of the natural electromagnetic fields used extends to an order of magnitude lower than can be produced artificially so that depths of investigation of several hundred metres are feasible.

9.5 Phase measuring systems

Tilt-angle methods such as VLF and AFMAG are widely used since the equipment is simple, relatively

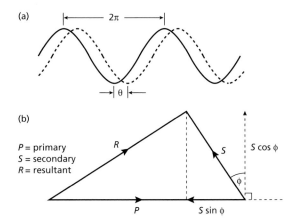

Fig. 9.8 (a) The phase difference θ between two waveforms. (b) Vector diagram illustrating the phase and amplitude relationships between primary, secondary and resultant electromagnetic fields.

cheap and the technique is rapid to employ. However, they provide little quantitative information on the conductor. More sophisticated EM surveying systems measure the phase and amplitude relationships between primary, secondary and resultant electromagnetic fields. The various types of system available are discussed in McCracken *et al.* (1986).

An alternating electromagnetic field can be represented by a sine wave with a wavelength of 2π (360°) (Fig. 9.8(a)). When one such wave lags behind another the waves are said to be out-of-phase. The phase difference can be represented by a phase angle θ corresponding to the angular separation of the waveforms. The phase relationships of electromagnetic waves can be represented on special vector diagrams in which vector length is proportional to field amplitude and the angle measured counterclockwise from the primary vector to the secondary vector represents the angular phase lag of the secondary field behind the primary.

The primary field P travels directly from transmitter to receiver above the ground and suffers no modification other than a small reduction in amplitude caused by geometric spreading. As the primary field penetrates the ground it is reduced in amplitude to a greater extent but remains in phase with the surface primary. The primary field induces an alternating voltage in a subsurface conductor with the same frequency as the primary but with a phase lag of $\pi/2$ (90°) according to the laws of electromagnetic induction. This may be represented on the vector diagram (Fig. 9.8(b)) by a vector $\pi/2$ counterclockwise to P.

The electrical properties of the conductor cause a further phase lag ϕ,

$$\phi = \tan^{-1}\left(\frac{2\pi f L}{r}\right) \tag{9.4}$$

where f is the frequency of the electromagnetic field, L the inductance of the conductor (its tendency to oppose a change in the applied field) and r the resistance of the conductor. For a good conductor ϕ will approach $\pi/2$ while for a poor conductor ϕ will be almost zero.

The net effect is that the secondary field S produced by the conductor lags behind the primary with a phase angle of $(\pi/2 + \phi)$. The resultant field R can now be constructed (Fig. 9.8(b)).

The projection of S on the horizontal (primary field) axis is $S\sin\phi$ and is an angle π out of phase with P. It is known as the *in-phase* or *real component* of S. The vertical projection is $S\cos\phi$, $\pi/2$ out of phase with P, and is known as the *out-of-phase*, *imaginary* or *quadrature* component.

Modern instruments are capable of splitting the secondary electromagnetic field into its real (Re) and imaginary (Im) components. The larger the ratio Re/Im, the better the conductor. Some systems, mainly airborne, simply measure the phase angle ϕ.

Classical phase measuring systems employed a fixed source, usually a very large loop of wire laid on the ground. These systems include the *two-frame*, *compensator* and *turam* systems. They are still in use but are more cumbersome than modern systems in which both transmitter and receiver are mobile. These latter systems are referred to as *twin-coil* or *slingram* systems.

A typical field set is shown in Fig. 9.9. The transmitter and receiver coils are about 1 m in diameter and are usually carried horizontally, although different orientations may be used. The coils are linked by a cable which carries a reference signal and also allows the coil separation to be accurately maintained at, normally, between 30 m and 100 m. The transmitter is powered by a portable AC generator. Output from the receiver coil passes through a compensator and decomposer (see below). The equipment is first read on barren ground and the compensator adjusted to produce zero output. By this means, the primary field is compensated so that the system subsequently responds only to secondary fields. Consequently, such EM methods reveal the presence of bodies of anomalous conductivity without providing information on absolute conductivity values. Over the survey area the decomposer splits the secondary field

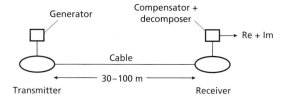

Fig. 9.9 Mobile transmitter–receiver EM field equipment.

into real and imaginary components which are usually displayed as a percentage of the primary field whose magnitude is relayed via the interconnecting cable. Traverses are generally made perpendicular to geological strike and readings plotted at the mid-point of the system. The maximum detection depth is about half the transmitter–receiver separation.

Fieldwork is simple and requires a crew of only two or three operators. The spacing and orientation of the coils is critical as a small percentage error in spacing can produce appreciable error in phase measurement. The coils must also be kept accurately horizontal and coplanar as small relative tilts can produce substantial errors. The required accuracy of spacing and orientation is difficult to maintain with large spacings and over uneven terrain.

Figure 9.10 shows a mobile transmitter–receiver EM profile across a sheet-like conductor in the Kankberg area of northern Sweden. A consequence of the coplanar horizontal coil system employed is that conducting bodies produce negative anomalies in both real and imaginary components with maximum amplitudes immediately above the conductor. The asymmetry of the anomalies is diagnostic of the inclination of the body, with the maximum gradient lying on the downdip side. In this case the large ratio of real to imaginary components over the ore body indicates the presence of a very good conductor, while a lesser ratio is observed over a sequence of graphite-bearing phyllites to the north.

9.6 Time-domain electromagnetic surveying

A significant problem with many EM surveying techniques is that a small secondary field must be measured in the presence of a much larger primary field, with a consequent decrease in accuracy. This problem is overcome in *time-domain electromagnetic surveying* (TDEM), sometimes called *pulsed* or *transient-field EM*, by using a primary field which is not continuous but consists of a series of pulses separated by periods when it is inactive. The

secondary field induced by the primary is only measured during the interval when the primary is absent. The eddy currents induced in a subsurface conductor tend to diffuse inwards towards its centre when the inducing field is removed and gradually dissipate by resistive heat loss. Within highly conductive bodies, however, eddy currents circulate around the boundary of the body and decay more slowly. Measurement of the rate of decay of the waning eddy currents thus provides a means of locating anomalously conducting bodies and estimating their conductivity. The analysis of the decaying secondary field is equivalent to analysing the response to a continuous EM wave at a number of frequencies. TDEM consequently bears the same relationship to continuous-wave EM as, for example, time-domain IP does to frequency-domain IP. INPUT® (Section 9.8.1) is an example of an airborne TDEM system.

In ground surveys, the primary pulsed EM field is generated by a transmitter that usually consists of a large rectangular loop of wire, several tens of metres across, which is laid on the ground. The transmitter loop can also be utilized as the receiver, or a second coil can be used for this purpose, either on the ground surface or down a borehole (Dyck & West 1984). The transient secondary field produced by the decaying eddy currents can last from less than a millisecond for poor conductors to more than 20 ms for good conductors. The decaying secondary field is quantified by measuring the temporal variation of the amplitude of the secondary at a number of fixed times (channels) after primary cut-off (Fig. 9.11). In good conductors the secondary field is of long duration and will register in most of the channels; in poor conductors the secondary field will only register in the channels recorded soon after the primary field becomes inactive. Repeated measurements can be stacked in a manner analogous to seismic waves (see Section 4.3) to improve the signal-to-noise ratio. The position and attitude of the conductor can be estimated from the change in amplitude from place to place of the secondary field in selected channels, while depth estimates can be made from the anomaly half-width. More quantitative interpretations can be made by simulation of the anomaly in terms of the computed response of simple geometric shapes such as spheres, cylinders or plates, or more simply by using the concept of equivalent current filaments which models the distribution of eddy currents in the conductor. Limited two-dimensional modelling (Oristaglio & Hohmann 1984) is also possible using a finite-difference approach.

A form of depth sounding can be made utilizing

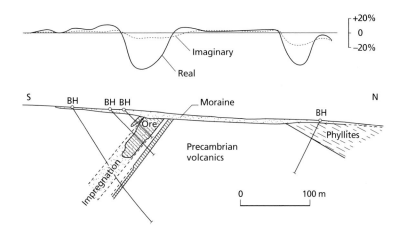

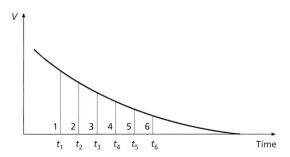

Fig. 9.10 Mobile transmitter–receiver profile, employing horizontal coplanar coils with a separation of 60 m and an operating frequency of 3.6 kHz, in the Kankberg area, north Sweden. Real and imaginary components are expressed as a percentage of the primary field. (After Parasnis 1973.)

Fig. 9.11 The quantification of a decaying TDEM response by measurement of its amplitude in a number of channels (1–6) at increasing times ($t_{1–6}$) after primary field cut-off. The amplitudes of the responses in the different channels are recorded along a profile.

TDEM (Frischnecht & Raab 1984). Only short offsets of transmitter and receiver are necessary and the array therefore crosses a minimum of geological boundaries such as faults and lithological contacts. By contrast, VES or continuous-wave EM methods are much more affected by near-surface conductivity inhomogeneities since long arrays are required. It is claimed that penetration of up to about 10 km can be achieved by TDEM sounding.

An example of a surface application of TDEM is presented in Fig. 9.12, which shows the results of a survey undertaken near Mount Minza, Northern Territory, Australia (Duckworth 1968, see also Spies 1976). The target, which had been revealed by other geophysical methods (Fig. 9.13), was a band of highly-conductive graphitic black shale, which has a conductivity in excess of 0.1 S m^{-1} in its pristine condition. In Fig. 9.12 the TDEM response is expressed in terms of the induced

voltage in the loop $e(t)$ normalized with respect to the current in the transmitter loop I. The response is shown for a number of different times after primary cut-off. The response persists into the latest channels, indicating the presence of a good conductor which corresponds to the graphitic shale. The asymmetry of the response curves and their variation from channel to channel allows the dip of the conductor to be estimated. The first channel, which logs the response to relatively shallow depths, peaks to the right. The maximum moves to the left in later channels, which give the response to progressively greater depths, indicating that the conductor dips in that direction.

An example of a survey using a borehole TDEM system is presented in Fig. 9.14, which shows results from the Single Tree Hill area, NSW, Australia (Boyd & Wiles 1984). Here semi-massive sulphides (pyrite and pyrrhotite), which occur in intensely sericitized tuffs with shale bands, have been penetrated by three drillholes. The TDEM responses at a suite of times after primary field cut-off, recorded as the receiver was lowered down the three drillholes, are shown. In hole PDS1, the response at early times indicates the presence of a conductor at a depth of 145 m. The negative response at later times at this depth is caused by the diffusion of eddy currents into the conductor past the receiver and indicates that the hole is near the edge of the conductor. In holes DS1 and DS2 the negative responses at 185 m and 225 m, respectively, indicate that the receiver passed outside, but near the edges of, the conductor at these depths. Also shown in the section is an interpretation of the TDEM data in terms of a model consisting of a rectangular current-carrying loop.

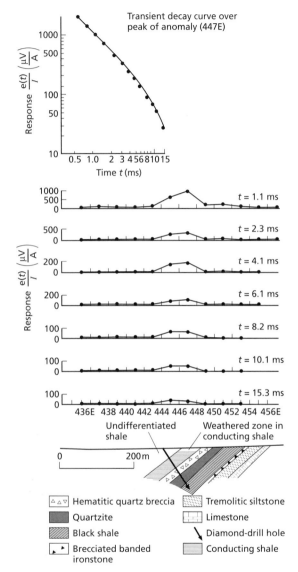

Fig. 9.12 TDEM profiles and geological section near Mount Minza, Northern Territory, Australia. (After Duckworth 1968.)

9.7 Non-contacting conductivity measurement

It is possible to obtain readings of ground conductivity by EM measurements (McNeill 1980). Measurements of this type can be made using standard resistivity methods (see Section 8.2), but, since these require the introduction of current into the ground via electrodes, they are labour intensive, slow and therefore costly. Moreover,

resistivity measurements are influenced by geological noise arising from near-surface resistivity variations which limit the resolution that can be achieved. The more recently developed non-contacting conductivity meters utilize EM fields and do not suffer from these drawbacks. No ground contact is required so that measurements can be made at walking pace and the subsurface volume sampled is averaged in such a way that resolution is considerably improved (Zalasiewicz et al. 1985).

The secondary EM field measured in a mobile transmitter–receiver survey (Section 9.5) is generally a complex function of the coil spacing s, the operating frequency f and the conductivity of the subsurface σ. However, it can be shown that if the product of s and the skin depth d (Section 9.2), known as the *induction number*, is much less than unity, the following relationship results:

$$\frac{H_s}{H_p} \approx \frac{i\omega\mu_0\sigma s^2}{4} \tag{9.5}$$

where H_s and H_p are the amplitudes of the secondary and primary EM fields, respectively, $\omega = 2\pi f$, μ_0 is the magnetic permeability of vacuum, and $i = \sqrt{(-1)}$, its presence indicating that the quadrature component is measured. Thus the ratio H_s/H_p is proportional to the ground conductivity σ. Since d depends on the product σf, estimation of the maximum probable value of σ allows the selection of f such that the above condition of low induction number is satisfied. The depth of penetration depends upon s and is independent of the conductivity distribution of the subsurface. Measurements taken at low induction number thus provide an apparent conductivity σ_a given by

$$\sigma_a = \frac{4}{\omega\mu_0 s^2}\frac{H_s}{H_p} \tag{9.6}$$

This relationship allows the construction of electromagnetic instruments which provide a direct reading of ground conductivity down to a predetermined depth. In one application the transmitter and receiver are horizontal dipoles mounted on a boom 3.7 m apart, providing a fixed depth of investigation of about 6 m. The instrument provides a rapid means of performing constant separation traversing (see Section 8.2.3) to a depth suitable for engineering and archaeological investigations. Where a greater depth of penetration is required, an

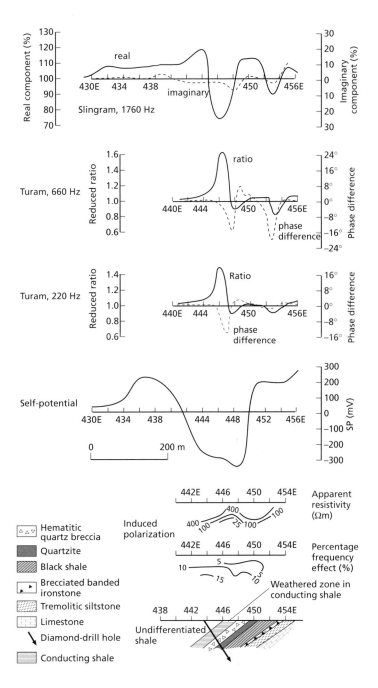

Fig. 9.13 Comparison of various geophysical methods over the same profile as shown in Fig. 9.12 near Mount Minza, Northern Territory, Australia. (After Duckworth 1968.)

instrument is used in which the transmitter and receiver, which usually take the form of vertical coplanar coils, are separate, so that their spacing is variable. Constant separation traversing (CST) can be performed with the subsurface energized to a desired depth, while vertical electrical sounding (see Section 8.2.3) can be undertaken by progressively increasing the transmitter–receiver separation.

A widely used instrument based on the above principles is the Geonics EM31.

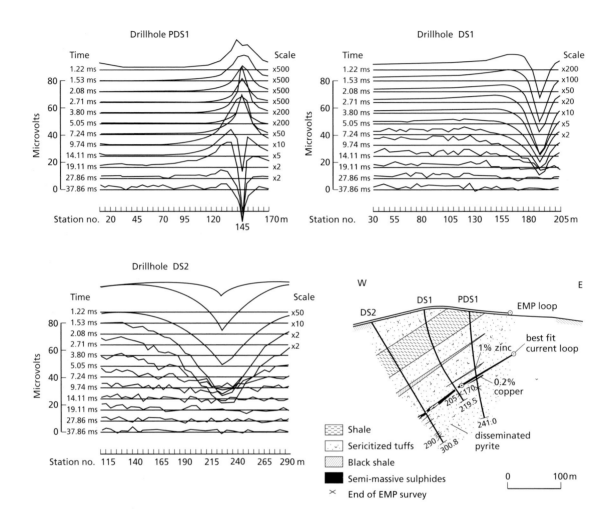

Fig. 9.14 Drillhole TDEM profiles and geological section over Single Tree Hill, NSW, Australia. (Redrawn from Boyd & Wiles 1984.)

9.8 Airborne electromagnetic surveying

Airborne EM techniques are widely used because of their speed and cost-effectiveness, and a large number of systems are available.

There is a broad division into *passive systems*, where only the receiver is airborne, and *active systems*, where both transmitter and receiver are mobile. Passive systems include airborne versions of the VLF and AFMAG methods. Independent transmitter methods can also be used with an airborne receiver, but are not very attractive as prior ground access to the survey area is required.

Active systems are more commonly used, as surveys can be performed in areas where ground access is difficult and provide more information than the passive tilt-

angle methods. They are, basically, ground mobile transmitter–receiver systems lifted into the air and interfaced with a continuous recording device. Certain specialized methods, described later, have been adopted to overcome the specific difficulties encountered in airborne work. Active systems comprise two main types, *fixed separation* and *quadrature*.

9.8.1 Fixed separation systems

In fixed separation systems the transmitter and receiver are maintained at a fixed separation, and real and imaginary components are monitored as in ground surveys. The coils are generally arranged to be vertical and either coplanar or coaxial. Accurate maintenance of separation and height is essential, and this is usually accomplished by mounting the transmitter and receiver either on the wings of an aircraft or on a beam carried beneath a helicopter. Compensating methods have to be employed to correct for minute changes in the relative positions of transmitter and receiver resulting from such factors as flexure of the wings, vibration and temperature changes. Since only a small transmitter–receiver separation is used to generate and detect an electromagnetic field over a relatively large distance, such minute changes in separation would cause significant distortion of the signal. Fixed-wing systems are generally flown at a ground clearance of 100–200 m, while helicopters can survey at elevations as low as 20 m.

Greater depth of penetration can be achieved by the use of two planes flying in tandem (Fig. 9.15), the rear plane carrying the transmitter and the forward plane towing the receiver mounted in a bird. Although the aircraft have to fly at a strictly regulated speed, altitude and separation, the use of a rotating primary field compensates for relative rotation of the receiver and transmitter. The rotating primary field is generated by a transmitter consisting of two orthogonal coils in the plane perpendicular to the flight direction. The coils are powered by the same AC source with the current to one coil shifted $\pi/2$ (90°) out-of-phase with respect to the other. The resulting field rotates about the flight line and is detected by a receiver with a similar coil configuration which passes the signals through a phase-shift network so that the output over a barren area is zero. The presence of a conductor is then indicated by non-zero output and the measured secondary field decomposed into real and imaginary components. Although penetration is increased and orientation errors are minimized, the method is relatively expensive and the interpretation of data is complicated by the complex coil system. It is possible to upward-continue airborne EM data. This diminishes variations caused by height fluctuations and anomalies of small, shallow sources.

Airborne TDEM methods, such as INPUT® (INduced PUlse Transient) (Barringer 1962), may be used to enhance the secondary field measurement. The discontinuous primary field shown in Fig. 9.16 is generated by passing pulses of current through a transmitter coil strung about an aircraft. The transient primary field induces currents within a subsurface conductor. These currents persist during the period when the primary field is shut off and the receiver becomes active. The exponential decay curve is sampled at several points and the signals displayed on a strip chart. The signal amplitude in successive sampling channels is, to a certain extent, diagnostic of the type of conductor present. Poor conductors produce a rapidly decaying voltage and only register on those channels sampling the voltage shortly after primary cut-off. Good conductors appear on all channels.

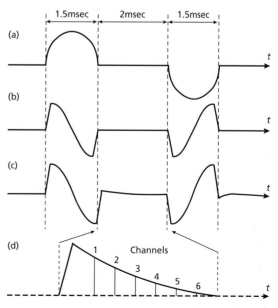

Fig. 9.16 Principle of the INPUT® system. (a) Primary field. (b) Receiver response to primary alone. (c) Receiver response in the presence of a secondary field. (d) Enlargement of the receiver signal during primary field cut-off. The amplitude of the decaying induced voltage is here sampled on six channels.

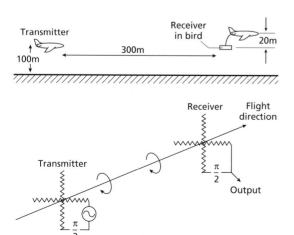

Fig. 9.15 The two-plane, rotary field, EM system.

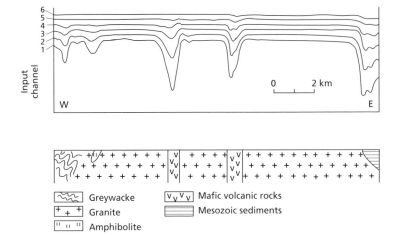

Greywacke

Granite

Amphibolite

Mafic volcanic rocks

Mesozoic sediments

Fig. 9.17 INPUT® profile across part of the Itapicuru Greenstone Belt, Brazil. (After Palacky 1981.)

INPUT® is more expensive than other airborne EM methods but provides greater depth penetration, possibly in excess of 100 m, because the secondary signal can be monitored more accurately in the absence of the primary field. It also provides a direct indication of the type of conductor present from the duration of the induced secondary field.

As well as being employed in the location of conducting ore bodies, airborne EM surveys can also be used as an aid to geological mapping. In humid and subtropical areas a weathered surface layer develops whose thickness and conductivity depend upon the local rock type. Figure 9.17 shows an INPUT® profile across part of the Itapicuru Greenstone Belt in Brazil, with sampling times increasing from 0.3 ms at channel 1 to 2.1 ms at channel 6. The transient response over mafic volcanic rocks and Mesozoic sediments is developed in all six channels, indicating that their weathered layer is highly conductive, while the response over greywacke is only apparent in channels 1–4, indicating a comparatively less conductive layer.

EM methods are being used increasingly in hydrogeological studies as they are more efficient than the resistivity methods classically used for this purpose. A series of case histories of the use of EM methods in groundwater studies is given in McNeill (1991).

9.8.2 Quadrature systems

Quadrature systems were the first airborne EM methods devised. The transmitter is usually a large aerial slung between the tail and wingtips of a fixed-wing aircraft and a nominally-horizontal receiver is towed behind the aircraft on a cable some 150 m long.

In quadrature systems the orientation and height of the receiver cannot be rigorously controlled as the receiver 'bird' oscillates in the slipstream. Consequently, the measurement of real and imaginary components is not possible as the strength of the field varies irregularly with movement of the receiver coil. However, the phase difference between the primary field and the resultant field caused by a conductor is independent of variation in the receiver orientation. A disadvantage of the method is that a given phase shift ϕ' may be caused by either a good or a poor conductor (Fig. 9.18). This problem is overcome by measuring the phase shift at two different primary frequencies, usually of the order of 400 and 2300 Hz. It can be shown that, if the ratio of low-frequency to high-frequency response exceeds unity, a good conductor is present.

Figure 9.19 shows a contour map of real component anomalies (in ppm of the primary field) over the Skellefteå orefield, northern Sweden. A fixed separation system was used, with vertical, coplanar coils mounted perpendicular to the flight direction on the wingtips of a small aircraft. Only contours above the noise level of some 100 ppm are presented. The pair of continuous anomaly belts in the southwest, with amplitudes exceeding 1000 ppm, corresponds to graphitic shales, which serve as guiding horizons in this orefield. The belt to the north of these is not continuous, and although in part related to sulphide ores, also results from a power cable. In the northern part of the area the three distinct anomaly centres all correspond to strong sulphide mineralization.

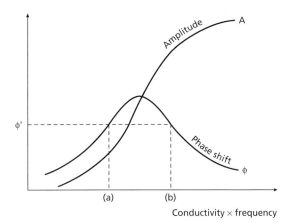

Fig. 9.18 The relationship between the phase/amplitude of a secondary electromagnetic field and the product of conductivity and frequency. A given phase shift ϕ' could result from a poor conductor (a) or a good conductor (b).

9.9 Interpretation of electromagnetic data

As with other types of geophysical data an indirect approach can be adopted in the interpretation of electromagnetic anomalies. The observed electromagnetic response is compared with the theoretical response, for the type of equipment used, to conductors of various shapes and conductivities. Theoretical computations of this type are quite complex and limited to simple geometric shapes such as spheres, cylinders, thin sheets and horizontal layers.

If the causative body is of complex geometry and variable conductivity, laboratory modelling may be used (Chakridi & Chouteau 1988). Because of the complexity of theoretical computations, this technique is used far more extensively in electromagnetic interpretation than in other types of geophysical interpretation. For example, to model a massive sulphide body in a well-conducting host rock, an aluminium model immersed in salt water may be used.

Master curves are available for simple interpretation of moving source–receiver data in cases where it may be assumed that the conductor has a simple geometric form. Figure 9.20 shows such a set of curves for a simple sheet-like dipping conductor of thickness t and depth d where the distance between horizontal, coplanar coils is a. The point corresponding to the maximum real and imaginary values, expressed as a percentage of the primary field, is plotted on the curves. From the curves coinciding with this point, the corresponding λ/a and d/a values

are determined. The latter ratio is readily converted into conductor depth. λ corresponds to $10^7/\sigma ft$, where σ is the conductivity of the sheet and f the frequency of the field. Since a and f are known, the product σt can be determined. By performing measurements at more than one frequency, σ and t can be computed separately.

Much electromagnetic interpretation is, however, only qualitative, particularly for airborne data. Contour maps of real or imaginary components provide information on the length and conductivity of conductors while the asymmetry of the profiles provides an estimate of the inclination of sheet-like bodies.

9.10 Limitations of the electromagnetic method

The electromagnetic method is a versatile and efficient survey technique, but it suffers from several drawbacks. As well as being caused by economic sources with a high conductivity such as ore bodies, electromagnetic anomalies can also result from non-economic sources such as graphite, water-filled shear zones, bodies of water and man-made features. Superficial layers with a high conductivity such as wet clays and graphite-bearing rocks may screen the effects of deeper conductors. Penetration is not very great, being limited by the frequency range that can be generated and detected. Unless natural fields are used, maximum penetration in ground surveys is limited to about 500 m, and is only about 50 m in airborne work. Finally, the quantitative interpretation of electromagnetic anomalies is complex.

9.11 Telluric and magnetotelluric field methods

9.11.1 Introduction

Within and around the Earth there exist large-scale, low-frequency, natural magnetic fields known as *magnetotelluric fields*. These induce natural alternating electric currents to flow within the Earth, known as *telluric currents*. Both of these natural fields can be used in prospecting.

Magnetotelluric fields are believed to result from the flow of charged particles in the ionosphere, as fluctuations in the fields correlate with diurnal variations in the geomagnetic field caused by solar emissions. Magnetotelluric fields penetrate the ground and there induce telluric currents to flow. The fields are of variable

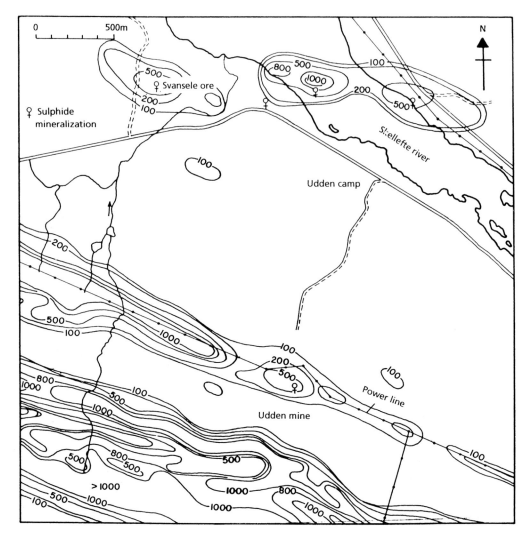

Fig. 9.19 Contour map of real component anomalies over part of the Skellefteå orefield, northern Sweden, obtained using an airborne system with vertical coplanar coils. Mean ground clearance 30 m, operating frequency 3.5 kHz. Contours in ppm of the primary field. (After Parasnis 1973.)

frequency, ranging from 10^{-5} Hz up to the audio range, and overlap the frequency range utilized in the AFMAG method (Section 9.4.3).

9.11.2 Surveying with telluric currents

Telluric currents flow within the Earth in large circular patterns that stay fixed with respect to the Sun. They normally flow in sheets parallel to the surface and extend to depths of several kilometres in the low frequencies. The telluric method is, in fact, the only electrical tech-

nique capable of penetrating to the depths of interest to the oil industry. Although variable in both their direction and intensity, telluric currents cause a mean potential gradient at the Earth's surface of about 10 mV km^{-1}.

Telluric currents are used in prospecting by measuring the potential differences they cause between points at the surface. Obviously no current electrodes are required and potential differences are monitored using non-polarizing electrodes or plates made of a chemically inert substance such as lead. Electrode spacing is typically 300–600 m in oil exploration and 30 m or less in

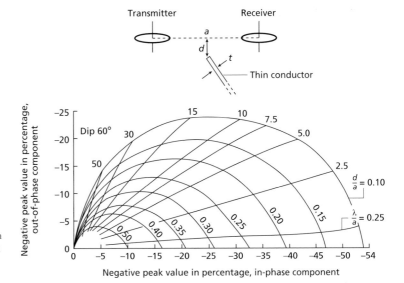

Fig. 9.20 Example of a vector diagram used in estimating the parameters of a thin dipping conductor from the peak real and imaginary component values. (Redrawn from Nair *et al.* 1968.)

mineral surveys. The potential electrodes are connected to an amplifier which drives a strip chart recorder or tape recorder.

If the electrical conductivity of the subsurface were uniform the potential gradient at the surface would be constant (Fig. 9.21(a)). Zones of differing conductivity deflect the current flow from the horizontal and cause distortion of the potential gradients measured at the surface. Figure 9.21(b) shows the distortion of current flow lines caused by a salt dome which, since it is a poor conductor, deflects the current lines into the overlying layers. Similar effects may be produced by anticlinal structures. Interpretation of anomalous potential gradients measured at the surface permits the location of subsurface zones of distinctive conductivity.

Telluric potential gradients are measured using orthogonal electrode pairs (Fig. 9.22(a)). In practice, the survey technique is complicated by temporal variation in direction and intensity of the telluric currents. To overcome this problem, one orthogonal electrode pair is read at a fixed base located on nearby barren ground and another moved over the survey area. At each observation point the potential differences between the pairs of electrodes at the base and at the mobile station are recorded simultaneously over a period of about 10 min. From the magnitude of the two horizontal components of the electrical field it is simple to find the variation in direction and magnitude of the resultant field at the two locations over the recording interval. The assumption is made that the ground is uniform beneath the base

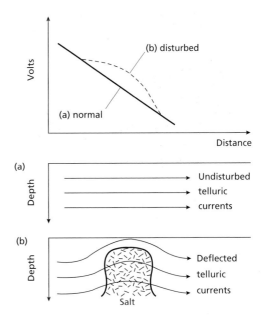

Fig. 9.21 The instantaneous potential gradient associated with telluric currents. (a) Normal, undisturbed gradient. (b) Disturbed gradient resulting from deflection of current flow by a salt dome.

electrodes so that the conductivity is the same in all directions. The resultant electrical field should also be constant in all directions and would describe a circle with time (Fig. 9.22(b)). To correct for variations in intensity of the telluric currents, a function is determined which,

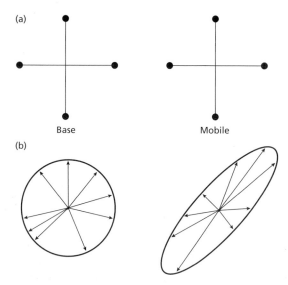

Fig. 9.22 (a) Base and mobile potential electrode sets used in telluric surveys. (b) The figure traced by the horizontal component of the telluric field over an undisturbed area (circle) and in the presence of a subsurface conductor (ellipse) after correction for temporal variations in telluric current intensity.

when applied to the base electrode results, constrains the resultant electric vector to describe a circle of unit radius. The same function is then applied to the mobile electrode data. Over an anomalous structure the conductivity of the ground is not the same in all directions and the magnitude of the corrected resultant electric field varies with direction. The resultant field vector traces an ellipse whose major axis lies in the direction of maximum conductivity. The relative disturbance at this point is conveniently measured by the ratio of the area of the ellipse to the area of the corresponding base circle. The results of a survey of this type over the Haynesville Salt Dome, Texas, USA are presented in Fig. 1.4. The solid circles represent locations where ellipse areas relative to a unit base circle have been computed. Contours of these values outline the known location of the dome with reasonable accuracy.

The telluric method is applicable to oil exploration as it is capable of detecting salt domes and anticlinal structures, both of which constitute potential hydrocarbon traps. As such, the method has been used in Europe, North Africa and the former Soviet Union. It is not widely used in the USA where oil traps tend to be too small in area to cause a significant distortion of telluric current flow. The telluric method can also be adapted to mineral exploration.

9.11.3 Magnetotelluric surveying

Prospecting using magnetotelluric fields is more complex than the telluric method as both the electric and magnetic fields must be measured. The technique does, however, provide more information on subsurface structure. The method is, for example, used in investigations of the crust and upper mantle (e.g. Hutton *et al.* 1980).

Telluric currents are monitored as before, although no base station is required. The magnetotelluric field is measured by its inductive effect on a coil about a metre in diameter or by use of a sensitive fluxgate magnetometer. Two orthogonal components are measured at each station.

The depth z to which a magnetotelluric field penetrates is dependent on its frequency f and the resistivity ρ of the substrate, according to equations of the form of (9.2) and (9.3), that is,

$$z = k\sqrt{\frac{\rho}{f}} \qquad (9.7)$$

where k is a constant. Consequently, depth penetration increases as frequency decreases. It can be shown that the amplitudes of the electric and magnetic fields, E and B, are related

$$\rho_a = \frac{0.2}{f}\left(\frac{E}{B}\right)^2 \qquad (9.8)$$

where f is in Hz, E in mV km^{-1} and B in nT. The apparent resistivity ρ_a thus varies inversely with frequency. The calculation of ρ_a for a number of decreasing frequencies thus provides resistivity information at progressively increasing depths and is essentially a form of vertical electrical sounding (see Section 8.2.3).

Interpretation of magnetotelluric data is most reliable in the case of horizontal layering. Master curves of apparent resistivity against period are available for two and three horizontal layers, vertical contacts and dykes, and interpretation may proceed in a similar manner to curve-matching techniques in the resistivity method (see Section 8.2.7). Routines are now available, however, which allow the computer modelling of two-dimensional structures.

9.12 Ground-penetrating radar

Ground-penetrating radar (GPR) (Davis & Annan 1989) is a technique of imaging the subsurface at high resolution. Although analogous in some ways to the seismic methods, it is included in this chapter as the propagation of radar waves through a medium is controlled by its electrical properties at high frequencies. A comprehensive account of modern advances in GPR is given by Reynolds (1997).

GPR is a non-destructive technique and can consequently be applied in urban and sensitive environments. GPR has many geological applications, such as imaging shallow soil and rock structure at high resolution, locating buried channels and mapping the water table. It also has several non-geological uses such as in archaeology, for the location of buried walls or cavities, and in forensic investigations, for the location of recently-disturbed ground where a burial has taken place.

GPR is similar in its principles to seismic reflection profiling (see Chapter 4) and sonar (see Section 4.15) surveying. A short radar pulse in the frequency band 10–1000 MHz is introduced into the ground. Radar velocities are controlled by the *dielectric constant (relative permittivity)* and conductivity of the subsurface.

The velocity of a radar wave (V) is given by:

$$V = \frac{c}{\sqrt{(\mu_r \varepsilon_r)}} \qquad (9.9)$$

where c is the velocity of light *in vacuo* ($3 \times 10^8 \, \mathrm{m\,s^{-1}}$), μ_r the relative magnetic permeability (Section 7.2), which is close to unity for non-magnetic rocks, and ε_r the relative dielectric permittivity.

In high resistivity rocks ($>10^2 \, \mathrm{ohm\,m}$) the propagation velocity of the pulse is mainly controlled by ε_r. Dielectric conduction takes place in such poor conductors and insulators, which have no free carriers, by the slight displacement of electrons with respect to their nuclei. Water has a dielectric constant of 80, whereas in most dry geological materials the dielectric constant is in the range 4–8. Consequently, the water content of materials exerts a strong influence on the propagation of a radar pulse.

A contrast in dielectric properties across an interface causes reflection of part of a radar pulse with a diminution of energy according to the reflection coefficient K, which is analogous to the seismic case (see Section 3.6.1),

$$K = \frac{(\sqrt{\varepsilon_{r2}} - \sqrt{\varepsilon_{r1}})}{(\sqrt{\varepsilon_{r2}} + \sqrt{\varepsilon_{r1}})} = \frac{(V_2 - V_1)}{(V_2 + V_1)} \qquad (9.10)$$

where ε_{r1} and ε_{r2} are the relative permittivities of the two media separated by the interface and V_1 and V_2 the radar velocities within them. Velocities of geological materials generally lie within the range 0.06–$0.175 \, \mathrm{m\,ns^{-1}}$.

Dielectric permittivity does not usually vary by more than a factor of 10 in most natural materials, so it is the more highly variable resistivity that controls the depth of penetration of a radar pulse. Generally, depth of penetration increases with increasing resistivity. Penetration is of the order of 20 m, although this may increase to 50 m under optimal conditions of low conductivity. As with seismic waves, there is a trade-off between depth of penetration and resolution, with the greater penetration achieved with the lower frequencies.

A transmitting antenna generates a wavetrain which comprises a pulse of radio waves with a frequency of about 50 kHz. This is transmitted into the subsurface. The arriving pulse is scanned at a fixed rate for a time adjusted to be of the order of the two-way travel time of the pulse. The pulse received by the receiving antenna is similarly a wavetrain, but differs from the transmitted wavetrain because of the modifications caused to it during its passage through the subsurface. The fact that the wavetrain comprises more than one wavelet complicates the subsequent interpretation. Since velocities of radar waves can be of the order of $0.3 \, \mathrm{m\,s^{-1}}$, accurate timing instrumentation is essential. The returned radar signals are amplified, digitized and recorded; the resulting data can be displayed on a radargram, which is very similar to a seismogram.

The depth of penetration of radio waves depends on their frequency and the nature of the material being surveyed. Figure 9.23 shows how the penetration varies in different materials over the frequency range 1–500 MHz. The permittivity of water is high compared to dry materials, so the water content and porosity are important controls on penetration.

There are three basic modes of deployment in GPR surveys, all of which have their seismic counterparts (Fig. 9.24):

1. Reflection profiling (Fig. 9.24(a)), in which the transmitter and antenna are kept at a small, fixed separation; this is often achieved by using the same antenna for transmission and reception.

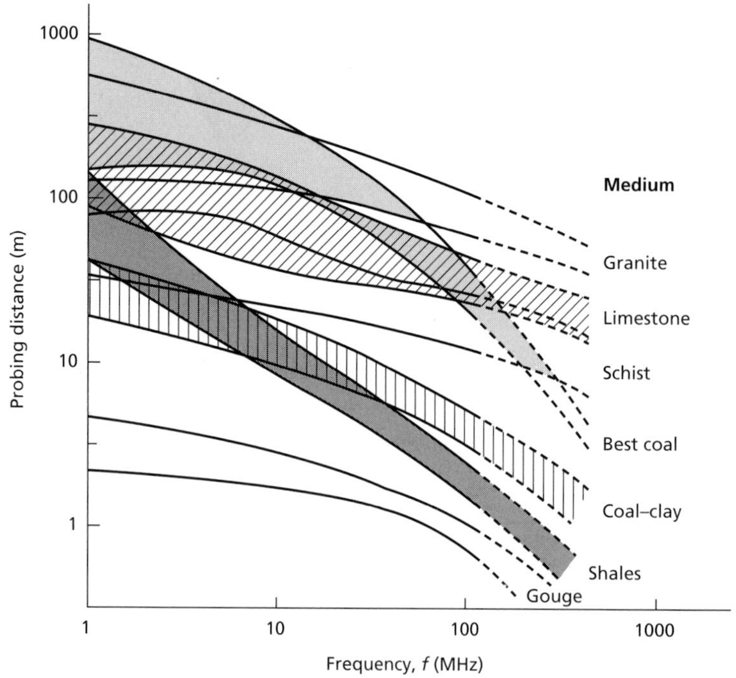

Fig. 9.23 The relationship between probing distance and frequency for different materials. (After Cook 1975.)

2. Velocity sounding (Fig. 9.24(b)), in which transmitter and antenna are moved apart about a fixed central point (the common depth point (CDP) method), or one kept stationary while the other is progressively moved away (the wide-angle reflection and refraction (WARR) method). The methods are designed to show how the radar velocity changes with depth. Without this information, velocities might be determined by correlating the radargram with a borehole section or with signals reflected from a body at known depth. In many cases, however, the velocities are guessed.

3. Transillumination (Fig. 9.24(c)), in which the transmitter and antenna are mounted on either side of the object of interest (e.g. a pillar in a mine). If it is arranged that there are many different configurations of transmitter and antenna, radar tomography can be carried out in a similar fashion to seismic (see Section 5.10) and resistivity (see Section 8.2.7) tomography.

Filtering of the radar signal can be applied during data acquisition, but is more conveniently performed on the digital output provided by modern instruments. The radar reflections can subsequently be enhanced by digital data-processing techniques very similar to those used in reflection seismology (see Section 4.8), of which migration is particularly important.

Interpretation of a radargram is commonly performed by interface mapping, similar to that used in the interpretation of seismograms. If amplitude fidelity has been retained in the radargram, zones of high attenuation can be recognized which represent high-conductivity areas such as are produced by clay accumulations. However, the identification of each band on a radargram as a distinct geological horizon would be incorrect because of the effects of multiples, interference with a previous reflection wavetrain, sideswipe (see Section 4.8), noise, etc. Processing of the radargram is simplified by deconvolution (see Section 4.8.2), which restores the shape of the downgoing wavetrain so that primary events can be recognized more easily. Migration is also particularly useful in that diffraction hyperbolae are removed and correct dips restored.

A GPR profile and its interpretation are shown in Fig. 9.25, which illustrates the detailed information provided by the technique.

Reflection

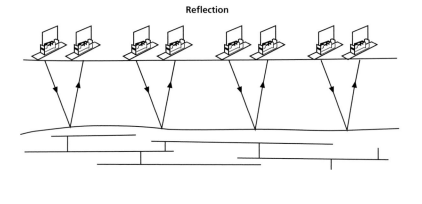

Transillumination

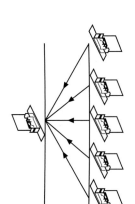

Common mid-point (CMP)

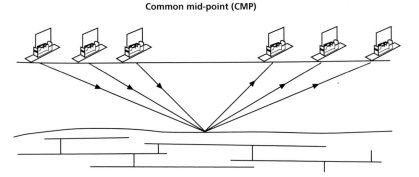

Fig. 9.24 The three basic modes of ground-penetrating radar.

9.13 Applications of electromagnetic surveying

The principal use of EM surveys is in the exploration for metalliferous mineral deposits, which differ significantly in their electrical properties from their host rocks. In spite of the limited depth of penetration, airborne techniques are frequently used in reconnaissance surveys, with aeromagnetic surveys often run in conjunction. EM methods are also used in the follow-up ground surveys which provide more precise information on the target area. Standard moving source–receiver methods (see Section 9.5) may be used for this purpose, although in rugged or forested terrain the VLF (see Section 9.4.2)

or AFMAG (see Section 9.4.3) methods may be preferred as no heavy equipment is required and there is no need to cut tracks for survey lines.

On a small scale, EM methods can be used in geotechnical and archaeological surveys to locate buried objects such as mine workings, pipes or treasure trove. The instruments used can take the form of metal detectors similar to the mine detectors used by army engineers, which have a depth of penetration of only a few centimetres and respond only to metal, or may be of the non-contacting conductivity meter-type described in Section 9.7, which have greater penetration and also respond to non-metallic resistivity anomalies.

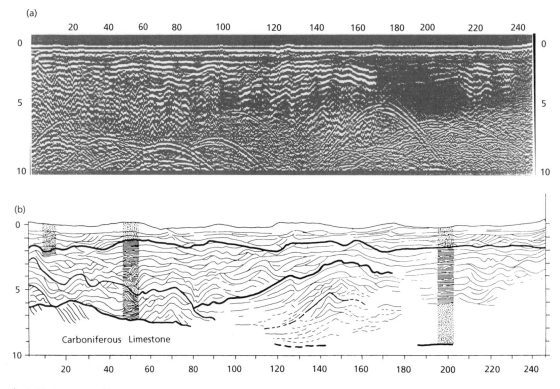

Fig. 9.25 (a) GPR profile. (b) Line drawing showing the interpretation of (a). Thick lines show truncation surfaces within a sequence of tufa overlying Carboniferous limestone. Data from three boreholes are also shown: dotted ornament = lime mud; horizontal ornament = sapropels; wavy lines = phytoherm framestones. All axes are in m. (After Pedley *et al.* 2000.)

Problems

1. Calculate the depth of penetration of electromagnetic fields with frequencies of 10, 500 and 2000 kHz in:

(a) wet sandstone with a conductivity of $10^{-1}\,\mathrm{S\,m^{-1}}$,

(b) massive limestone with a conductivity of $2.5 \times 10^{-4}\,\mathrm{S\,m^{-1}}$,

(c) granite with a conductivity of $10^{-6}\,\mathrm{S\,m^{-1}}$.

2. Figure 9.26(a) shows four profiles obtained during a tilt-angle EM survey near Uchi Lake, Ontario, the horizontal axes being displayed in their correct relative geographical positions. The survey was performed using transmitter and receiver in the form of vertical loops kept at a fixed separation of 120 m. Sketch in the location of the subsurface conductor and comment on its geometry. Figure 9.26(b) shows a repeat of profile 3 using a fixed transmitter and a mobile receiver operated at frequencies of 480 and 1800 Hz. Where was the transmitter located and what form did it take? What additional information is provided by this profile?

3. During a phase measuring EM survey, the resultant EM field was observed to have an amplitude 78% of that of the primary field and lagged behind it with an angular phase difference of 22°. Determine the amplitude of the secondary field of the subsurface conductor and of its real and imaginary components, all expressed as a percentage of the primary. What do these results reveal about the nature of the conductor?

Continued

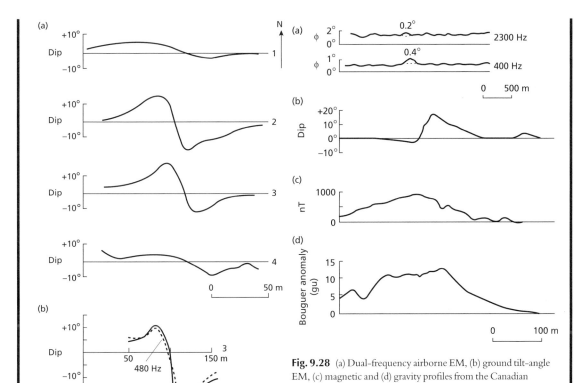

Fig. 9.28 (a) Dual-frequency airborne EM, (b) ground tilt-angle EM, (c) magnetic and (d) gravity profiles from the Canadian Shield. See Question 5. (After Paterson 1967.)

Fig. 9.26 (a) Tilt-angle profiles from an EM survey near Uchi Lake, Ontario. (b) Profile 3 repeated with dual-frequency EM equipment. See Question 2. (After Telford et al. 1990.)

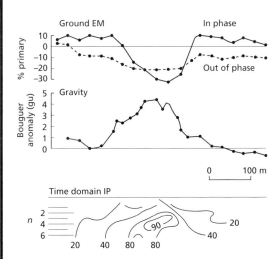

Fig. 9.27 Ground EM profile, Bouguer gravity profile and chargeability pseudosection representing results from a double-dipole IP electrode spread, all from a survey in Bahia, Brazil. See Question 4. (After Palacky & Sena 1979.)

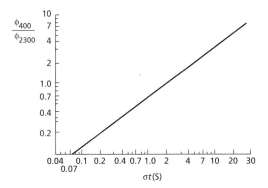

Fig. 9.29 Characteristic curve for an airborne EM system over a half-plane. ϕ_{400}/ϕ_{2300} is the ratio of peak responses at 400 Hz and 2300 Hz respectively, σ and t are the conductivity in $S\,m^{-1}$ and thickness of the conductor in metres, respectively. See Question 5. (After Paterson 1967.)

Continued

4. Figure 9.27 shows various ground geophysical measurements taken over volcanic terrain in Bahia, Brazil. The EM survey was conducted with a system using horizontal, coplanar coils 100 m apart and a frequency of 444 Hz. The time-domain IP survey used a double-dipole array with a basic electrode separation of 25 m. Interpret these data as fully as possible. What further information would be necessary before an exploratory borehole were sunk?

5. Figure 9.28 shows the results of airborne and ground geophysical surveys over an area of the Canadian Shield. The airborne EM survey used a quadrature system with measurements of phase angle taken at 2300 and 400 Hz. The ground tilt-angle EM survey was undertaken with a vertical-loop system using a local transmitter. Interpret and comment upon these results. Figure 9.29 can be used to estimate the product of conductivity and conductor thickness from the airborne data.

6. Which geophysical methods are particularly suitable for archaeological applications?

Further reading

Boissonas, E. & Leonardon, E.G. (1948) Geophysical exploration by telluric currents with special reference to a survey of the Haynesville Salt Dome, Wood County, Texas. *Geophysics*, **13**, 387–403.

Cagniard, L. (1953) Basic theory of the magnetotelluric method of geophysical prospecting. *Geophysics*, **18**, 605–35.

Davis, J.L. & Annan, A.P. (1989) Ground-penetrating radar for high-resolution mapping of soil and rock stratigraphy. *Geophys. Prosp.*, **37**, 531–51.

Dobrin, M.B. & Savit, C.H. (1988) *Introduction to Geophysical Prospecting*, 4th edn. McGraw-Hill, New York.

Jewell, T.R. & Ward, S.H. (1963) The influence of conductivity inhomogeneities upon audiofrequency magnetic fields. *Geophysics*, **28**, 201–21.

Keller, G.V. & Frischnecht, F.C. (1966) *Electrical Methods in Geophysical Prospecting*. Pergamon, Oxford.

Milsom, J. (1989) *Field Geophysics*. Open University Press, Milton Keynes.

Parasnis, D.S. (1973) *Mining Geophysics*. Elsevier, Amsterdam.

Parasnis, D.S. (1996) *Principles of Applied Geophysics*, 5th edn. Chapman & Hall, London.

Pedley, H.M., Hill, I. & Brasington, J. (2000) Three dimensional modelling of a Holocene tufa system in the Lathkill Valley, N. Derbyshire, using ground penetrating radar. *Sedimentology*, **47**, 721–37.

Reynolds, J.M. (1997) *An Introduction to Applied and Environmental Geophysics*. Wiley, Chichester.

Telford, W.M., Geldart, L.P. & Sheriff, R.E. (1990) *Applied Geophysics*, 2nd edn. Cambridge University Press, Cambridge.

Wait, J.R. (1982) *Geo-Electromagnetism*. Academic Press, New York.

10 Radiometric surveying

10.1 Introduction

Surveying for radioactive minerals has become important over the last few decades because of the demand for nuclear fuels. Radiometric surveying is employed in the search for deposits necessary for this application, and also for non-radioactive deposits associated with radioactive elements such as titanium and zirconium. Radiometric surveys are of use in geological mapping as different rock types can be recognized from their distinctive radioactive signature (Moxham 1963, Pires & Harthill 1989). There are in excess of 50 naturally occurring radioactive isotopes, but the majority are rare or only very weakly radioactive. The elements of principal interest in radiometric exploration are uranium (^{238}U), thorium (^{232}Th) and potassium (^{40}K). The latter isotope is widespread in potassium-rich rocks which may not be associated with concentrations of U and Th. Potassium can thus obscure the presence of economically important deposits and constitutes a form of geological 'noise' in this type of surveying. Figure 10.1 shows a ternary diagram illustrating the relative abundances of ^{238}U, ^{232}Th and ^{40}K in different rock types.

Radiometric surveys are less widely used than the other geophysical methods as they seek a very specific target. Probably the most common application of radiometric techniques is in geophysical borehole logging (see Section 11.7).

10.2 Radioactive decay

Elements whose atomic nuclei contain the same number of protons but different numbers of neutrons are termed isotopes. They are forms of the same element with different atomic weights. A conventional notation for describing an element A in terms of its atomic number n and atomic weight w is $^{w}_{n}A$. Certain isotopes are unstable and may disintegrate spontaneously to form other elements.

The disintegration is accompanied by the emission of radioactivity of three possible types.

Alpha particles are helium nuclei $^{4}_{2}He$ which are emitted from the nucleus during certain disintegrations:

$$^{w}_{n}A \rightarrow \, ^{w-4}_{n-2}B + \, ^{4}_{2}He$$

Beta particles are electrons which may be emitted when a neutron splits into a proton and an electron during certain disintegrations. The proton remains within the nucleus so that the atomic weight remains the same but the atomic number increases by one to form a new element:

$$^{w}_{n}A \rightarrow \, ^{w}_{n+1}B + e^{-}$$

Gamma rays are pure electromagnetic radiation released from excited nuclei during disintegrations. They are characterized by frequencies in excess of about 10^{16} Hz and differ from X-rays only in being of higher energy.

In addition to these emissions, a further process occurs in some radioactive elements which also releases energy in the form of gamma rays. This is known as K capture and takes place when an electron from the innermost (K) shell enters the nucleus. The atomic number decreases and a new element is formed:

$$^{w}_{n}A + e^{-} \rightarrow \, ^{w}_{n-1}B$$

Radioactive decay may lead to the formation of a stable element or a further radioactive product which itself undergoes decay. The rate of decay is exponential so that

$$N = N_{0}e^{-\lambda t}$$

where N is the number of atoms remaining after time t from an initial number N_{0} at time $t = 0$. λ is a decay constant characteristic of the particular element. The half-life of an element is defined as the time taken for N_{0} to decrease by a half. Half-lives vary from 10^{-7} s for $^{212}_{84}Po$ to about 10^{13} Ma for $^{204}_{82}Pb$. The fact that decay constants

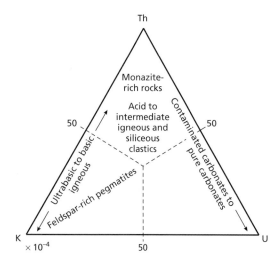

Fig. 10.1 Relative abundances of radioactive elements in different rock types. Also shown are the relative radioactivities of the radioelements. (After Wollenberg 1977.)

are accurately known and unaffected by external conditions such as temperature, pressure and chemical composition forms the basis of radiometric dating.

The radioactive emissions have very different pene-

trating properties. Alpha particles are effectively stopped by a sheet of paper, beta particles are stopped by a few millimetres of aluminium and gamma rays are only stopped by several centimetres of lead. In air, alpha particles can travel no more than a few centimetres, beta particles only a few decimetres and gamma rays several hundreds of metres. Alpha particles thus cannot be detected in radiometric surveying and beta particles only in ground surveys. Only gamma rays can be detected in airborne surveys.

There are three radioactive series of uranium and thorium whose parents are $^{235}_{92}U$, $^{238}_{92}U$ and $^{232}_{90}Th$. These all decay eventually to stable isotopes of lead via intermediate, daughter radioisotopes. About 89% of ^{40}K decays by beta emission to ^{40}Ca and 11% to ^{40}Ar by K-capture.

10.3 Radioactive minerals

There is a large number of radioactive minerals (for a full list see Durrance 1986), but the more common are given in Table 10.1 with their modes of occurrence.

The nature of the mineral in which the radioisotope is found is irrelevant for detection purposes as the prospecting techniques locate the element itself.

Table 10.1 Radioactive minerals. (From Telford *et al.* 1990.)

Potassium

Mineral	(i)	Orthoclase and microcline feldspars [KAlSi$_3$O$_8$]
	(ii)	Muscovite [H$_2$KAl(SiO$_4$)$_3$]
	(iii)	Alunite [K$_2$Al$_6$(OH)$_{12}$SiO$_4$]
	(iv)	Sylvite, carnallite [KCl, MgCl$_2$·6H$_2$O]
Occurrence	(i)	Main constituents in acid igneous rocks and pegmatites
	(ii)	Main constituents in acid igneous rocks and pegmatites
	(iii)	Alteration in acid volcanics
	(iv)	Saline deposits in sediments

Thorium

Mineral	(i)	Monazite [ThO$_2$ + rare earth phosphate]
	(ii)	Thorianite [(Th,U)O$_2$]
	(iii)	Thorite, uranothorite [ThSiO$_4$ + U]
Occurrence	(i)	Granites, pegmatites, gneiss
	(ii), (iii)	Granites, pegmatites, placers

Uranium

Mineral	(i)	Uraninite [oxide of U, Pb, Ra + Th, rare earths]
	(ii)	Carnotite [K$_2$O.2UO$_3$.V$_2$O$_5$.2H$_2$O]
	(iii)	Gummite [uraninite alteration]
Occurrence	(i)	Granites, pegmatites and with vein deposits of Ag, Pb, Cu, etc.
	(ii)	Sandstones
	(iii)	Associated with uraninite

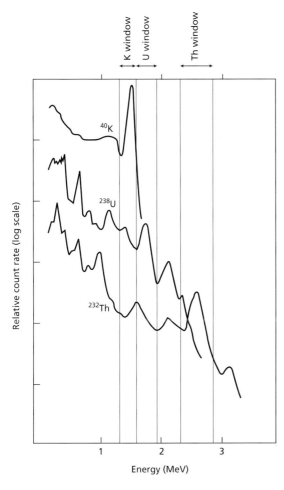

Fig. 10.2 Energy spectra of ^{40}K, ^{238}U and ^{232}Th and their measurement windows.

10.4 Instruments for measuring radioactivity

Several types of detector are available for radiometric surveys, results being conventionally displayed as the number of counts of emissions over a fixed period of time. Radioactive decay is a random process following a Poisson distribution with time so that adequate count times are important if the statistical error in counting decay events is to be kept at an acceptable level.

The standard unit of gamma radiation is the Roentgen (R). This corresponds to the quantity of radiation that would produce 2.083×10^{15} pairs of ions per cubic metre at standard temperature and pressure. Radiation anomalies are usually expressed in μR per hour.

10.4.1 Geiger counter

The *Geiger* (or *Geiger–Müller*) *counter* responds primarily to beta particles. The detecting element is a sealed glass tube containing an inert gas, such as argon, at low pressure plus a trace of a quenching agent such as water vapour, alcohol or methane. Within the tube a cylindrical cathode surrounds a thin axial anode and a power source maintains a potential difference of several hundred volts between them. Incoming beta particles ionize the gas and the positive ions and electrons formed are accelerated towards the electrodes, ionizing more gas en route. These cause discharge pulses across an anode resistor which, after amplification, may be registered as clicks, while an integrating circuit displays the number of counts per minute. The quenching agent suppresses the secondary emission of electrons resulting from bombardment of the cathode by positive ions.

The Geiger counter is cheap and easy to use. However, since it only responds to beta particles, its use is limited to ground surveys over terrain with little soil cover.

10.4.2 Scintillation counter

The *scintillation counter* or *scintillometer* is used to measure gamma radiation based on the phenomenon that certain substances such as thallium-treated sodium iodide and lithium-drifted germanium convert gamma rays to light; that is, they *scintillate*. Photons of light impinging upon a semi-transparent cathode of a photomultiplier cause the emission of electrons. The photomultiplier amplifies the electron pulse before its arrival at the anode where it is further amplified and integrated to provide a display in counts per minute.

The scintillation counter is more expensive than the Geiger counter and less easy to transport, but it is almost 100% efficient in detecting gamma rays. Versions are available which can be mounted in ground transport or aircraft.

10.4.3 Gamma-ray spectrometer

The *gamma-ray spectromet* is an extension of the scintillation counter that enables the source element to be identified. This is possible as the spectra of gamma rays from ^{40}K, ^{232}Th and ^{238}U contain peaks which represent stages in the decay series. Since the higher the frequency of gamma radiation, the higher its contained energy, it is customary to express the spectrum in terms of energy levels. A form of windowing whereby the energy levels between predetermined upper and lower levels are

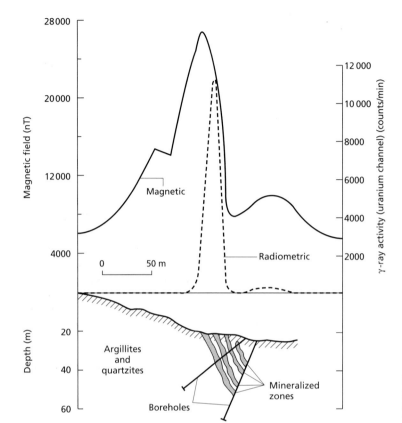

Fig. 10.3 Radiometric and magnetic profiles over pitchblende–magnetite mineralization in Labrador. (After Telford *et al.* 1990.)

monitored then provides a diagnostic means of discriminating between different sources. Figure 10.2 shows the gamma ray spectra of ^{238}U, ^{232}Th and ^{40}K and it is apparent that measurements at 1.76, 2.62 and 1.46 MeV, respectively, provide a discrimination of the source (1 MeV = 10^6 electronvolts, 1 electronvolt being the energy acquired by a particle of unit charge falling through a potential of 1 volt). These devices are sometimes termed *pulse-height analysers* as the intensity of the scintillation pulses is approximately proportional to the original gamma ray energy.

Gamma ray spectrometers for airborne use are often calibrated by flying over an area of known radioisotope concentration or by positioning the aircraft on a concrete slab fabricated with a known proportion of radio-isotopes. The actual concentrations of ^{238}U, ^{232}Th and ^{40}K in the field can then be estimated from survey data.

10.4.4 Radon emanometer

Radon is the only gaseous radioactive element. Being a noble gas it does not form compounds with other elements and moves freely through pores, joints and faults in the subsurface either as a gas or dissolved in groundwater. It is one of the products of the ^{238}U decay series, with a half-life of 3.8 days, and the presence of ^{222}Rn at the surface is often an indication of buried uranium concentrations.

The *radon emanometer* samples air drawn from a shallow drillhole. The sample is filtered, dried and passed to an ionization chamber where alpha particle activity is immediately monitored to provide a count rate.

The emanometer is relatively slow to use in the field. It does, however, represent a means of detecting deeper deposits of uranium than the other methods described above, since spectrometers will only register gamma rays originating in the top metre or so of the subsurface (Telford 1982). Because of its high mobility, radon can have travelled a considerable distance from the source of uranium before being detected. The emanometer has also been used to map faults, which provide channels for the transport of radon generated at depth (Abdoh & Pilkington 1989). This technique is advantageous when there is no great difference in rock properties across the fault that could be detected by other geophysical methods.

10.5 Field surveys

As previously stated, Geiger counter investigations are limited to ground surveys. Count rates are noted and their significance assessed with respect to background effects resulting from the potassium content of the local rocks, nuclear fallout and cosmic radiation. An appreciable anomaly would usually be in excess of three times the background count rate.

Scintillation counters may also be used in ground surveys and are usually sited on rock exposures. The ground surface should be relatively flat so that radioactive emissions originate from the half-space below the instrument. If this condition does not obtain, a lead collimator can be used to ensure that radioactive emissions do not arrive from elevated areas flanking the instrument.

Most radiometric surveying is carried out from the air, employing larger scintillation sensors than in ground instruments, with a consequent increase in measurement sensitivity. Instruments are interfaced with strip recorders and position fixing is by means of the methods discussed in Section 7.8. Radiometric measurements are normally taken in conjunction with magnetic and electromagnetic readings, so providing additional datasets at minimal extra cost. In surveying for relatively small deposits the slow speed of helicopters is often advantageous and provides greater discrimination and amplitude of response. Flight altitude is usually less than 100 m and, because of the weak penetrative powers of radioactive emissions, the information obtained relates only to the top metre or so of the ground.

The interpretation of radiometric data is mainly qualitative, although characteristic curves are available for certain elementary shapes which provide the parameter: (surface area) × (source intensity).

10.6 Example of radiometric surveying

Figure 10.3 shows a ground magnetic and gamma-ray profile across a zone of uranium mineralization in Labrador. This was obtained from contour maps of a small area identified from a regional airborne survey. There are strong coincident magnetic and radiometric anomalies, the source of which was investigated by two boreholes. The anomalies arise from magnetite and pitchblende, located immediately beneath the anomaly maxima, in an argillaceous and quartzitic host. Pitchblende is a variety of massive, botryoidal or colloform uraninite.

Further reading

Durrance, E.M. (1986) *Radioactivity in Geology.* Ellis Horwood, Chichester.

Milsom, J. (1989) *Field Geophysics.* Open University Press, Milton Keynes.

Telford, W.M. (1982) Radon mapping in the search for uranium. *In:* Fitch, A.A. (ed.), *Developments in Geophysical Exploration Methods.* Applied Science, London, 155–94.

Telford, W.M., Geldart, L.P. & Sheriff, R.E. (1990) *Applied Geophysics*, 2nd edn. Cambridge University Press, Cambridge.

Wollenberg, H.A. (1977) Radiometric methods. *In:* Morse, J.G. (ed.), *Nuclear Methods in Mineral Exploration and Production.* Elsevier, Amsterdam, 5–36.

11 Geophysical borehole logging

11.1 Introduction to drilling

Shallow boreholes may be excavated by percussion drilling, in which rock fragments are blown out of the hole by air pressure. Most boreholes, however, are sunk by rotary drilling in which the detritus produced by rotating teeth on a rock bit drilling head is flushed to the surface by a drilling fluid (or 'mud'), which holds it in suspension. The drilling fluid also lubricates and cools the bit and its density is carefully controlled so that the pressure it exerts is sufficient to exceed that of any pore fluids encountered so as to prevent blowouts. Rotary drilling using a core drill in place of a rock bit to obtain core samples is not so widely applied because of its higher costs and slower rate of progress.

The deposition of particles held in suspension in the drilling fluid seals porous wall rocks to form a *mudcake* (Fig. 11.1). Mudcakes up to several millimetres thick can build up on the borehole wall and, since the character of the mudcake is determined by the porosity and permeability of the wallrock in which it is developed, investigation of the mudcake properties indirectly provides insight into these 'poroperm' properties. The drilling fluid filtrate penetrates the wallrock and completely displaces indigenous fluids in a 'flushed zone' which can be several centimetres thick (Fig. 11.1). Beyond lies an *annulus of invasion* where the proportion of filtrate gradually decreases to zero. This zone of invasion is a few centimetres thick in rock such as shale, but can be up to a few metres wide in more permeable and porous rocks.

Casing may be introduced into borehole sections immediately after drilling to prevent collapse of the wallrock into the hole. Cased holes are lined with piping, the voids between wallrock and pipe being filled with cement. Boreholes with no casing are termed *open holes*.

11.2 Principles of well logging

The fragments of rock flushed to the surface during drilling are often difficult to interpret as they have been mixed and leached by the drilling fluid and often provide little information on the intrinsic physical properties of the formations from which they derive. Geophysical borehole logging, also known as *downhole geophysical surveying* or *wire-line logging*, is used to derive further information about the sequence of rocks penetrated by a borehole. Of particular value is the ability to define the depth to geological interfaces or beds that have a characteristic geophysical signature, to provide a means of correlating geological information between boreholes and to obtain information on the *in situ* properties of the wallrock. Potentially, any of the normal geophysical surveying techniques described in previous chapters may be adapted for use in borehole logging, but in practice the most useful and widely-applied methods are based on electrical resistivity, electromagnetic induction, self-potential, natural and induced radioactivity, sonic velocity and temperature.

These methods and some other specialized logging techniques, such as gravity and magnetic logging, are described below. In addition, several other types of subsurface geophysical measurements may be taken in a borehole environment. Of these, perhaps the most important and widely used is vertical seismic profiling, as discussed in Section 4.13.

The instrumentation necessary for borehole logging is housed in a cylindrical metal tube known as a *sonde*. Sondes are suspended in the borehole from an armoured multicore cable. They are lowered to the base of the section of the hole to be logged, and logging is carried out as the sonde is winched back up through the section. Logging data are commonly recorded on a paper strip chart and also on magnetic tape in analogue or digital form for subsequent computer processing. The surface

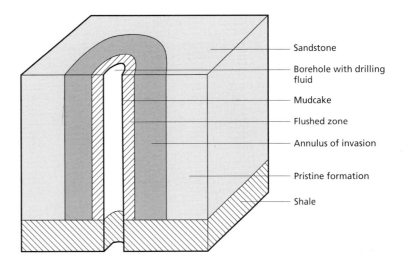

Sandstone

Borehole with drilling
fluid

Mudcake

Flushed zone

Annulus of invasion

Pristine formation

Shale

Fig. 11.1 The borehole environment.

instrumentation, including recorders, cable drums and winches, is usually installed in a special recording truck located near the wellhead. Sondes normally contain combinations of logging tools that do not mutually interfere, so that a wide suite of geophysical logs may be obtained from a limited number of logging runs.

Several techniques of borehole logging are used together to overcome the problems of mudcake and drilling fluid filtrate invasion so as to investigate the properties of the pristine wallrock. Open holes can be surveyed with the full complement of logging tools. Casing prevents the use of logging methods based on electrical resistivity and distorts measurement of seismic velocities. Consequently only a few of the logging methods, such as those based on radioactivity, can be used in cased holes.

Logging techniques are very widely used in the investigation of boreholes drilled for hydrocarbon exploration, as they provide important *in situ* properties of possible reservoir rocks. They are also used in hydrogeological exploration for similar reasons. A review of the methodology and applications of borehole logging at sea is given in Goldberg (1997). Some modern case histories and reviews of recent developments are given in Lovell *et al.* (1999).

permeability, proportion of water and/or hydrocarbon saturation, stratal dip and temperature.

Formation thickness and lithology are normally determined by comparison of borehole logs with the log of a cored hole. The most useful logs are those based on resistivity (Section 11.4), self-potential (Section 11.6), radioactivity (Section 11.7) and sonic velocity (Section 11.8), and these are often used in combination to obtain an unambiguous section. The calliper log, which measures changes in borehole diameter, also provides information on the lithologies present. In general, larger diameters reflect the presence of less cohesive wallrocks which are easily eroded during drilling.

Porosity estimates are usually based on measurements of resistivity, sonic velocity and radioactivity. In addition, porosity estimates may be obtained by gamma-ray density logging (Section 11.7.2), neutron–gamma-ray logging (Section 11.7.3) and nuclear magnetic resonance logging (Section 11.10). The methodology is described in the relevant sections which follow. Permeability and water and hydrocarbon saturation are derived from resistivity measurements. Stratal dip and temperature are determined by their own specialized logs.

11.3 Formation evaluation

The geological properties obtainable from borehole logging are: formation thickness and lithology, porosity,

11.4 Resistivity logging

In this chapter the symbol R is used for resistivity to avoid confusion with the symbol ρ used for density.

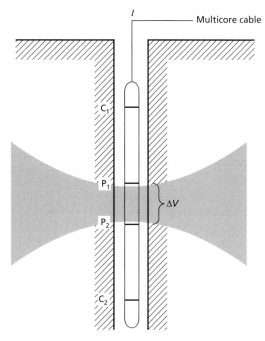

Fig. 11.2 The general form of electrode configuration in resistivity logging. The shaded area represents the effective region energized by the system.

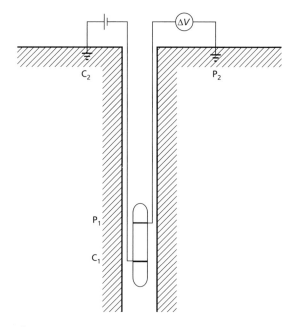

Fig. 11.3 The normal log.

The general equation for computing apparent resistivity R_a for any downhole electrode configuration is

$$R_a = \frac{4\pi\Delta V}{I\left\{\left(\dfrac{1}{C_1P_1} - \dfrac{1}{C_2P_1}\right) - \left(\dfrac{1}{C_1P_2} - \dfrac{1}{C_2P_2}\right)\right\}} \quad (11.1)$$

where C_1, C_2 are the current electrodes, P_1, P_2 the potential electrodes between which there is a potential difference ΔV, and I is the current flowing in the circuit (Fig. 11.2). This is similar to equation (8.9) but with a factor of 4 instead of 2, as the current is flowing in a full space rather than the halfspace associated with surface surveying.

Different electrode configurations are used to give information on different zones around the borehole. Switching devices allow the connection of different sets of electrodes so that several types of resistivity log can be measured during a single passage of the sonde.

The region energized by any particular current electrode configuration can be estimated by considering the equipotential surfaces on which the potential electrodes lie. In a homogeneous medium, the potential

difference between the electrodes reflects the current density and resistivity in that region. The same potential difference would be obtained no matter what the position of the potential electrode pair. The zone energized is consequently the region between the equipotential surfaces on which the potential electrodes lie. Figure 11.2 shows the energized zone in a homogeneous medium.

11.4.1 Normal log

In the *normal log*, only one potential and current electrode are mounted on the sonde, the other pair being grounded some distance from the borehole (Fig. 11.3). By substitution in equation (11.1)

$$R_a = 4\pi C_1 P_1 \frac{\Delta V}{I} \quad (11.2)$$

Since C_1P_1 and I are constant, R_a varies with ΔV, and the output can be calibrated directly in ohm m. The zone energized by this configuration is a thick shell with an inner radius C_1P_1 and a large outer radius. However, the current density decreases rapidly as the separation of C_1

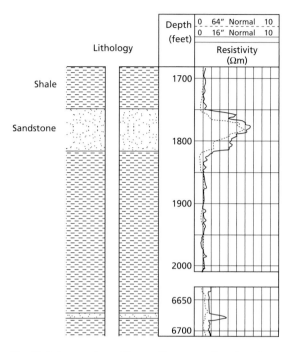

Fig. 11.4 A comparison of short and long normal logs through a sequence of sandstone and shale. (After Robinson & Çoruh 1988.)

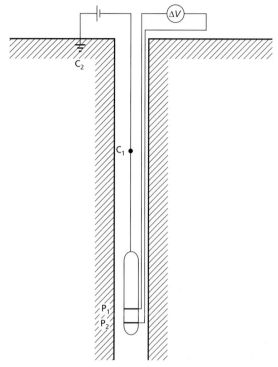

Fig. 11.5 The lateral log.

and P_2 increases, so that measurements of resistivity correspond to those in a relatively thin spherical shell. The presence of drilling fluid and resistivity contrasts across lithological boundaries cause current refractions so that the zone tested changes in shape with position in the hole.

It is possible to correct for the invasion of drilling fluid by using the results of investigations with different electrode separations (*short normal log* 16 in (406 mm), *long normal log* 64 in (1626 mm)) which give different penetration into the wallrock. Comparison of these logs with standard correction charts (known as departure curves) allows removal of drilling-fluid effects.

The normal log is characterized by smooth changes in resistivity as lithological boundaries are traversed by the sonde because the zone of testing precedes the sonde and the adjacent bed controls the apparent resistivity. Examples of short and long normal logs are given in Fig. 11.4.

11.4.2 Lateral log

In the *lateral log* the in-hole current electrode C_1 is a con-

siderable distance above the potential electrode pair, and is usually mounted on the wire about 6 m above a short sonde containing P_1 and P_2 about 800 mm apart (Fig. 11.5). For this electrode configuration

$$R_a = \frac{4\pi\Delta V}{I\left(\dfrac{1}{C_1P_1} - \dfrac{1}{C_1P_2}\right)} \tag{11.3}$$

An alternative configuration uses C_1 mounted below the potential electrode pair.

The measured potential difference varies in proportion to the resistivity, so the output can be calibrated directly in ohm m. The zone energized extends much farther into the wallrock than with normal logs, and the apparent resistivity thus approaches the pristine wallrock value more closely.

The electrode configuration causes asymmetry in the apparent resistivity signature as the potential electrode pair descends through one bed while the current electrode may be moving through another. Thin beds

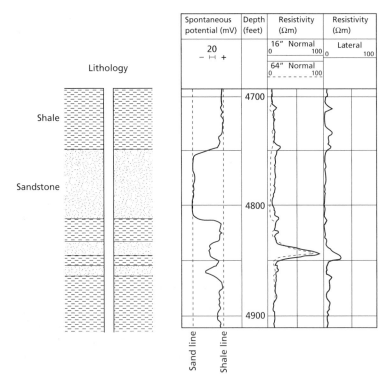

Lithology	Spontaneous potential (mV)	Depth (feet)	Resistivity (Ωm)	Resistivity (Ωm)

Shale

Sandstone

Sand line   Shale line

20
− ⊢ +

16" Normal
0   100

64" Normal
0 _ _ _ _ _ _ _ 100

Lateral
0   100

4700

4800

4900

Fig. 11.6 The lateral log compared with normal and self-potential logs. (After Guyod 1974.)

produce spurious peaks below them. The lateral log does, however, give a clear indication of the lower boundary of a formation. An example of a lateral log and comparison with normal and self-potential logs is given in Fig. 11.6. As with the normal logs, corrections for the effects of invasion can be applied by making use of standard charts.

11.4.3 Laterolog

The normal and lateral logs described above have no control on the direction of current flow through the wallrock. By contrast, the *laterolog* (or *guard log*) is a focused log in which the current is directed horizontally so that the zone tested has the form of a circular disc. This may be achieved by the use of a short electrode 75–300 mm long between two long (guard) electrodes about 1.5 m long (Fig. 11.7). The current supply to the electrodes is automatically adjusted so as to maintain them all at the same potential. Since no potential difference exists between the electrodes, the current flows outwards horizontally, effectively energizing the wallrock to a depth of about three times the length of the guard electrodes. The use of a fixed potential has the consequence that the current in the central electrode varies in proportion to

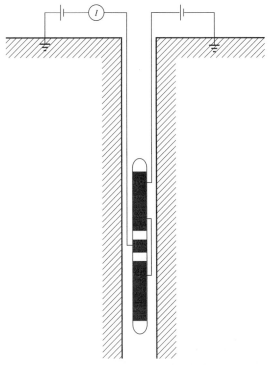

Fig. 11.7 The laterolog.

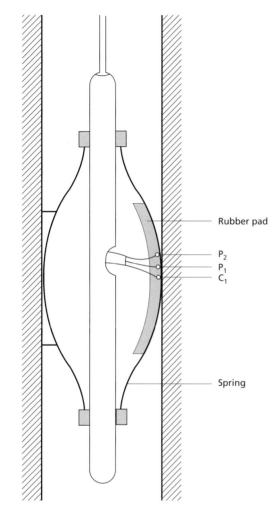

Fig. 11.8 The microlog.

insulating pad pressed firmly against the wallrock by a power-driven expansion device (Fig. 11.8). The depth of penetration is typically about 100 mm. Different electrode arrangements allow the measurement of micronormal, microlateral and microlaterolog apparent resistivities that are equivalent to normal, lateral and laterolog measurements with much smaller electrode spacings. The log has to be moved very slowly and it is normally used only in short borehole sections which are of particular interest.

As the electrode spacing is so small, the effects of the borehole diameter, drilling fluid and adjacent beds are negligible. Very thin beds register sharply, but the main use of the microlog is to measure the resistivities of the mudcake and zone of invasion, which are needed to convert log measurements into true resistivities.

11.4.5 Porosity estimation

Porosity is defined as the fractional volume of pore spaces in a rock. The method of *porosity estimation* is based on the relationship between *formation factor F* and porosity ϕ discovered by Archie (1942). *F* is a function of rock texture and defined as

$$F = \frac{R_{\mathrm{f}}}{R_{\mathrm{w}}} \tag{11.4}$$

where R_{f} and R_{w} are the resistivities of the saturated formation and pore fluid, respectively (Section 8.2.2). Porosity and formation factor are related by

$$\phi = aF^{-m} \tag{11.5}$$

where *a* is an empirical constant specific to the rocks of the area of interest, and *m* a constant known as the *cementation factor* which depends on the grain size and complexity of the paths between pores (*tortuosity*). Normal limits on *a* and *m*, derived experimentally, are given by

$$0.62 < a < 1.0, \qquad \text{and} \qquad 2.0 < m < 3.0$$

the apparent resistivity so that the output can be calibrated in ohm m.

The focusing of the log makes it sensitive to thin beds down to the same thickness as the length of the central electrode. The zone of invasion has a pronounced effect which can be estimated from the results of normal and lateral logging and corrected using standard charts.

11.4.4 Microlog

The *microlog* (or *wall-resistivity log*) makes measurements at very small electrode spacings by using small, button-shaped electrodes 25–50 mm apart mounted on an

11.4.6 Water and hydrocarbon saturation estimation

Natural pore water is generally a good conductor of electricity because of the presence of dissolved salts.

Hydrocarbons, however, are poor conductors and cause an increase in the measured resistivity of a rock relative to that in which water is the pore fluid. Hydrocarbons displace pore water and cause it to be reduced to an irreducible minimum level. Archie (1942) described a method of estimating the proportion of pore water present (the *water saturation S*) based on laboratory measurements of the resistivities of sandstone cores containing varying proportions of hydrocarbons and pore water of fixed salinity. If R_f and R_h are the resistivities of (matrix + pore water) and (matrix + pore water + hydrocarbons), respectively, then

$$S = \left(\frac{R_f}{R_h} \right)^{1/n} \tag{11.6}$$

where n is the *saturation exponent*. The experimentally determined limits of n are $1.5 < n < 3.0$, although n is usually assumed to be 2 where there is no evidence to the contrary.

Combining equations (11.4) and (11.6) gives an alternative expression for S

$$S = \left(\frac{FR_w}{R_h} \right)^{1/n} \tag{11.7}$$

R_f is determined in parts of the borehole which are known to be saturated with water.

11.4.7 Permeability estimation

Permeability (k) is a measure of the capacity of a formation to transmit fluid under the influence of a pressure gradient. It is dependent upon the degree of interconnection of the pores, the size of the pore throats and the active capillary forces. It is estimated from the minimum pore water remaining after displacement of the rest by hydrocarbons (the *irreducible water saturation* S_{irr}), which in turn is estimated from resistivity measurements in parts of the formation where irreducible saturation obtains:

$$k = \left(\frac{c\phi^3}{S_{irr}} \right)^2 \tag{11.8}$$

where ϕ is determined as in Section 11.4.5 and c is a constant dependent on the lithology and grain size of the

formation. Large errors in determining the parameters from which k is derived render permeability the most difficult reservoir property to estimate.

k is commonly expressed in darcies, a unit corresponding to a permeability which allows a flow of $1\,\mathrm{mm\,s^{-1}}$ of a fluid of viscosity $10^{-3}\,\mathrm{Pa\,s}$ through an area of $100\,\mathrm{mm^2}$ under a pressure gradient of $0.1\,\mathrm{atm\,mm^{-1}}$. Reservoirs commonly exhibit values of permeability from a few millidarcies to 1 darcy.

11.4.8 Resistivity dipmeter log

The sonde of the *dipmeter log* contains four equally-spaced microresistivity electrodes at the same horizontal level, which allow the formation dip and strike to be estimated. The orientation of the sonde is determined by reference to a magnetic compass and its deviation from the vertical by reference to a spirit level or pendulum. The four electrodes are mounted at right angles to each other round the sonde. If the beds are horizontal, identical readings are obtained at each electrode. Non-identical readings can be used to determine dip and strike. In fact the four electrodes can be used to make four three-point dip calculations as a control on data quality. Dipmeter results are commonly displayed on *tadpole plots* (Fig. 11.9).

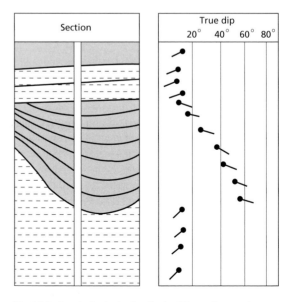

Fig. 11.9 A typical tadpole plot obtained from a dipmeter log.

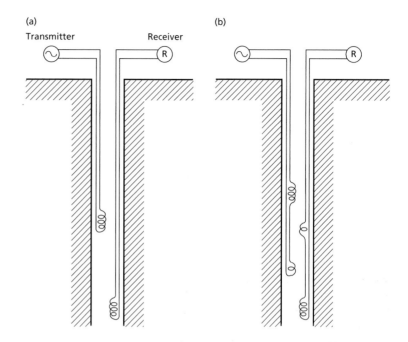

Fig. 11.10 (a) A simple induction log.
(b) A focused induction log.

11.5 Induction logging

The *induction log* is used in dry holes or boreholes that contain non-conductive drilling fluid which electrically insulates the sonde. The wallrock is energized by an electromagnetic field, typically of about 20 kHz, which generates eddy currents in the wallrock by electromagnetic induction. The secondary EM field created is registered at a receiver which is compensated for direct coupling with the primary field and which allows a direct estimate of apparent resistivity to be made. The set-up is thus similar to the surface moving coil-receiver EM system described in Section 9.5.

The two-coil system shown in Fig. 11.10(a) is unfocused and the induced EM field flows in circular paths around the borehole, with a depth of investigation of about 75% of the transmitter–receiver separation. Lithological boundaries show up as gradual changes in apparent resistivity as they are traversed. When combined with information from other logs, corrections for invasion can be made from standard charts.

Clearer indications of lithological contacts can be obtained using a focused log such as that shown in Fig. 11.10(b), in which two extra coils are mounted near the receiver and transmitter and wired in series with them. Such an arrangement provides a depth of penetration of

about twice the transmitter–receiver separation. This particular focused system has the disadvantage that spurious apparent resistivities are produced at boundaries, but this effect may be compensated by employing additional coils.

See Section 9.6 for the application of time-domain electromagnetic techniques in borehole surveys.

11.6 Self-potential logging

In the *self-potential* (*SP*) *log*, measurements of potential difference are made in boreholes filled with conductive drilling fluid between an electrode on the sonde and a grounded electrode at the surface (Fig. 11.11).

The SP effect (Section 8.4.2) originates from the movement of ions at different speeds between two fluids of differing concentration. The effect is pronounced across the boundary between sandstone and shale, as the invasion of drilling mud filtrate is greater into the sandstone. Near the borehole there is a contact between mud filtrate in the sandstone and pore fluid of different salinity in the shale. The movement of ions necessary to nullify this difference is impeded by the membrane polarization effect (Section 8.3.2) of the clay minerals in the shale. This causes an imbalance of charge across the

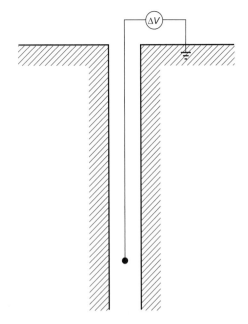

Fig. 11.11 The self-potential log.

boundary and generates a potential difference of a few tens to a few hundreds of millivolts.

In sequences of sandstone and shale, the sandstone anomaly is negative with respect to the shale. This SP effect provides a sharper indication of the boundary than resistivity logs. In such sequences it is possible to draw a 'shale line' through the anomaly maxima and a 'sand line' through the minima (see Fig. 11.6). The proportion of sand to shale at intermediate anomalies can then be estimated by interpolation.

The main applications of SP logging are the identification of boundaries between shale horizons and more porous beds, their correlation between boreholes, and the determination of the volume of shale in porous beds. They have also been used to locate coal seams. In hydrocarbon-bearing zones the SP log has less deflection than normal and this 'hydrocarbon suppression' can be an indicator of their presence.

11.7 Radiometric logging

Radiometric logs make use of either the natural radioactivity produced by the unstable elements ^{238}U, ^{232}Th and ^{40}K (Section 10.2), or radioactivity induced by the bombardment of stable nuclei with gamma rays or neutrons. Gamma rays are detected by a scintillation counter (Section 10.4.2) or occasionally by a Geiger–Müller counter (Section 10.4.1) or an ionization chamber. Radioactivity in borehole measurements is usually expressed in API (American Petroleum Institute) units, which are defined according to reference levels in a test pit at the University of Houston.

11.7.1 Natural gamma radiation log

Shales usually contain small quantities of radioactive elements, in particular ^{40}K which occurs in micas, alkali feldspars and clay minerals, and trace amounts of ^{238}U and ^{232}Th. These produce detectable gamma radiation from which the source can be distinguished by spectrometry; that is, measurements in selected energy bands (Section 10.4.3). The *natural gamma radiation log* consequently detects shale horizons and can provide an estimate of the clay content of other sedimentary rocks. Potassium-rich evaporites are also distinguished. An example of this type of log is shown in Fig. 11.12.

The natural gamma radiation log (or *gamma log*) measures radioactivity originating within a few decimetres of the borehole. Because of the statistical nature of gamma-ray emissions, a recording time of several seconds is necessary to obtain a reasonable count, so the sensitivity of the log depends on the count time and the speed with which the hole is logged. Reasonable results are obtained with a count time of 2 s and a speed of 150 mm s^{-1}. Measurements can be made in cased wells, but the intensity of the radiation is reduced by about 30%.

11.7.2 Gamma-ray density log

In the *gamma-ray density* (or *gamma–gamma*) *log*, artificial gamma rays from a ^{60}Co or ^{137}Cs source are utilized. Gamma-ray photons collide 'elastically' with electrons and are reduced in energy, a phenomenon known as *Compton scattering*. The number of collisions over any particular interval of time depends upon the abundance of electrons present (the *electron density index*), which in turn is a function of the density of the formation. Density is thus estimated by measuring the proportion of gamma radiation returned to the detector by Compton scattering.

The relationship between the formation density ρ_f and electron density index ρ_e depends upon the elements present

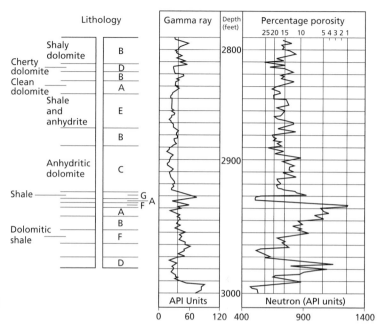

Fig. 11.12 Natural gamma and neutron logs over the same sequence of dolomite and shale. (After Wood *et al.* 1974.)

$$\rho_f = \frac{\rho_e w}{2 \sum N} \qquad (11.9)$$

where w is the molecular weight of the constituents of the formation and N is the atomic number of the elements present, which specifies the number of electrons.

The sonde has a plough-shaped leading edge which cuts through the mudcake, and is pressed against the wallrock by a spring. Most of the scattering takes place within about 75 mm of the sonde. A modern version of the sonde uses long and short spacings for the detectors which are sensitive to material far from and near to the sonde, respectively.

Porosity ϕ may be estimated from the density measurements. For a rock of formation density ρ_f, matrix density ρ_m and pore fluid density ρ_w

$$\rho_f = \phi \rho_w + (1 - \phi) \rho_m \qquad (11.10)$$

Thus

$$\phi = \frac{(\rho_m - \rho_f)}{(\rho_m - \rho_w)} \qquad (11.11)$$

11.7.3 Neutron–gamma-ray log

In the neutron–gamma-ray (or neutron) log, non-radioactive elements are bombarded with neutrons and, as a result of neutron capture by the nuclei, they are stimulated to emit gamma rays which provide information on porosity. The sonde contains a neutron source, consisting of a small quantity of a radioactive substance such as Pu–Be, and a scintillation counter (Section 10.4.2) a fixed distance apart.

The neutrons collide with atomic nuclei in the wallrock. Most nuclei are much more massive than neutrons, which rebound 'elastically' with very little loss of kinetic energy. However, a hydrogen ion has almost the same mass as a neutron, so collision transfers considerable kinetic energy and slows the neutron to the point at which it can be absorbed by a larger nucleus. This neutron capture, which normally occurs within 600 mm of the borehole, gives rise to gamma radiation, a proportion of which impinges on the scintillation counter. The intensity of the radiation is controlled by how far it has travelled from the point of neutron capture. This distance depends mainly on the hydrogen-ion concentration: the higher the concentration, the closer the neutron capture to the borehole and the higher the level of radiation.

In sandstone and limestone all hydrogen ions are present in pore fluids or hydrocarbons, so the hydrogen ion concentration is entirely dependent upon the porosity. In shales, however, hydrogen can also derive from micas and clay minerals. Consequently, the lithology must be determined by other logs (e.g. gamma log) before porosity estimates can be made in this way. Similar count times and logging speeds to other radiometric methods are used. The method is suitable for use in both cased and uncased boreholes. An example is given in Fig. 11.12.

11.8 Sonic logging

The *sonic log*, also known as the *continuous velocity* or *acoustic log*, determines the seismic velocities of the formations traversed. The sonde normally contains two receivers about 300 mm apart and an acoustic source some 900–1500 mm from the nearest receiver (Fig. 11.13(a)). The source generates ultrasonic pulses at a frequency of 20–40 kHz.

Since the wallrock invariably has a greater velocity than the drilling fluid, part of the sonic pulse is critically refracted in the wallrock and part of its energy returns to the sonde as a head wave (Section 3.6.3). Each sonic pulse activates a timer so that the differential travel time between the receivers can be measured. If the sonde is tilted in the well, or if the well diameter varies, different path lengths result. This problem is overcome, in a borehole-compensated log, by using a second source on the other side of the receivers (Fig. 11.13(b)) so that the tilt effect is self-cancelling when all four travel paths are considered.

Porosity ϕ may be estimated from the sonic measurements (see Section 3.4). For a rock whose matrix velocity (the velocity of its solid components) is V_m and pore fluid velocity is V_w, the formation velocity V_f is given by

$$\frac{1}{V_f} = \frac{\phi}{V_w} + \frac{1-\phi}{V_m} \qquad (11.12)$$

The velocity of the matrix can be determined from cuttings and that of the fluid from standard values.

Sondes of the dimensions described above have transmission path lengths that lead to penetrations of only a few centimetres into the wallrock and allow the discrimination of beds only a few decimetres in thickness. However, they are greatly affected by drilling damage to the wallrock and, to overcome this, longer sondes with source–geophone spacings of 2.1–3.7 m may be used. In addition to providing porosity estimates, sonic logs may be used for correlation between boreholes and are also used in the interpretation of seismic reflection data by providing velocities for the conversion of reflection times into depths. An example is given in Fig. 11.14.

Sonic logs can also provide useful attenuation information, usually from the first P-wave arrival. Attenuation (Section 3.5) is a function of many variables including wavelength, wave type, rock texture, type and nature of pore fluid and the presence of fractures and fissures. However, in a cased well, the attenuation is at a minimum when the casing is held in a thick annulus of cement and at a maximum when the casing is free. This forms the basis of the cement bond log (or cement evaluation probe) which is used to investigate the effectiveness of the casing. Other techniques make use of both P- and S-wave travel times to estimate the *in situ* elastic moduli (Section 5.11). See also the description of vertical seismic profiling in Section 4.13.

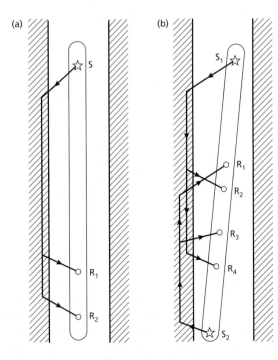

Fig. 11.13 (a) A simple sonic log. (b) A borehole-compensated sonic log.

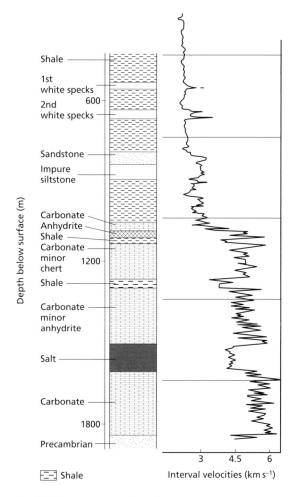

Fig. 11.14 A continuous velocity log. (After Grant & West 1965.)

is the *thermal conductivity* of the relevant wallrock, which is usually determined by laboratory measurement.

Temperature gradients within about 20 m of the Earth's surface are strongly affected by diurnal and seasonal changes in solar heating and do not provide reliable estimates of heat flux. Porous strata can also strongly influence temperature gradients by the ingress of connate water and because their contained pore fluids act as a thermal sink. Heat flux measurements are commonly made to assess the potential of an area for geothermal energy utilization.

11.10 Magnetic logging

11.10.1 Magnetic log

The normal *magnetic log* has only limited application. The magnetic field is either measured with a downhole flux-gate or with a proton magnetometer (Section 7.6) or a susceptibility meter is utilized. Anomalous readings indicate the presence of magnetic minerals.

11.10.2 Nuclear magnetic resonance log

The *nuclear magnetic resonance* (or *free fluid index*) *log* is used to estimate the hydrogen ion concentration in formation fluids and, hence, to obtain a measure of porosity. The method of measurement resembles that of the proton magnetometer, but with the formation fluid taking the place of the sensor. A pulsed magnetic field causes the alignment of some of the hydrogen ions in a direction different from the Earth's field. A receiver measures the amplitude and decay rate of the precession of the protons as they realign in the geomagnetic field direction when the polarizing field is inactive. The amplitude measurements provide an estimate of the amount of fluid in the pore spaces and the rate of decay is diagnostic of the type of fluid present.

11.11 Gravity logging

In situations where density is a function of depth only, the strata being substantially horizontal, stepwise measurement of the vertical gravity gradient with a *gravity log* can be used to estimate mean densities according to the calculation given in Section 6.9.

A specialized borehole gravimeter of LaCoste and Romberg type (Section 6.4) is used for gravity logging.

11.9 Temperature logging

Temperature gradients may be measured through a borehole section using a sonde on which a number of closely-spaced thermistor probes are mounted. The vertical *heat flux H* is estimated by

$$H = k_z \frac{d\theta}{dz} \tag{11.13}$$

where $d\theta/dz$ is the vertical temperature gradient and k_z

The instrument has a diameter of about 100 mm, an accuracy of ±5 microgal, and is capable of operation in temperatures up to 120°C and pressures up to 80 MPa. The normal vertical spacing of observations is about 6 m and, if depths are determined to ±50 mm, densities can be estimated to ±0.01 Mg m^{-3}, which corresponds to an accuracy of porosity estimation of about ±1%. The density applies to the part of the formation lying within about five times the spacing between observations. This is more accurate than other methods of measuring density in boreholes and can be used in cased holes. It is, however, time consuming as each reading can take 10–20 min and the meter is so costly that it can only be risked in boreholes in excellent condition.

Problems

1. A sandstone, when saturated with water of resistivity 5 ohm m, has a resistivity of 40 ohm m. Calculate the probable range of porosity for this rock.

2. On a sonic log, the travel time observed in a sandstone was 568 μs over a source–receiver distance of 2.5 m. Given that the seismic velocities of quartz and pore fluid are 5.95 and 1.46 km s^{-1}, respectively, calculate the porosity of the sandstone. What would be the effect on the observed travel time and velocity of the sandstone if the pore fluid were methane with a velocity of 0.49 km s^{-1}?

3. During the drilling of an exploratory borehole, the rock chippings flushed to the surface indicated the presence of a sandstone–shale sequence. The lateral log revealed a discontinuity at 10 m depth below which the resistivity decreased markedly. The SP log showed no deflection at this depth and recorded consistently low values. The gamma-ray density log indicated an increase in density with depth across the interface from 2.24 to 2.35 Mg m^{-3}.

(a) Infer, giving your reasons, the nature of the discontinuity.

(b) What porosity information is provided by the data?

4. Figure 11.15 shows the SP and short normal (including a partial expanded scale version), long normal and lateral resistivity logs of a borehole penetrating a sedimentary sequence. Interpret the logs as fully as possible.

5. Figure 11.16 shows the SP, induction, laterolog, sonic, calliper and gamma logs of a bore-

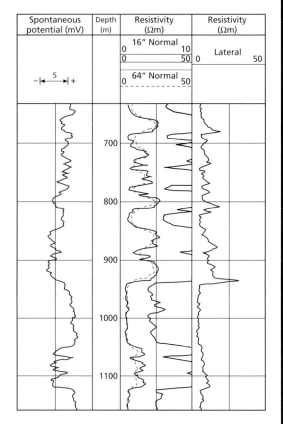

Fig. 11.15 SP and resistivity logs pertaining to Question 4. (After Desbrandes 1985.)

hole in a sequence of shale and sandstone. Interpret the logs as fully as possible.

6. Two gravity readings in a borehole, 100 m apart vertically, reveal a measured gravity difference of 107.5 gu. What is the average density of the rocks between the two observation levels?

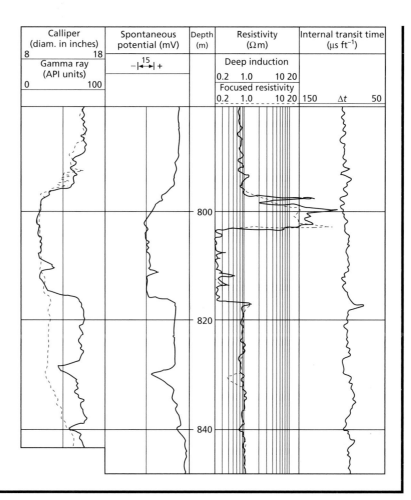

Calliper (diam. in inches) 8 18	Spontaneous potential (mV)	Depth (m)	Resistivity (Ωm)	Internal transit time (μs ft^{-1})

Fig. 11.16 SP, induction, resistivity, sonic, calliper and gamma-ray logs pertaining to Question 5. (After Ellis 1987.)

Further reading

Asquith, G.B. & Gibson, C.R. (1982) *Basic Well Log Analysis for Geologists.* Am. Assoc. Petroleum Geologists, Tulsa.

Chapellier, D. (1992) *Well Logging in Hydrogeology.* A.A. Balkema Publishers, Brookfield, Vermont.

Desbrandes, R. (1985) *Encyclopedia of Well Logging.* Graham & Trotman, London.

Dyck, A.V. & Young, R.P. (1985) Physical characterization of rock masses using borehole methods. *Geophysics,* **50**, 2530–41.

Ellis, D.V. (1987) *Well Logging for Earth Scientists.* Elsevier, Amsterdam.

Hearst, J.R. & Nelson, P.H. (1985) *Well Logging for Physical Properties.* McGraw-Hill, New York.

Hurst, A., Lovell, M.A. & Morton A.C. (eds) (1990) *Geological Applications of Wireline Logs.* Spec. Pub. Geol. Soc. Lond. **48**.

Labo, J. (1986) *A Practical Introduction to Borehole Geophysics.* Soc. Econ. Geologists, Tulsa.

Pirson, S.J. (1977) *Geologic Well Log Analysis* (2nd edn). Gulf, Houston.

Rider, M.H. (1986) *The Geological Interpretation of Well Logs.* Blackie, London.

Robinson, E.S. & Çoruh, C. (1988) *Basic Exploration Geophysics.* John Wiley & Sons, New York.

Segesman, F.F. (1980) Well logging method. *Geophysics,* **45**, 1667–84.

Serra, O. (1984) *Fundamentals of Well-log Interpretation, 1. The Acquisition of Logging Data.* Elsevier, Amsterdam.

Serra, O. (1986) *Fundamentals of Well-log Interpretation, 2. The Interpretation of Logging Data.* Elsevier, Amsterdam.

Snyder, D.D. & Fleming, D.B. (1985) Well logging – a 25 year perspective. *Geophysics,* **50**, 2504–29.

Tittman, J. (1987) Geophysical well logging. *In*: Samis, C.G. & Henyey, T.L. (eds), *Methods of Experimental Physics, Vol. 24, Part B – Field Measurements.* Academic Press, Orlando, 441–615.

Appendix: SI, c.g.s. and Imperial (customary USA) units and conversion factors

Quantity	SI name	SI symbol	c.g.s. equivalent	Imperial (USA) equivalent
Mass	kilogram	kg	10^3 g	2.205 lb
Time	second	s	s	s
Length	metre	m	10^2 cm	39.37 in
				3.281 ft
Acceleration	metre s^{-2}	m s^{-2}	10^2 cm s^{-2} = 10^2 gal	39.37 in s^{-2}
Gravity	gravity unit	gu = μm s^{-2}	10^{-1} milligal (mgal)	3.937×10^{-5} in s^{-2}
Density	megagram m^{-3}	Mg m^{-3}	g cm^{-3}	3.613×10^{-2} lb in^{-3}
				62.421 lb ft^{-3}
Force	newton	N	10^5 dyne	0.2248 lb (force)
Pressure	pascal	Pa = N m^2	10 dyne cm^{-2} = 10^{-5} bar	1.45×10^{-4} lb in^{-2}
Energy	joule	J	10^7 erg	0.7375 ft lb
Power	watt	W = J s^{-1}	10^7 erg s^{-1}	0.7375 ft lb s^{-1}
				1.341×10^{-3} hp
Temperature	T	°C*	°C	$(1.8T + 32)$°F
Current	ampere	A	A	A
Potential	volt	V	V	V
Resistance	ohm	Ω = V A^{-1}	Ω	Ω
Resistivity	ohm m	Ω m	10^2 Ω cm	3.281 ohm ft
Conductance	siemens	S = Ω^{-1}	mho	mho
Conductivity	siemens m^{-1}	S m^{-1}	10^{-2} mho cm^{-1}	0.3048 mho ft^{-1}
Dielectric constant	dimensionless			
Magnetic flux	weber	Wb = V s	10^8 maxwell	
Magnetic flux density (B)	tesla	T = Wb m^{-2}	10^4 gauss (G)	
Magnetic anomaly	nanotesla	nT = 10^{-9} T	gamma (γ) = 10^{-5} G	
Magnetizing field (H)	ampere m^{-1}	A m^{-1}	$4\pi \times 10^{-3}$ oersted (Oe)	
Inductance	henry	H = Wb A^{-1}	10^9 emu (electromagnetic unit)	
Permeability of vacuum (μ_0)	henry m^{-1}	$4\pi \times 10^{-7}$ H m^{-1}	1	
Susceptibility	dimensionless	k	4π emu	
Magnetic pole strength	ampere m	A m	10 emu	
Magnetic moment	ampere m^2	A m^2	10^3 emu	
Magnetization (J)	ampere m^{-1}	A m^{-1}	10^{-3} emu cm^{-3}	

* Strictly, SI temperatures should be stated in kelvin (K = 273.15 + °C). In this book, however, temperatures are given in the more familiar Centigrade (Celsius) scale.

References

Abdoh, A. & Pilkington, M. (1989) Radon emanation studies of the Ile Bizard Fault, Montreal. *Geoexploration*, **25**, 341–54.

Al-Chalabi, M. (1972) Interpretation of gravity anomalies by non-linear optimisation. *Geophys. Prosp.*, **20**, 1–16.

Ali, J. & Hill, I.A. (1991) Reflection seismics for shallow geological applications: a case study from Central England. *J. Geol. Soc. London*, **148**, 219–22.

Al-Sadi, H.N. (1980) *Seismic Exploration*. Birkhauser Verlag, Basel.

Anstey, N.A. (1965) Wiggles. *J. Can. Soc. Exploration Geophysicists*, **1**, 13–43.

Anstey, N.A. (1977) *Seismic Interpretation: The Physical Aspects*. IHRDC, Boston.

Apostolopoulos, G., Louis, I. & Lagios, E. (1997) The SP method in the geothermal exploration of Greece. *Geophysics*, **62**, 1715–23.

Archie, G.E. (1942) The electrical resistivity log as an aid in determining some reservoir characteristics. *Trans. Am. Inst. Mining Met. Eng.*, **146**, 54–62.

Arnaud Gerkens, J.C. d' (1989) *Foundations of Exploration Geophysics*. Elsevier, Amsterdam.

Arzi, A.A. (1975) Microgravimetry for engineering applications. *Geophys. Prosp.*, **23**, 408–25.

Aspinall, A. & Walker, A.R. (1975) The earth resistivity instrument and its application to shallow earth surveys. *Underground Services*, **3**, 12–15.

Ates, A. & Kearey, P. (1995) A new method for determining magnetization direction from gravity and magnetic anomalies: application to the deep structure of the Worcester Graben. *J. Geol. Soc. Lond.*, **152**, 561–6.

Baeten, G., Fokkema, J. & Ziolkowski, A. (1988) The marine vibrator source. *First Break*, **6**(9), 285–94.

Bamford, D., Nunn, K., Prodehl, C. & Jacob, B. (1978) LISPB-IV. Crustal structure of northern Britain. *Geophys. J. R. astr. Soc.*, **54**, 43–60.

Baranov, V. & Naudy, H. (1964) Numerical calculation of the formula of reduction to the magnetic pole (airborne). *Geophysics*, **29**, 67–79.

Barazangi, M. & Brown, L. (eds) (1986) *Reflection Seismology: The Continental Crust*. AGU Geodynamics Series, 14. American Geophysical Union, Washington.

Barker, R.D. (1981) The offset system of electrical resistivity sounding and its use with a multicore cable. *Geophys. Prosp.*, **29**, 128–43.

Barker, R.D. (1997) Electrical imaging and its application in engineering investigations. *In*: McCann, D.M., Eddleston, M.,

Fenning, P.J. & Reeves, G.M. (eds), *Modern Geophysics in Engineering Geology*. Geological Society Engineering Geology Special Publication **12**, 37–43.

Barraclough, D.R. & Malin, S.R.C. (1971) *Synthesis of International Geomagnetic Reference Field Values*. Inst. Geol. Sci. Rep. No. 71/1.

Barringer, A.R. (1962) A new approach to exploration: the INPUT airborne electrical pulse prospecting system. *Mining Congress J.*, **48**, 49–52.

Barton, P.J. (1986) Comparison of deep reflection and refraction structures in the North Sea. *In*: Barazangi, M. & Brown, L. (eds), *Reflection Seismology: a Global Perspective*. Geodynamics Series, **13**, 297–300, American Geophysical Union, Washington DC.

Bayerly, M. & Brooks, M. (1980) A seismic study of deep structure in South Wales using quarry blasts. *Geophys. J. R. astr. Soc.*, **60**, 1–19.

Beamish, D. (1998) Three-dimensional modelling of VLF data. *J. Appl. Geophys.*, **63**, 63–76.

Bell, F.G. (1993) *Engineering Geology*. Blackwell Science, Oxford.

Bell, R.E. & Watts, A.B. (1986) Evaluation of the BGM-3 sea gravity meter system onboard R/V Conrad. *Geophysics*, **51**, 1480–93.

Bell, R.E., Childers, V.A., Arko, R.A., Blankenship, D.D. & Brozena, J.M. (1999) Airborne gravity and precise positioning for geologic applications. *J. Geophys. Res.*, **104**, 15281–92.

Beltrão, J.F., Silver, J.B.C. & Costa, J.C. (1991) Robust polynomial fitting method for regional gravity estimation. *Geophysics*, **56**, 80–9.

Berry, M.J. & West, G.F. (1966) An interpretation of the first-arrival data of the Lake Superior experiment by the time term method. *Bull. Seismol. Soc. Am.*, **56**, 141–71.

Bhattacharya, P.K. & Patra, H.P. (1968) *Direct Current Electrical Sounding*. Elsevier, Amsterdam.

Birch, F. (1960) The velocity of compressional waves in rocks to ten kilobars, Part 1. *J. Geophys. Res.*, **65**, 1083–102.

Birch, F. (1961) The velocity of compressional waves in rocks to ten kilobars, Part 2. *J. Geophys. Res.*, **66**, 2199–224.

Blakely, R.J. & Connard, G.G. (1989) Crustal structures using magnetic data. *In*: Pakiser, L.C. & Mooney, W.D. (eds), *Geophysical Framework of the Continental United States*. Geol. Soc. America Mem. **172**, 45–60.

Boissonnas, E. & Leonardon, E.G. (1948) Geophysical exploration by telluric currents with special reference to a survey of the Salt Dome, Wood County, Texas. *Geophysics*, **13**, 387–403.

Bolt, B.A. (1982) *Inside the Earth*. Freeman, San Francisco.

Bott, M.H.P. (1982) *The Interior of the Earth*. Edward Arnold, London.

Bott, M.H.P. & Scott, P. (1964) Recent geophysical studies in southwest England. *In*: Hosking, K.F.G. & Shrimpton, G.H. (eds), *Present Views of Some Aspects of the Geology of Devon and Cornwall*. Royal Geological Society of Cornwall.

Bott, M.H.P., Day, A.A. & Masson-Smith, D. (1958) The geological interpretation of gravity and magnetic surveys in Devon and Cornwall. *Phil. Trans. R. Soc.*, **215A**, 161–91.

Bowin, C., Aldrich, T.C. & Folinsbee, R.A. (1972) VSA gravity meter system: Tests and recent developments. *J. Geophys. Res.*, **77**, 2018–33.

Bowker, A. (1991) Quantitative interpretation of three-dimensional mise-à-la-masse data. A case history from Gairloch, northwest Scotland. *Geoexploration*, **28**, 1–22.

Boyd, G.W. & Wiles, C.J. (1984) The Newmont drill-hole EMP system – Examples from eastern Australia. *Geophysics*, **49**, 949–56.

Brewer, J.A. (1983) Profiling continental basement: the key to understanding structures in the sedimentary cover. *First Break*, **1**, 25–31.

Brewer, J.A. & Oliver, J.E. (1980) Seismic reflection studies of deep crustal structure. *Annu. Rev. Earth planet. Sci.*, **8**, 205–30.

Brigham, E.O. (1974) *The Fast Fourier Transform*. Prentice-Hall, New Jersey.

Brooks, M. & Ferentinos, G. (1984) Tectonics and sedimentation in the Gulf of Corinth and the Zakynthos and Kefallinia Channels, western Greece. *Tectonophysics*, **101**, 25–54.

Brooks, M., Doody, J.J. & Al-Rawi, F.R.J. (1984) Major crustal reflectors beneath SW England. *J. Geol. Soc. Lond.*, **141**, 97–103.

Brooks, M., Mechie, J. & Llewellyn, D.J. (1983) Geophysical investigations in the Variscides of southwest Britain. *In*: Hancock, P.L. (ed.), *The Variscan Fold Belt in the British Isles*. Hilger, Bristol, 186–97.

Brown, A.R. (1986) *Interpretation of Three-Dimensional Seismic Data*. AAPG Memoir 42, American Association of Petroleum Geologists, Tulsa.

Brown, J.M., Niebauer, T.M., Richter, B. *et al.* (1999) Miniaturized gravimeter may greatly improve measurements. *EOS (Trans. A.G.U.), Electronic Supplement* at http://www.agu.org/eos_elec/

Bugg, S.F. & Lloyd, J.W. (1976) A study of freshwater lens configuration in the Cayman Islands using resistivity methods. *Q. J. Eng. Geol.*, **9**, 291–302.

Cady, J.W. (1980) Calculation of gravity and magnetic anomalies of finite-length right polygonal prisms. *Geophysics*, **45**, 1507–12.

CamposEnriquez, J.O., MoralesRodriguez, H.F., Dominguez-Mendez, F. & Birch, F.S. (1998) Gauss's theorem, mass deficiency at Chicxulub crater (Yucatan, Mexico), and the extinction of the dinosaurs. *Geophysics*, **63**, 1585–94.

Cardarelli, E. & de Nardis, R. (2001) Seismic refraction, isotropic and anisotropic seismic tomography on an ancient monument (Antonino and Faustina temple AD141). *Geophys. Prosp.*, **49**, 228–41.

Carpenter, P.J., Calkin, S.F. & Kaufmann, R.S. (1991) Assessing a fractured landfill cover using electrical resistivity and seismic refraction techniques. *Geophysics*, **50**, 1896–904.

Cassell, B. (1984) Vertical seismic profiles – an introduction. *First Break*, **2**(11), 9–19.

Castagna, J.P. & Bachus, M.M. (1993) *Offset-Dependent Reflectivity: Theory and Practice of AVO Analysis. Investigations in Geophysics*. Society of Exploration Geophysicists, Tulsa.

Casten, U. & Gram, C. (1989) Recent developments in underground gravity surveys. *Geophys. Prosp.*, **37**, 73–91.

Cerveny, V. & Ravindra, R. (1971) *Theory of Seismic Head Waves*. University of Toronto Press, Toronto.

Cerveny, V., Langer, J. & Psencik, I. (1974) Computation of geometric spreading of seismic body waves in laterally inhomogeneous media with curved interfaces. *Geophys. J. R. astr. Soc.*, **38**, 9–19.

Chakridi, R. & Chouteau, M. (1988) Design of models for electromagnetic scale modelling. *Geophys. Prosp.*, **36**, 537–50.

Christensen, N.I. & Fountain, D.M. (1975) Constitution of the lower continental crust based on experimental studies of seismic velocities in granulites. *Bull. Geol. Am.*, **86**, 227–36.

Clark, A.J. (1986) Archaeological geophysics in Britain. *Geophysics*, **51**, 1404–13.

Cogbill, A.H. (1990) Gravity terrain corrections calculated using digital elevation models. *Geophysics*, **55**, 102–6.

Cook, J.C. (1975) Radar transparencies of mine and tunnel rocks. *Geophysics*, **40**, 865–85.

Cook, K.L. & Van Nostrand, R.G. (1954) Interpretation of resistivity data over filled sinks. *Geophysics*, **19**, 761–90.

Cooper, G.R.J. (1997) GravMap and PFproc: software for filtering geophysical map data. *Comput. Geosci.*, **23**, 91–101.

Corry, C.E. (1985) Spontaneous polarization associated with porphyry sulfide mineralization. *Geophysics*, **50**, 1020–34.

Cox, M. (2001) *Static Corrections for Seismic Reflection Surveys*. Society of Exploration Geophysicists, Tulsa.

Cunningham, A.B. (1974) Refraction data from single-ended refraction profiles. *Geophysics*, **39**, 292–301.

Daily, W. & Owen, E. (1991) Cross-borehole resistivity tomography. *Geophysics*, **56**, 1228–35.

Davis, J.L. & Annan, A.P. (1989) Ground-penetrating radar for high-resolution mapping of soil and rock stratigraphy. *Geophys. Prosp.*, **37**, 531–51.

Davis, J.L., Prescott, W.H., Svarc, J.L. & Wendt, K.J. (1989) Assessment of Global Positioning System measurements for studies of crustal deformation. *J. Geophys. Res.*, **94**, 13635–50.

Desbrandes, R. (1985) *Encyclopedia of Well Logging*. Graham & Trotman, London.

Dey, A. & Morrison, H.F. (1979) Resistivity modelling for arbitrarily shaped two-dimensional structures. *Geophys. Prosp.*, **27**, 106–36.

Dix, C.H. (1955) Seismic velocities from surface measurements. *Geophysics*, **20**, 68–86.

Dobrin, M.B. & Savit, C.H. (1988) *Introduction to Geophysical Prospecting* (4th edn). McGraw Hill, New York.

Duckworth, K. (1968) *Mount Minza Area Experimental Geophysical Surveys, Northern Territory 1966 and 1967*. Record No. 1968/107 of the B.M.R., Australia.

Duncan, P.M. & Garland, G.D. (1977) A gravity study of the Saguenay area, Quebec. *Can. J. Earth Sci.*, **14**, 145–52.

Durrance, E.M. (1986) *Radioactivity in Geology*. Ellis Horwood, Chichester.

Dyck, A.V. & West, G.F. (1984) The role of simple computer models in interpretations of wide-band, drillhole electromagnetic surveys in mineral exploration. *Geophysics*, **49**, 957–80.

Ebraheem, A.M., Hamburger, M.W., Bayless, E.R. & Krothe, N.C. (1990) A study of acid mine drainage using earth resistivity measurements. *Groundwater*, **28**, 361–8.

Ellis, D.V. (1987) *Well Logging for Earth Scientists*. Elsevier, New York.

Fountain, D.K. (1972) Geophysical case-histories of disseminated sulfide. *Geophysics*, **37**, 142–59.

Fournier, C. (1989) Spontaneous potentials and resistivity surveys applied to hydrogeology in a volcanic area: case history of the Chaîne des Puys (Puy de Dôme, France). *Geophys. Prosp.*, **37**, 647–68.

Frischknecht, F.C. & Raab, P.V. (1984) Time-domain electromagnetic soundings at the Nevada Test Site, Nevada. *Geophysics*, **49**, 981–92.

Gardner, G.H.F., Gardner, L.W. & Gregory, A.R. (1974) Formation velocity and density – the diagnostic basics for stratigraphic traps. *Geophysics*, **39**, 770–80.

Garland, G.D. (1951) Combined analysis of gravity and magnetic anomalies. *Geophysics*, **16**, 51–62.

Garland, G.D. (1965) *The Earth's Shape and Gravity*. Pergamon, Oxford.

Gilbert, D. & Galdeano, A. (1985) A computer program to perform transformations of gravimetric and aeromagnetic surveys. *Comput. Geosci.*, **11**, 553–88.

Goldberg, D. (1997) The role of downhole measurements in marine geology and geophysics. *Rev. Geophys.*, **35**, 315–42.

Gondwe, E. (1991) Saline water intrusion in southeast Tanzania. *Geoexploration*, **27**, 25–34.

Götze, H.-J. & Lahmeyer, B. (1988) Application of three-dimensional interactive modelling in gravity and magnetics. *Geophysics*, **53**, 1096–108.

Grant, F.S. & West, G.F. (1965) *Interpretation Theory in Applied Geophysics*. McGraw-Hill, New York.

Griffiths, D.H. & King, R.F. (1981) *Applied Geophysics for Geologists and Engineers*. Pergamon, Oxford.

Gunn, P.J. (1975) Linear transformations of gravity and magnetic fields. *Geophys. Prosp.*, **23**, 300–12.

Gunn, P. (1998) Aeromagnetics locates prospective areas and prospects. *The Leading Edge*, **17**, 67–9.

Guyod, H. (1974) Electrolog. *In: Log Review 1*. Dresser Industries, Houston.

Habberjam, G.M. (1979) *Apparent Resistivity and the Use of Square Array Techniques*. Gebrüder Borntraeger, Berlin.

Hagedoorn, J.G. (1959) The plus–minus method of interpreting seismic refraction sections. *Geophys. Prosp.*, **7**, 158–82.

Halpenny, J.F. & Darbha, D.M. (1995) Airborne gravity tests over Lake Ontario. *Geophysics*, **60**, 61–5.

Hatton, L., Worthington, M.H. & Makin, J. (1986) *Seismic Data Processing*. Blackwell Scientific Publications, Oxford.

Heirtzler, J.R., Dickson, G.O., Herron, E.M., Pitman, W.C. & Le Pichon, X. (1968) Marine magnetic anomalies, geomagnetic field reversals, and motions of the ocean floor and continents. *J. Geophys. Res.*, **73**, 2119–36.

Hjelt, S.E., Kaikkonen, P. & Pietilaä, R. (1985) On the interpretation of VLF resistivity measurements. *Geoexploration*, **23**, 171–81.

Hood, P.J. & Teskey, D.J. (1989) Aeromagnetic gradiometer program of the Geological Survey of Canada. *Geophysics*, **54**, 1012–22.

Hooper, W. & McDowell, P. (1977) Magnetic surveying for buried mineshafts and wells. *Ground Engineering*, **10**, 21–3.

Hubbert, M.K. (1934) Results of Earth-resistivity survey on various geological structures in Illinois. *Trans. Am. Inst. Mining Met. Eng.*, **110**, 9–29.

Hutton, V.R.S., Ingham, M.R. & Mbipom, E.W. (1980) An electrical model of the crust and upper mantle in Scotland. *Nature, Lond.*, **287**, 30–3.

IAG (International Association of Geodesy) (1971) *Geodetic Reference System 1967*. Pub. Spec. No. 3 du Bulletin Géodésique.

Ivansson, S. (1986) Seismic borehole tomography – Theory and computational methods. *Proc. IEEE*, **74**, 328–38.

Jack, I. (1997) *Time Lapse Seismic in Reservoir Management*. Society of Exploration Geophysicists, Short Course Notes, Society of Exploration Geophysicists, Tulsa.

Johnson, S.H. (1976) Interpretation of split-spread refraction data in terms of plane dipping layers. *Geophysics*, **41**, 418–24.

Kanasewich, E.R. (1981) *Time Sequence Analysis in Geophysics* (3rd edn). University of Alberta.

Kearey, P. & Allison, J.R. (1980) A geomagnetic investigation of Carboniferous igneous rocks at Tickenham, County of Avon. *Geol. Mag.*, **117**, 587–93.

Kearey, P. & Vine, F.J. (1990) *Global Tectonics*. Blackwell Scientific Publications, Oxford.

Kearey, P. & Vine, F.J. (1996) *Global Tectonics* (2nd edn). Blackwell Science, Oxford.

Keating, P.B. (1998) Weighted Euler deconvolution of gravity data. *Geophysics*, **63**, 1595–603.

Kilty, K.T. (1984) On the origin and interpretation of self-potential anomalies. *Geophys. Prosp.*, **32**, 51–62.

Klemperer, S.L. and the BIRPS group (1987) Reflectivity of the crystalline crust: hypotheses and tests. *Geophys. J. R. astr. Soc.*, **89**, 217–22.

Knopoff, L. (1983) The thickness of the lithosphere from the dispersion of surface waves. *Geophys. J. R. astr. Soc.*, **74**, 55–81.

Koefoed, O. (1979) *Geosounding Principles, 1 – Resistivity Sounding Measurements*. Elsevier, Amsterdam.

Labson, V.F., Becker, A., Morrison, H.F. & Conti, U. (1985) Geophysical exploration with audiofrequency natural magnetic fields. *Geophysics*, **50**, 656–64.

LaFehr, T.R. (1991) Standardization in gravity reduction. *Geophysics*, **56**, 1170–8.

Langore, L., Alikaj, P. & Gjovreku, D. (1989) Achievements in copper sulphide exploration in Albania with IP and EM methods. *Geophys. Prosp.*, **37**, 975–91.

Lavergne, M. (1989) *Seismic Methods*. Editions Technip, Paris.

Lee, M.K., Pharaoh, T.C. & Soper, N.J. (1990) Structural trends in central Britain from images of gravity and aeromagnetic fields. *J. Geol. Soc. Lond.*, **147**, 241–58.

Le Tirant, P. (1979) *Seabed Reconnaissance and Offshore Soil Mechanics*. Editions Technip, Paris.

Lines, L.R., Schultz, A.K. & Treitel, S. (1988) Cooperative inversion of geophysical data. *Geophysics*, **53**, 8–20.

Logn, O. (1954) Mapping nearly vertical discontinuities by Earth resistivities. *Geophysics*, **19**, 739–60.

Loke, M.H. & Barker, R.D. (1995) Least-squares inversion of apparent resistivity pseudosections by a quasi-Newton method. *Geophysics*, **60**, 1682–90.

Loke, M.H. & Barker, R.D. (1996) Rapid least-squares deconvolution of apparent resistivity pseudosections. *Geophys. Prosp.*, **44**, 131–52.

Lovell, M.A., Williamson, G. & Harvey, P.K. (1999) Borehole Images: Applications and Case Histories. *Spec. Pub. Geol. Soc., London* **159**, 235 pp.

March, D.W. & Bailey, A.D. (1983) Two-dimensional transform and seismic processing. *First Break*, **1**(1), 9–21.

Mason, R.G. & Raff, R.D. (1961) Magnetic survey off the west coast of North America, 32°N to 42°N. *Bull. Geol. Soc. Am.*, **72**, 1259–66.

McCann, D.M., Eddleston, M., Fenning, P.J. & Reeves, G.M. (eds) (1997) *Modern Geophysics in Engineering Geology*. Geological Society Engineering Geology Special Publication **12**.

McCracken, K.G., Oristaglio, M.L. & Hohmann, G.W. (1986) A comparison of electromagnetic exploration systems. *Geophysics*, **51**, 810–8.

McNeill, J.D. (1980) *Electromagnetic Terrain Conductivity Measurement at Low Induction Numbers*. Technical Note TN-6, Geonics, Mississauga.

McNeill, J.D. (1991) Advances in electromagnetic methods for groundwater studies. *Geoexploration*, **27**, 43–54.

Meckel, L.D. & Nath, A.K. (1977) Geologic considerations for stratigraphic modelling and interpretation. *In*: Payton, C.E. (ed.), *Seismic Stratigraphy – Applications to Hydrocarbon Exploration*. AAPG Memoir 26, 417–38.

Merkel, R.H. (1972) The use of resistivity techniques to delineate acid mine drainage in groundwater. *Groundwater*, **10**, No. 5, 38–42.

Mittermayer, E. (1969) Numerical formulas for the Geodetic Reference System 1967. *Bolletino di Geofisca Teorica ed Applicata*, **11**, 96–107.

Morelli, C., Gantor, C., Honkasalo, T. *et al.* (1971) *The International Gravity Standardisation Net*. Pub. Spec. No. 4 du Bulletin Géodésique.

Moxham, R.M. (1963) Natural radioactivity in Washington County, Maryland. *Geophysics*, **28**, 262–72.

Nafe, J.E. & Drake, C.L. (1963) Physical properties of marine sediments. *In*: Hill, M.N. (ed.), *The Sea, Vol. 3*. Interscience Publishers, New York, 794–815.

Nair, M.R., Biswas, S.K. & Mazumdar, K. (1968) Experimental studies on the electromagnetic response of tilted conducting half-planes to a horizontal-loop prospective system. *Geoexploration*, **6**, 207–44.

Neidell, N.S. & Poggiagliolmi, E. (1977) Stratigraphic modelling and interpretation – geophysical principles. *In*: Payton, C.E. (ed.), *Seismic Stratigraphy – Applications to Hydrocarbon Exploration*.

Memoir 26, American Association of Petroleum Geologists, Tulsa, 389–416.

Nettleton, L.L. (1976) *Gravity and Magnetics in Oil Exploration*. McGraw-Hill, New York.

O'Brien, P.N.S. (1974) Aspects of seismic research in the oil industry. *Geoexploration*, **12**, 75–96.

Ogilvy, R.D., Cuadra, A., Jackson, P.D. & Monte, J.L. (1991) Detection of an air-filled drainage gallery by the VLF resistivity method. *Geophys. Prosp.*, **39**, 845–60.

Orellana, E. & Mooney, H.M. (1966) *Master Tables and Curves for Vertical Electrical Sounding Over Layered Structures*. Interciencia, Madrid.

Orellana, E. & Mooney, H.M. (1972) *Two and Three Layer Master Curves and Auxiliary Point Diagrams for Vertical Electrical Sounding Using Wenner Arrangement*. Interciencia, Madrid.

Oristaglio, M.L. & Hohmann, G.W. (1984) Diffusion of electromagnetic fields into a two-dimensional earth: A finite-difference approach. *Geophysics*, **49**, 870–94.

Palacky, G.J. (1981) The airborne electromagnetic method as a tool of geological mapping. *Geophys. Prosp.*, **29**, 60–88.

Palacky, G.J. & Sena, F.O. (1979) Conductor identification in tropical terrains – case histories from the Itapicuru greenstone belt, Bahia, Brazil. *Geophysics*, **44**, 1941–62.

Palmer, D. (1980) *The Generalised Reciprocal Method of Seismic Refraction Interpretation*. Society of Exploration Geophysicists, Tulsa.

Panissod, C., Dabas, M., Hesse, A. *et al.* (1998) Recent developments in shallow-depth electrical and electrostatic prospecting using mobile arrays. *Geophysics*, **63**, 1542–50.

Parasnis, D.S. (1973) *Mining Geophysics*. Elsevier, Amsterdam.

Parker, R.L. (1977) Understanding inverse theory. *Annu. Rev. Earth planet. Sci.*, **5**, 35–64.

Paterson, N.R. (1967) Exploration for massive sulphides in the Canadian Shield. *In*: Morley, L.W. (ed.), *Mining and Groundwater Geophysics*. Econ. Geol. Report No. 26, Geol. Surv. Canada, 275–89.

Paterson, N.R. & Reeves, C.V. (1985) Applications of gravity and magnetic surveys: the state-of-the-art in 1985. *Geophysics*, **50**, 2558–94.

Payton, C.E. (ed.) (1977) *Seismic Stratigraphy – Applications to Hydrocarbon Exploration*. Memoir 26, American Association of Petroleum Geologists, Tulsa.

Peddie, N.W. (1983) International geomagnetic reference field – its evolution and the difference in total field intensity between new and old models for 1965–1980. *Geophysics*, **48**, 1691–6.

Pedley, H.M., Hill, I. & Brasington, J. (2000) Three dimensional modelling of a Holocene tufa system in the Lathkill Valley, N. Derbyshire, using ground penetrating radar. *Sedimentology*, **47**, 721–37.

Peters, J.W. & Dugan, A.F. (1945) Gravity and magnetic investigations at the Grand Saline Salt Dome, Van Zandt Co., Texas. *Geophysics*, **10**, 376–93.

Pires, A.C.B. & Harthill, N. (1989) Statistical analysis of airborne gamma-ray data for geologic mapping purposes: Crixas-Itapaci area, Goias, Brazil. *Geophysics*, **54**, 1326–32.

Rayner, J.N. (1971) *An Introduction to Spectral Analysis*. Pion, England.

Reid, A.B., Allsop, J.M., Granser, H., Millett, A.J. & Somerton,

I.W. (1990) Magnetic interpretation in three dimensions using Euler deconvolution. *Geophysics*, **55**, 80–91.

Reilly, W.I. (1972) Use of the International System of Units (SI) in geophysical publications. *N.Z.J. Geol. Geophys.*, **15**, 148–58.

Reynolds, J.M. (1997) *An Introduction to Applied and Environmental Geophysics*. Wiley, Chichester.

Robinson, E.A. & Treitel, S. (1980) *Geophysical Signal Analysis*. Prentice-Hall, London.

Robinson, E.A. & Treitel, S. (2000) *Geophysical Signal Analysis*. Society of Exploration Geophysicists, Tulsa, USA.

Robinson, E.S. & Çoruh, C. (1988) *Basic Exploration Geophysics*. Wiley, New York.

Sakuma, A. (1986) *Second Internal Comparison of Gravimeters*. Internal Report, Bureau International des Poids et Mesures.

Sato, M. & Mooney, H.H. (1960) The electrochemical mechanism of sulphide self potentials. *Geophysics*, **25**, 226–49.

Schramm, M.W., Dedman, E.V. & Lindsey, J.P. (1977) Practical stratigraphic modelling and interpretation. *In*: Payton, C.E. (ed.), *Seismic Stratigraphy – Applications to Hydrocarbon Exploration*. Memoir 26, American Association of Petroleum Geologists, Tulsa, 477–502.

SEG (1997) Digital Tape Standards (SEG-A, SEG-B, SEG-C, SEG-Y and SEG-D formats plus SEG-D rev 1&2). Compiled by SEG Technical Standards Committee. Society of Exploration Geophysicists, Tulsa, USA.

Seigel, H.O. (1967) The induced polarisation method. *In*: Morley, L.W. (ed.), *Mining and Groundwater Geophysics*. Econ. Geol. Report No. 26, Geol. Survey of Canada, 123–37.

Seigel, H.O. & McConnell, T. (1998) Regional surveys using a helicopter-suspended gravimeter. *Leading Edge*, **17**, 47–9.

Sharma, P. (1976) *Geophysical Methods in Geology*. Elsevier, Amsterdam.

Sheriff, R.E. (1973) *Encyclopedic Dictionary of Exploration Geophysics*. Society of Exploration Geophysicists, Tulsa.

Sheriff, R.E. (1978) *A First Course in Geophysical Exploration and Interpretation*. IHRDC, Boston.

Sheriff, R.E. (1980) *Seismic Stratigraphy*. IHRDC, Boston.

Sheriff, R.E. & Geldart, L.P. (1982) *Exploration Seismology Vol. 1: History, Theory and Data Acquisition*. Cambridge University Press, Cambridge.

Sheriff, R.E. & Geldart, L.P. (1983) *Exploration Seismology Vol. 2: Data-Processing and Interpretation*. Cambridge University Press, Cambridge.

Smith, R.A. (1959) Some depth formulae for local magnetic and gravity anomalies. *Geophys. Prosp.*, **7**, 55–63.

Snyder, D.B. & Flack, C.A. (1990) A Caledonian age for reflections within the mantle lithosphere north and west of Scotland. *Tectonics*, **9**, 903–22.

Spector, A. & Grant, F.S. (1970) Statistical models for interpreting aeromagnetic data. *Geophysics*, **35**, 293–302.

Spies, B.R. (1976) The transient electromagnetic method in Australia. *B.M.R. J. Austral. Geol. & Geophys.*, **1**, 23–32.

Spies, B.R. (1989) Depth of investigation in electromagnetic sounding methods. *Geophysics*, **54**, 872–88.

Sprenke, K.F. (1989) Efficient terrain corrections: A geostatistical analysis. *Geophysics*, **54**, 1622–8.

Stacey, R.A. (1971) Interpretation of the gravity anomaly at Darnley Bay, N.W.T. *Can. J. Earth Sci.*, **8**, 1037–42.

Stoffa, P.L. & Buhl, P. (1979) Two-ship multichannel seismic experiments for deep crustal studies: expanded spread and constant offset profiles. *J. Geophys. Res.*, **84**, 7645–60.

Sundararajan, N., Rao, P.S. & Sunitha, V. (1998) An analytical method to interpret self-potential anomalies caused by 2-D inclined sheets. *Geophysics*, **63**, 1551–5.

Talwani, M. & Ewing, M. (1960) Rapid computation of gravitational attraction of three-dimensional bodies of arbitrary shape. *Geophysics*, **25**, 203–25.

Talwani, M., Le Pichon, X. & Ewing, M. (1965) Crustal structure of the mid-ocean ridges 2. Computed model from gravity and seismic refraction data. *J. Geophys. Res.*, **70**, 341–52.

Talwani, M., Worzel, J.L. & Landisman, M. (1959) Rapid gravity computations for two-dimensional bodies with applications to the Mendocino submarine fracture zones. *J. Geophys. Res.*, **64**, 49–59.

Taner, M.T. & Koehler, F. (1969) Velocity spectra – digital computer derivation and applications of velocity functions. *Geophysics*, **34**, 859–81.

Telford, W.M. (1982) Radon mapping in the search for uranium. *In*: Fitch, A.A. (ed.), *Developments in Geophysical Exploration Methods*. Applied Science, London, 155–94.

Telford, W.M., Geldart, L.P., Sheriff, R.E. & Keys, D.A. (1976) *Applied Geophysics*. Cambridge University Press, Cambridge.

Telford, W.M., Geldart, L.P. & Sheriff, R.E. (1990) *Applied Geophysics*, 2nd edn. Cambridge University Press, Cambridge.

Thomas, M.D. & Kearey, P. (1980) Gravity anomalies, block-faulting and Andean-type tectonism in the eastern Churchill Province. *Nature, Lond.*, **283**, 61–3.

Thornburgh, H.R. (1930) Wave-front diagrams in seismic interpretation. *Bull. Am. Assoc. Petrol. Geol.*, **14**, 185–200.

Van Overmeeren, R.A. (1975) A combination of gravity and seismic refraction measurements, applied to groundwater explorations near Taltal, Province of Antofagasta, Chile. *Geophys. Prosp.*, **23**, 248–58.

Van Overmeeren, R.A. (1989) Aquifer boundaries explored by geoelectrical measurements in the coastal plain of Yemen: a case of equivalence. *Geophysics*, **54**, 38–48.

Vine, F.J. & Matthews, D.H. (1963) Magnetic anomalies over oceanic ridges. *Nature, Lond.*, **199**, 947–9.

Vogelsang, D. (1995) *Environmental Geophysics, a Practical Guide*. Springer-Verlag, Berlin.

Waters, H. (1978) *Reflection Seismology – a Tool for Energy Resource Exploration*. Wiley, New York.

Webb, J.E. (1966) The search for iron ore, Eyre Peninsula, South Australia. *In*: *Mining Geophysics, Vol. 1*. Society of Exploration Geophysicists, Tulsa, 379–90.

Westbrook, G.K. (1975) The structure of the crust and upper mantle in the region of Barbados and the Lesser Antilles. *Geophys. J. R. astr. Soc.*, **43**, 201–42.

Whitcomb, J.H. (1987) Surface measurements of the Earth's gravity field. *In*: Samis, C.G. & Henyey, T.L. (eds), *Methods of Experimental Physics, Vol. 24, Part B – Field Measurements*. Academic Press, Orlando, 127–61.

White, P.S. (1966) Airborne electromagnetic survey and ground follow-up in north-western Quebec. *In: Mining Geophysics, Vol. 1.* Society of Exploration Geophysicists, Tulsa, 252–61.

Willmore, P.L. & Bancroft, A.M. (1960) The time-term approach to refraction seismology. *Geophys. J. R. astr. Soc.*, **3**, 419–32.

Wold, R.J. & Cooper, A.K. (1989) Marine magnetic gradiometer – a tool for the seismic interpreter. *The Leading Edge*, **8**, 22–7.

Wollenberg, H.A. (1977) Radiometric methods. *In:* Morse, J.G. (ed.), *Nuclear Methods in Mineral Exploration and Production.* Elsevier Science, Amsterdam, 5–36.

Wong, J., Bregman, N., West, G. & Hurley, P. (1987) Crosshole seismic scanning and tomography. *The Leading Edge*, **6**, 36–41.

Wood, R.D., Wichmann, P.A. & Watt, H.B. (1974) Gamma ray – neutron log. *In: Log Review 1.* Dresser Industries, Houston.

Wright, C., Barton, T., Goleby, B.R., Spence, A.G. & Pester, D. (1990) The interpretation of expanding spread profiles: examples from central and eastern Australia. *Tectonophysics*, **173**, 73–82.

Yilmaz, O. (1987) *Seismic Data Processing.* Society of Exploration Geophysicists, Tulsa.

Yilmaz, O. (2001) *Seismic Data Analysis: Processing, Inversion and Analysis of Seismic Data.* (2 Vols). Society of Exploration Geophysicists, Tulsa.

Yüngül, S. (1954) Spontaneous polarisation survey of a copper deposit at Sariyer, Turkey. *Geophysics*, **19**, 455–58.

Zalasiewicz, J.A., Mathers, S.J. & Cornwell, J.D. (1985) The application of ground conductivity measurements to geological mapping. *Q. J. Eng. Geol. Lond.*, **18**, 139–48.

Zoeppritz, K. (1919) Uber reflexion und durchgang seismischer wellen durch Unstetigkerlsflaschen. Berlin, Uber Erdbebenwellen VII B, Nachrichten der Koniglichen Gesellschaft der Wissensschaften zu Gottingen, math-phys. Kl. pp. 57–84.

Zohdy, A.A.R. (1989) A new method for the automatic interpretation of Schlumberger and Wenner sounding curves. *Geophysics*, **54**, 245–53.

Index